Immunity in Evolution

**WITHDRAWN**

## Immunity in Evolution

John J. Marchalonis

Harvard University Press
Cambridge, Massachusetts
1977

Library of Congress Cataloging in Publication Data
Marchalonis, John J     1940-
  Immunity in evolution.

  "Publication no. 1900 from the Walter and Eliza Hall
Institute of Medical Research."
  Bibliography:  p.
  Includes index.
  1.  Immunotaxonomy.  2.  Evolution.  I.  Walter and
Eliza Hall Institute of Medical Research, Melbourne.

II.  Title.  [DNLM:  1.  Evolution.  2.  Immunity.
QW504  M315i]
QR183.8.M37        596'.02'9        75-37719
ISBN 0-674-44445-0

To Anne

# Acknowledgments

I thank Sir F. M. Burnet, Professor G. J. V. Nossal, Dr. P. Kincade, and Dr. A. Szenberg for helpful advice in the preparation of this manuscript. Original work discussed was supported in part by the American Heart Association, the Anna Fuller Fund, the Australian Research Grants Committee, the Damon Runyon Memorial Fund, the National Health and Medical Research Council of Australia, and the United States Public Health Service. This is publication number 1900 from The Walter and Eliza Hall Institute of Medical Research.

J. J. M.

# Contents

# Tables

# Figures

# Foreword

*by F. M. Burnet*

Immunology began, in the hands of Pasteur and his followers, as the search for an answer to one of the most urgent of human problems: the high mortality from infectious disease, particularly in the young. But all through its history the best workers in the field have been as much interested in a scholarly understanding of the processes of immunity as in preventing or curing disease. Today most of the diseases that are, from their character, capable of being prevented by immunological means have either vanished from most parts of the world or are as effectively controlled as practical considerations allow. There will always be work to do in perfecting vaccines and improving the logistics of applying them when and where they are needed, but the growing edge of immunology is elsewhere. Its real concern, as I see it, is to use the uniquely interdisciplinary position of the phenomena of immunology to throw light on the whole range of biological fields that it impinges upon: biochemistry, genetics, cytology, medicine, and even ecology. In even broader terms, our objective is to weld the findings and theories of immunology into the pattern of general biology.

Currently, research seems to be concentrated on a rather limited number of major topics: (1) the chemical structure of antibody and the genetic origin of immune diversity, (2) the functions of the cells concerned in immunity and the nature of their interactions, and (3) immunological function in the lower vertebrates and analogous activities in invertebrates, i.e. comparative immunology, with, of course, special interest in the evolution of the immune response.

To discuss immunity in an evolutionary context requires, in addition to wide reading, some firsthand experience in all three of the fields I have mentioned. I have been close enough to Dr. Marchalonis's work to know that he has such a background, and I believe that in this book he has produced a valuable and critical study of the whole field. It may even emerge as the first definitive statement of the evolutionary origins and development of vertebrate immunity.

Since his arrival in Melbourne, Dr. Marchalonis has made particularly good use of our living fossils of the southern hemisphere: lung fish, tuatara, and echidna. He is an excellent biologist, but, in addition, he has been able to make use of all the more refined biochemical and radiochemical techniques which my own generation lacked. I believe that the result is an important contribution to general biology as well as to immunology. It should be particularly useful to workers who are interested in using less familiar experimental animals to elucidate some of the general aspects of immunology for which they may offer special advantages.

# 1 Mammalian Immunity:
## Guidelines and Problems for
## Evolutionary Studies

"All our reasonings concerning matter of fact are founded
on a species of Analogy, which leads us to expect from any
cause the same events, which we have observed to result
from similar causes. Where the causes are entirely similar,
the analogy is perfect, and the inference, drawn from it,
is regarded as certain and conclusive: nor does any man
ever entertain a doubt, where he sees a piece of iron, that it
will have weight and cohesion of parts; as in all other instances, which have ever fallen under his observation. But
where the objects have not so exact a similarity, the analogy is less perfect, and the inference is less conclusive;
though still it has some force, in proportion to the degree
of similarity and resemblance. The anatomical observations, formed upon one animal are, by this species of
reasoning, extended to all animals; and it is certain, that
when the circulation of the blood, for instance, is clearly
proved to have place in one creature, as a frog, or fish, it
forms a strong presumption, that the same principle has
place in all."

David Hume, 1777.

Although studies with mammals provide the preponderance of our
knowledge of immunological phenomena, many investigators, motivated
by a desire to understand the basic mechanisms governing these important biological reactions, have attempted to trace immunity to its
origins in lower vertebrates and even invertebrates. Metchnikoff (1884,

1905) clearly recognized the value of studies employing nonmammalian species. His pioneering observations of a phagocytosis, a phenomenon ubiquitous among living organisms, were made using the water flea *Daphnia*. Noguchi (1903), working in the first "golden age of immunology," endeavored to place the theoretical concepts of Bordet, Ehrlich, and others within the context of evolutionary biology. Much of the work of these early leaders as well as that of Widal and Sicard (1897) and Cantacuzene (1923) is definitive and incisive, but more precise methodological approaches have refuted some of it. However, the aims of recent investigators into the phylogenetic origins of immunity remain essentially the same as those of earlier workers, and can be stated as follows: first, to determine the generality of the mechanisms of immune recognition, processing of antigen, and antibody formation or other immune effector functions of lymphocytes, and second, to disclose simplified systems which will serve as useful models in attempts to elucidate the molecular basis of immune function. For example, it is conceivable that certain organisms might be capable of exhibiting some immune reactions but might lack the full repertoire characteristic of mammalian immunity.

Modern research, like the initial studies, is directed toward the application of the principles and techniques of immunology to infectious disease and to other clinical problems such as destruction of tumors and transplanted organs. Furthermore, immune reactions must now be analyzed in terms of present knowledge of cell biology and molecular biology in order to clarify the mechanisms regulating recognition of antigens, immune differentiation, and immune effector functions. Evolutionary studies of immunity make it possible to compare the process of antibody evolution with that of other proteins such as hemoglobins and cytochromes. Experiments can be performed to assess the origins of cellular functions necessary in immunity using animals which either are primitive immunologically (e.g., lampreys) or possess developmental stages amenable to direct experimental manipulation (e.g., larval frogs). In this book, I will review our present knowledge of immune phenomena as they occur in lower vertebrates and will attempt to interpret them and their phylogenetic distribution as problems in evolutionary biology and cell biology. Prior to considering studies on lower vertebrates and invertebrates in detail, one must first introduce the key problems of the immune response of mammals which will serve as a foundation for such a discussion.

## The biology of mammalian immunity

The immune response of vertebrates comprises a diverse collection of biological phenomena which share three properties: the response is induced by some material that is usually foreign to the animal; the reaction is specific for the challenging agent; and the response is mediated by blood cells of the lymphoid series. Two broad categories of immune reaction have been distinguished operationally. In one case, the capacity to respond in an immune manner can be conferred upon an animal that has not been challenged by the foreign antigen merely by transfer of blood serum from an immune animal. This type of reaction is termed humoral immunity and is defined by the presence of specific antibodies in the serum. The second broad class of immune response, cellular immunity, can be conferred upon a native animal by transfer of living cells but not by infusion of immune serum.

A classical example of a humoral immune response is the protection given by immunization with an altered form of a bacterial toxin that has been rendered harmless but is still capable of inducing circulating antibodies. The neurotoxin that causes tetanus, produced by the bacillus *Clostridium tetani*, can be inactivated by heating or treatment with fixatives such as formalin and injected into persons who might run a risk of contracting the disease. This inactivated toxin, or toxoid, serves as the foreign antigen which elicits a sequence of cellular events culminating in the formation of circulating proteins termed antibodies, which combine specifically with the challenging agent. If the immunized person were infected with growing tetanus bacilli, these antibodies, or antitoxin, would combine in the blood with the toxin produced by the bacilli and render it inactive (Landsteiner 1962). Moreover, complexes of toxin with antitoxin would be readily removed from circulation by phagocytic cells.

An example of cellular immunity that has been studied in detail is the phenomenon of delayed response to contact sensitization (Chase 1959). Contact of small organic molecules, such as the catechol from poison ivy, with skin may sensitize an individual so that a second application results in specific inflammation at the site of contact within 24 to 48 hours. This response meets the criteria for vertebrate immunity but does not require the presence of circulating antibodies. A second example of cell-mediated immunity which has recently received considerable attention is the response of the body to organs transplanted from another member of the same species. The rejection of such transplants is thought to be mediated largely by cellular immunity. In this case, paradoxically, circulating antibodies may have the opposite effect and protect the graft from lymphocytes activated to destroy it (World Health Organization 1969a).

The chemical properties of mammalian antibodies have recently been elucidated in great detail (Edelman et al. 1969; Putnam et al. 1972, 1973). This body of information provides a firm background upon which to analyze the structures of antibody proteins synthesized by lower vertebrates and renders feasible direct comparison between blood proteins of invertebrates and antibodies of vertebrate species. Although circulating antibodies are not involved in cell mediated immunity, the lymphocytes which carry out these functions possess receptor sites on their surfaces which are antibodies (Ehrlich 1900; Greaves and Hogg 1971; Marchalonis and Cone 1973a). Comparative analysis of the structure of vertebrate antibodies provides both information on the genetic events governing the evolution of a protein and a criterion for determining the emergence of vertebrate immunity.

In the pages that follow I shall consider antibody production as a problem of evolution and differentiation with the major emphasis on studies of the physical and chemical characteristics of antibodies. These techniques provide a direct means of assessing similarities in immune responses among widely divergent animal species. Cellular aspects of antibody formation will be described and correlated with the evolution of basic immunological mechanisms; for example, the phenomenon of immunological memory or enhanced response following a second stimulation with antigen will be discussed. A second basic immunological phenomenon is that of tolerance or specific suppression of an antibody response because of prior treatment with the challenging agent. The mechanisms regulating these phenomena are not completely understood at present, and their phylogenetic emergence in certain primitive vertebrates may serve as yet another means of elucidating them. Although immunophylogeny has only been subjected to extensive investigation in the past ten years, sufficient evidence has accrued to develop a consistent scheme for the evolution of antibodies and mechanisms involved in antibody formation. The present text will attempt to analyze and integrate these recent data on the phylogenetic origins of antibody structure and synthesis.

## Antibody formation

The general phenomenology of antibody formation is illustrated in Figure 1. In this case it is a graph of amount of antibody evoked by a bacteriophage, a viral antigen, with which an animal has been injected. It can be seen from this figure that there is a short lag phase during which no detectable viral inactivation by the serum is observed. This lag is followed by an exponential rise in antibody activity and a final plateau. In a mammal this plateau would probably be reached within two weeks. With viral antigen, the plateau may last indefinitely, sometimes for months or years, but in general, sometime after the plateau

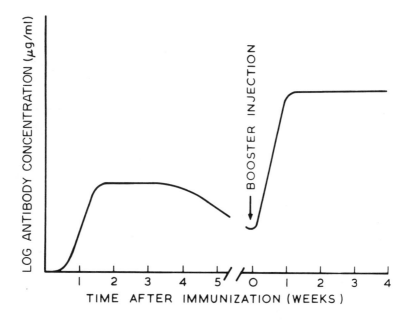

Fig. 1. Generalized kinetics of appearance of antibodies in mammals immunized to bacteriophage.

has been reached there is a decline. This figure also shows the effect of a second injection of antigen, indicated by the arrow. This booster injection brings about immunological anamesis, or memory. Antibody production again occurs, but in this case the appearance of antibody in the serum is more rapid and the plateau is attained more quickly than in the primary response. In addition, the final level of peak titer of antibody in the secondary response is significantly higher than it is in the primary. These are the most general parameters by which to gauge antibody production. In a time course such as is shown here, different types of antibodies may appear, and their sequence of emergence is a function of time after antigen given.

Extensive studies by a variety of workers, particularly Uhr and his colleagues (Uhr, Finkelstein, and Franklin 1962; Uhr 1964), have shown that in the situation where the antigen is of viral origin, the antibody which appears in the mammal's serum early after immunization is the immune macroglobulin or IgM antibody. The molecular structure of this immunoglobulin and the others will be described in detail in Chapter 2. Later, after a single injection of virus, antibody activity can be detected in another type of immunoglobulin, the IgG class of antibodies. Eventually the preponderance of antibody activity is found in these molecules. Following secondary injection, the vast bulk of antibody activity occurs in the IgG immunoglobulins. The nature of the mechanisms underlying the succession of IgM by IgG provides an example of gene activation in antibody formation. The succession of IgM by IgG type antibodies was thought to be a general phenomenon, but in practice, it is not always so. As will be seen later in this text, many lower animals provide an excellent system for the study of antibody production in which antibody activity is restricted to molecules of the IgM class.

Essentially any macromolecule can serve as an antigen; that is, if it is injected into an animal to which it is foreign, it will elicit the production of antibody specific to it. These macromolecules can be carbohydrates, proteins, lipids, or even nucleic acids. The actual antigenic determinant which elicits antibody production and to which the antibody binds specifically need not be the entire macromolecule. It is usually a quite small portion of the macromolecule and may be about the size of six amino acids or less. Antibodies can also be formed to small, very simple, organic chemical compounds, for example, para-aminobenzoic acid, which Landsteiner used in his early critical studies (Landsteiner 1962). This compound, called a hapten, will not alone induce the formation of circulating antibodies; but if it is coupled to a macromolecular carrier, the complex is immunogenic, and antibodies which bind to the hapten are synthesized. The combination of antibody with hapten is stereospecific and has been likened to the complementary fit of a key into a lock.

## Cellular aspects of immunity

The cellular aspects of antibody production have been subjected to intensive study in recent years. It is now generally accepted that members of lymphocyte populations within an individual animal are phenotypically restricted in their capacity to respond to antigens. Burnet (1959), in his clonal selection theory of immunologically competent cells, argued that the population of lymphocytes in the body was analogous to a population of microbes, where individual cells spontaneously differ from one another in the capacity to respond to new environmental stimuli. In a bacterial population, individual change occurs by means of spontaneous mutation, and it is well established that bacteria possess the capacity to survive a vast array of toxic chemicals because large populations contain certain individuals which are resistant to the chemical. In lymphocyte populations, each individual cell is considered to be capable of making only one immunoglobulin. This molecule has a combining site which is sterically complementary to one antigen, but would probably bind also to antigens structurally similar to the major one. Although each cell is restricted in the range of antibodies that it can produce, the population expresses the potential to form antibodies of a great variety because each cell differs from other cells in the antibody that it expresses. Clonal restriction of lymphocytes is a spontaneous process that does not require the presence of antigen. A variety of possible mechanisms, including somatic mutation, somatic recombination of inherited genes, and differential expression of germ line genes might account for clonal selection. These will be discussed in more detail later.

The second major aspect of the clonal selection theory is that each cell expresses its particular antibody on its surface and that this molecule serves as its receptor for antigen. Contact of a cell receptor with an antigen for which it exhibits steric complementarity would initiate division and differentiation of the cell into one that was actively secreting antibody. This concept is illustrated in Figure 2 where cells 2 and 1024 exhibit receptors which can bind to antigen x, albeit with different affinity constants. These cells then proliferate, forming clones of progeny that secrete antibodies to this antigen. The other cells are not activated by this antigen because their immune receptors do not combine with it.

The cellular basis of mammalian immunity has recently become complicated by the demonstration that two distinct types of lymphocytes

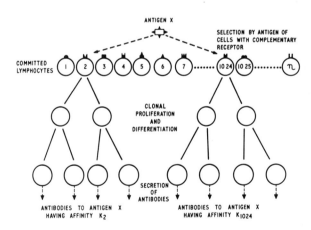

Fig. 2. Diagram of the cellular events involved in the clonal selection of specifically reactive lymphocytes by antigen.

function in immune responses (Claman and Chaperon 1969; Miller and Mitchell 1969). Figure 3 presents a schematic diagram depicting the distribution of lymphocytes in the body and their respective immunological roles. Hemopoietic stem cells arise in the yolk sac of the embryo and localize in the bone marrow. In the adult mammal, stem cells ancestral to lymphocytes are generated in the bone marrow and migrate to so-called primary lymphoid organs, such as the thymus or the bursa of Fabricius in birds or its equivalent in mammals, where they differentiate respectively into thymus-derived lymphocytes (T cells) and bone marrow-derived lymphocytes (B cells). Both cell types possess specific receptors for antigen, but their reactions to antigen are quite distinct. Present evidence indicates that both T cells and B cells are clonally restricted in the specificity of the receptors which they express on their surfaces.

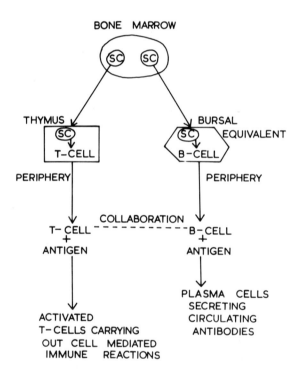

Fig. 3. Diagram of the origins and functions of thymus-derived lymphocytes (T cells) and bone-marrow-derived lymphocytes (B cells). SC signifies lymphopoietic stem cells. Since this pattern depicts the situation in mammals, the source of B cells is designated the "bursal equivalent." In chickens, B cells are generated within a discrete lymphoid organ termed the bursa of Fabricius.

T cells do not synthesize easily detectable amounts of antibody but divide after contact with antigen and differentiate into cells capable of destroying tumors or allografts (killer cells), or into cells able to collaborate with B cells in the elaboration of antibody formation to antigens such as erythrocytes or hapten-carrier complexes (helper cells). T cells are the lymphocytes responsible for the mediation of cellular immune reactions. Following stimulation by antigen, such cells do not differentiate into plasma cells, but become large lymphocytes.

B cells represent the line that differentiates into plasma cells, which are the major producers of circulating antibody. B cells can differentiate into antibody-forming plasma cells after contact with so-called thymus independent antigens, such as the polymerized flagellin of *Salmonella adelaide* bacteria. Thymus-independent antigens are usually large polymeric structures, which can interact directly with the B lymphocyte. The B lymphocyte itself is capable of making small amounts of antibody, but the plasma cell is a much more efficient factory for the production of immunoglobulin. Recent evidence has shown that the *in vitro* reaction of mouse B lymphocytes to flagellin polymer requires neither macrophages nor thymus-bone marrow interactions (Diener, O'Callaghan, and Kraft 1971). Antibody formation to other antigens, however, is a much more complex process. Figure 4 illustrates possible mechanisms for responses to thymus-dependent antigens, such as dinitrophenyl-albumin, which require the obligatory presence of both macrophages and T lymphocytes in addition to B lymphocytes. This figure illustrates the complexity of the interactions and the difficulty of determining a single mechanism to explain collaborative interactions among these three cell types. The general scheme represented here includes both antigen specific and nonspecific factors produced by T cells which influence the activation and proliferation of B cells. Evidence presently exists for both types of factors (Feldmann and Nossal 1972; Katz and Benacerraf 1972; Gorczinsky, Miller, and Phillips 1973). On the basis of the involvement of these cell types, antigen, one antigen-specific factor, and one nonspecific factor, at least 720 or 6! simple models are possible.

The diagram depicts a possible model for cooperation between T cells and B cells in which a specific collaborative factor is required. Since antibodies show exact specificity for antigen, a minimal hypothesis would be that T cell immunoglobulin functions as the collaborative factor in this series of cellular interactions. In the mechanism illustrated here the macrophage may serve two roles: it can phagocytize and process antigens so that immunogenic determinants are readily exposed (this phase may be necessary when the antigen is a large particle such as a foreign erythrocyte) or the macrophage can serve a direct role in cell collaboration by concentrating released T cell receptor immunoglobulin or com-

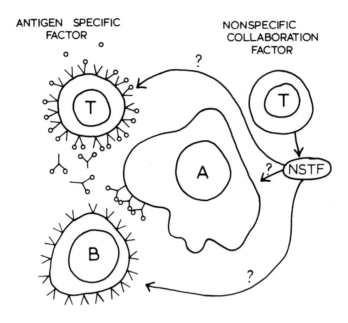

ANTIGEN SPECIFIC
FACTOR

NONSPECIFIC
COLLABORATION
FACTOR

Fig. 4. Schematic illustration of possible modes of collaboration among T cells, B cells, and macrophages in the elaboration of an immune response. The left side of the diagram represents antigen-specific cooperation, which is mediated by complexes of antigen and T-cell receptor immunoglobulins that are concentrated on the macrophage (A) surface. The right side shows possible collaborations involving a T-cell factor which does not show specificity for antigen (NSTF) but induces B cells to proliferate and differentiate. Such nonspecific factors might affect any of the cells involved, and the depiction of only one nonspecific factor here is purely arbitrary. As illustrated in Fig. *53* in Chapter 11, direct contact between specific T and B cells has also been postulated to account for some collaborative effects.

plexes of T cell immunoglobulin antigen upon its surface. These receptor molecules combined with antigen might present the antigenic determinants to specific B lymphocytes as an array comparable to the polymeric state characteristic of thymus-independent antigens (Feldmann and Nossal 1972). Other mechanisms of interaction between T cells and B cells are possible. One mechanism, which has been the subject of considerable attention, is the proposal that T lymphocytes concentrate antigen on their surfaces and then interact directly with B cells to trigger the latter into division and differentiation (Mitchison, Rajewsky, and Taylor 1970). This mechanism will be considered in Chapter 11, where data obtained from studies of the immune responses of lower vertebrates to hapten-carrier complexes will be analyzed. Another possibility is that activated T cells secrete some molecule distinct from immuno-globulin, which is not specific for antigen, but acts as a stimulator of B cells (Katz and Benacerraf 1972). These models are not mutually exclusive. I would emphasize that collaboration between these two cell types is an established fact but that the precise mechanisms of the inter-action remain to be determined. The response to thymus-dependent antigens in the whole animal might in fact consist of an orchestration of the sorts of mechanisms described here.

The cellular schemes for antibody production illustrated above are presented as these processes might occur under *in vitro* cell culture conditions. The mechanisms of antigen processing and antibody forma-tion are more involved in the whole animal because the macro-phages, lymphocytes, and plasma cells are localized in organs of the lymphatic system. Furthermore, the particular cell types are restricted to certain structural areas within the lymphoid organs. The spleen and lymph nodes are the major antibody-producing organs in mammals, and the thymus is necessary in early development to potentiate the immuno-logical capacities of the other lymphoid organs (Good and Papermaster 1964; Miller and Mitchell 1969). Structurally, these organs are divided into a cortical region, which is rich in lymphocytes, and a medullary region containing large numbers of macrophages. The sequence for antibody production in the intact animal proceeds as follows: injected antigen is taken up by macrophages in the nearest lymph node draining the region of entry and partially degraded antigen is transferred to antigen-reactive lymphoid cells within the lymph node. Following the events of T- and B-cell cooperation in the case of thymus-dependent antigens or direct contact with antigen in the case of thymus-independent antigens, the B lymphocytes which have been stimulated with antigen then undergo the process of division and differentiation into plasma cells. The plasma cells are restricted to lymphoid organs in normal ani-mals, but lymphocytes are free to wander throughout the blood stream and tissues. Specific patterns of retention of antigen in lymphoid tissues and their possible relationships to the induction of immunity and toler-ance will be considered in a later chapter.

## Cellular differentiation in antibody formation

Antibody production is a process which involves specific activation of cells. A small lymphocyte is a resting cell, having a diameter of less than 10 $\mu$. Approximately 90-95% of its volume is occupied by the nucleus. The cell possesses very little in the way of specialized cytoplasmic structures such as endoplasmic reticulum or Golgi apparatus. It is activated by contact with antigen to become a large blast cell through a process involving nucleic acid synthesis, protein synthesis, and, in addition, processes of activation which may be common to all differentiation (Hirschhorn 1967). Very little is known about the sequence of events following contact of antigen with the cell surface which brings about the specific activation of immunoglobulin genes. Present data indicate that the receptor on the lymphocyte is antibody (Greaves and Hogg 1971; Marchalonis and Cone 1973a; Warner 1974) and that contact with the antigen triggers differentiation and antibody production. In general, activation of cells by contact with chemical inducers is a problem not only of immunology but of many areas of cell biology.

The process of activation is followed by one of differentiation. This process includes morphological differentiation as illustrated in the electron micrographs shown in Figures 5, 6, and 7. The activated cell becomes larger and develops new cytoplasmic structures. By the plasma cell stage (Figure 7), huge numbers of polysomes are present in the form of rough endoplasmic reticulum, and a prominent Golgi apparatus has appeared. In short, a resting cell has differentiated into a cell which has the appearance of one specialized for the synthesis and secretion of protein. This pattern of development obtains for the B-lymphocyte line. Small lymphocytes of the T-cell series show morphological characteristics very similar to the small lymphocyte shown here (Figure 5). After activation, the T lymphocytes progress to cells resembling the large lymphocyte or blast illustrated in Figure 6. These cells do not differentiate into antibody-secreting plasma cells. It is clear that the antigen-directed differentiation of lymphocytes is both a morphological and a biochemical process.

## Relevance of mammalian immunity to evolutionary studies

Despite the complexity of the immune response in mammals, it is possible to assign definitive criteria to vertebrate immune phenomena. These consist of inducibility, specificity, and the requirement that the response be mediated by lymphoid cells. Although some would contest this proposition (Crone, Koch, and Simonsen 1972), it is most likely that T lymphocytes as well as B lymphocytes possess receptor molecules consisting of antibodies on their surface (Feldmann 1972; Hogg and Greaves 1972; Warner 1972, 1974; Marchalonis and Cone 1973a; Roelants et al. 1974; Roelants, Forni, and Pernis 1973; Lawrence, Spiegelberg, and Weigle 1973; Ashman and Raff 1973; Goldschneider and Cogen 1973; Moroz and Hahn 1973; Rieber and Riethmuller 1974a). The presence of protein molecules homologous to antibodies of higher vertebrates in the serum or on cells therefore provides an indication of verte-

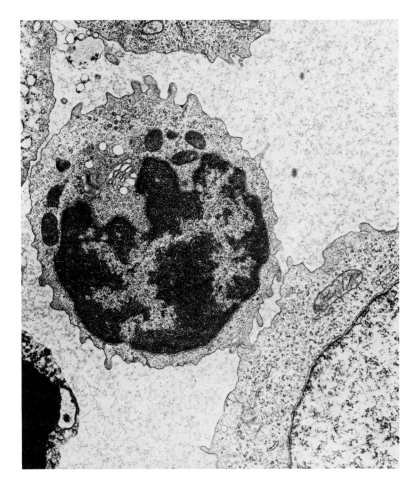

Fig. 5. Electron micrograph of a typical small lymphocyte from the mouse
(x 9,900). Courtesy of Dr. T. Mandel.

brate-type immunity. Other workers, especially Hildemann (1972), have
stressed functional criteria, such as the development of immunological
memory, as prerequisites for immunity. However, situations exist in
mice (Bernacerraf and McDevitt 1972) and toads (Diener and Marchalonis 1970) where a clear-cut primary immune response is generated but
neither positive nor negative memory is demonstrated upon second challenge with antigen. In addition, attempts to delineate precisely the origins of vertebrate-type immunity face a serious difficulty in definition
because some cellular immune (T cell) reactions of mammals have been
classed as quasi-immunological (Hildemann 1974). Phenomena such as

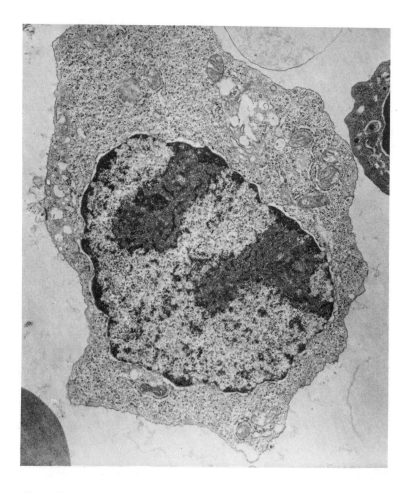

Fig. 6. Large lymphocyte or lymphoblast of the mouse. Note the presence of clusters of polysomes and the presence of small amounts of rough endoplasmic reticulum (x 4,100). Photograph courtesy of Dr. T. Mandel.

the mixed-lymphocyte reaction, in which lymphocytes bearing certain receptors proliferate specifically in response to lymphocytes carrying a particular surface marker, may not involve immunoglobulin receptors (Crone, Koch, and Simonsen 1972). Moreover the degree of proliferation in these responses is inconsistent with present concepts of the clonal restriction of immunologically competent cells (see Chapter 13), and immune-type memory may not exist in such reactions (Hildemann and Reddy 1973). Until the molecular nature of the recognition systems involved in quasi-immunological reactions of mammals is known, the comparison with potentially similar systems in invertebrates is difficult.

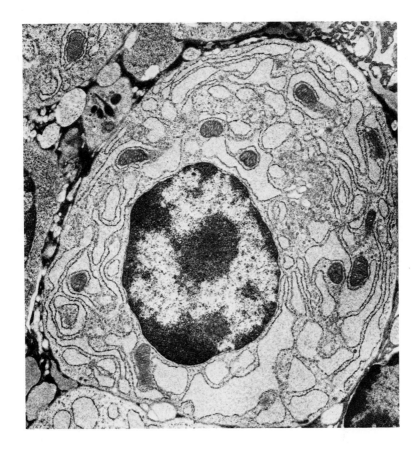

Fig. 7. Typical plasma cell of the mouse. Note large amounts of rough endoplasmic reticulum and the prominent Golgi apparatus (x 10,300).
Courtesy of Dr. T. Mandel.

In terms of the information considered thus far, we can pose three
questions to serve as a guide in analysis of the evolution of immunity:
(1) When did lymphocytes and plasma cells emerge phylogenetically?
(2) What are the minimal structural requirements of the lymphoid system
which will allow cellular and humoral immunity to function?
(3) When does the distinction between T and B lymphocytes arise?
A corollary question might also be asked; namely, do phylogenetic
studies provide any insight into the structure, function and evolutionary
emergence of T- and B-cell receptors for antigen?

These questions appear direct and straightforward, but I must emphasize that attempts to recognize homology of function in comparing
cells of different species contain an element of uncertainty. The classifi-

cation of cells as either lymphocytes or plasma cells is reasonably clear-cut when they possess morphological properties characteristic of such cells in mammals and express or secrete immunoglobulins, respectively. These criteria can be best applied to lymphocytes of the B-lymphocyte series. The origin of T cells is a more complicated problem which must be based upon largely uncharacterized functions of such cells in responses taken to be thymus-dependent in mammals. The interpretation of possible homologies of lymphoid structures which appear in primitive vertebrates is also unclear, but partial answers to these guideline questions now exist and will be discussed below.

Having considered some of the basic chemical and cellular aspects of antibody formation, I will next consider a more detailed description of the structures of human antibodies. This information is required in order to provide a basis for comparison when studies of antibodies of lower vertebrates and presumptive antibody precursors of invertebrates are analyzed.

# 2 Structure of Mammalian Immunoglobulins:
## Basis for
## Phylogenetic Comparisons

*The structure of antibodies*

Antibodies belong to a group of serum proteins termed immunoglobulins. Among the definitive characteristics of immunoglobulins are the possession of antibody activity, serological relatedness, and the solubility and electrophoretic properties of $\gamma$-globulins. Serological relatedness means that if one type of human immunoglobulin, IgG for example, is used as an antigen to immunize rabbits, the antibody which is induced will cross-react with human IgM as well as the other human immunoglobulin classes. The reasons for this will be considered below. Molecules which are structurally and/or antigenically related to immunoglobulins often appear in serum of patients suffering from certain diseases. Although such proteins might be incapable of binding known antigens, they must be considered part of the immunoglobulin family of molecules (see Chapter 16). Beginning in 1959 (Edelman 1959; Porter 1959), immunoglobulin structure has been investigated with the incisive techniques of protein chemistry, and remarkable progress has been made in this area. The complete covalent structures of a human IgG myeloma immunoglobulin (Edelman and Gall 1969; Edelman 1970) and an IgM molecule (Putnam et al. 1973) have been determined. Since studies of the structure of human and murine immunoglobulins have recently been reviewed in great detail (Edelman and Gall 1969; Gally and Edelman 1972; Natvig and Kunkel 1973), I shall limit my comments to a brief consideration of those aspects of direct relevance to evolutionary comparisons.

The major structural properties of mammalian immunoglobulins are illustrated in Figure 8, which presents a diagrammatic representation of human IgG or 7S immunoglobulin. These proteins are composed of two types of polypeptide chains, termed light chains and heavy chains, which are usually linked covalently via disulfide bonds. The basic pattern of immunoglobulin structure is a unit consisting of two identical pairs of light and heavy chains. Each light chain is paired with a heavy chain, and the combining site for antigen is formed by interaction of the two chains. The combining site consists of a cavity 15Å wide × 12Å deep × 20Å long in which the hypervariable regions of the light chain and the heavy chain contribute contact residues (Padlan et al. 1974).

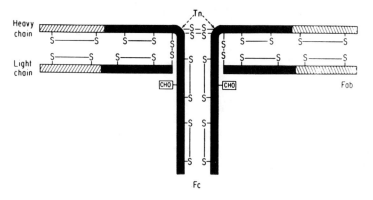

Fig. 8. Schematic representation of the IgG immunoglobulin molecule. Variable regions are indicated by the striped pattern. Constant regions are solid black. S-S indicates the positions of disulfide bonds; Tn, site of cleavage by trypsin to form Fc and Fab pieces; CHO, site of attachment of carbohydrate to the gamma chain. (From Gally and Edelman 1970).

Immunoglobulins have been resolved into their constituent polypeptide chains by reduction to cleave interchain disulfide bonds followed by alkylation to protect the newly formed sulfhydryl groups. The chains can then be separated by gel filtration (Fleischman, Pain, and Porter 1962) or ion exchange chromatography (Edelman and Poulik 1961) in dissociating solvents, which are required to disrupt the noncovalent interactions that stabilize the union of light chains and heavy chains. The molecules have also been subjected to limited treatment with proteolytic enzymes, such as papain, which cleave the molecule into fragments (Porter 1959). One fragment, the Fab, retains the capacity to bind antigen. The other, the Fc, can be readily crystallized and represents the portion of the heavy chain to which complement components bind. These fractionation patterns are illustrated in Figure 9. This scheme will prove

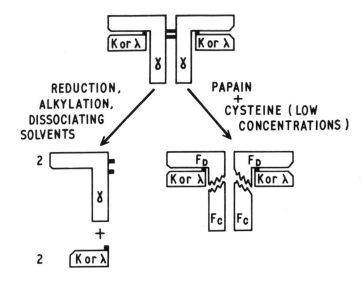

Fig. 9. Simplified diagram illustrating the fragmentation of IgG immunoglobulin by cleavage of disulfide bonds (reduction) or by proteolytic
enzymes (papain). The location(s) of inter-heavy-chain disulfide bonds
after proteolysis in the presence of cysteine is not shown. The enzy-
matically-produced fragment containing the light chain and a piece of
the gamma chain (Fd fragment) is termed the Fab fragment. It contains
the combining site for antigen.

useful in subsequent chapters which describe the isolation of polypeptide
chains of immunoglobulins of nonmammalian species.

The properties described so far are those of the polypeptide portions
of immunoglobulins. Although light chains usually do not contain
oligosaccharide units, heavy chains are glycoproteins, which may contain
over 10% carbohydrate in the case of IgM molecules. The carbohydrate
moiety associated with immunoglobulins consists of fucose, mannose,
galactose, glucosamine and sialic acid. Its function is unknown, but it
apparently plays no role in combining with antigens.

## Immunoglobulin classes

Immunoglobulins within a mammalian species constitute a family
of proteins. As illustrated in Table 1, man possesses five types of serum
immunoglobulins that are serologically related because they all share the
same type of light chains. Two antigenic types of light chains, which also
possess characteristic amino acid sequences, are found in mammals.
These have been termed κ and λ chains. Immunoglobulin classes are

TABLE 1: Some properties of immunoglobulin classes of man.

| Class | Serum conc. (mg/100 ml) | M.W. | Sedimentation coefficient | Cross the placenta? | Light chains[a] |
|-------|-------------------------|------|---------------------------|---------------------|-----------------|
| IgG | 800-1,680 | 150,000 | 6.7S | Yes | $\kappa, \lambda$ |
| IgA | 140-420 | 150,000-400,000 | 7-11S | No | $\kappa, \lambda$ |
| IgM | 50-100 | 950,000 | 18-20S | No | $\kappa, \lambda$ |
| IgD | 3-40 | 180,000 | 6.2-6.8S | No | $\kappa, \lambda$ |
| IgE | 0.01-0.14 | 190,000 | 8.2S | No | $\kappa, \lambda$ |

defined by the presence of class-distinctive heavy chains that differ in a variety of properties, including molecular weight and amino acid sequence. They have been given Greek letter designations, and immunoglobulin composition can be expressed in molecular formulae by analogy with hemoglobulins (World Health Organization 1964). For example, the structural formula of IgG is $\kappa_2\gamma_2$ or $\lambda_2\gamma_2$ whereas that of IgM is usually $(\kappa_2\mu_2)_5$, indicating that five units, of the form 2 light chains—2 heavy chains, are linked to form the intact molecule. The IgM pentamer (M. W. 900,000) readily dissociates to form the $\kappa_2\mu_2$ or $\lambda_2\mu_2$ subunit (M. W. 180,000) upon treatment with relatively low concentrations of reducing agents. The capacity to aggregate in a specific fashion reflects the properties of the particular heavy chains present in the molecule. A simplified planar diagram of the IgM pentamer is shown in Figure 10.

The type of heavy chains also determines which functional properties the immunoglobulin class possesses. IgA and IgE proteins, for example, do not fix complement in the classical sense. IgE, however, can carry out the unique role of fixing to mast cells and initiating anaphylactic reactions. If this IgE binds to antigen, the combination causes the mast cell to lyse, thereby releasing pharmacologically active agents such as histamine, slow reactive substance, serotonin, and bradykinin. The eventual result of this process is an asthmatic reaction (Bennich and Johansson 1971). All types of immunoglobulins can bind antigens.

Certain immunoglobulin classes can be further differentiated into

| Heavy chains | Heavy chain M.W. | Molecular formula[d] | Carbohydrate content (%) | Antibody activity | Complement fixation[e] |
|---|---|---|---|---|---|
| $\gamma$ | 53,000 | $\gamma_2 \kappa_2$ $\gamma_2 \lambda_2$ | 2 | Yes | Yes |
| $a$ | 64,000 $(52,000)$[b] | $(a_2\kappa_2)_n$ $(a_2\lambda_2)_n$ | 6-10 | Yes | No |
| $\mu$ | 70,000 | $(\mu_2\kappa_2)_5$ $(\mu_2\lambda_2)_5$ | 9-12 | Yes | Yes |
| $\delta$ | 60,000 $(70,000)$[c] | $\delta_2\kappa_2$ $\delta_2\lambda_2$ | 12.7[c] | Yes | ? |
| $\varepsilon$ | 72,500 | $\varepsilon_2\kappa_2$ $\varepsilon_2\lambda_2$ | 11.7 | Yes | ? |

Source: Unless stated otherwise, the data presented here were compiled by Edelman and Gall 1969.
a. All light chains have a molecular weight of 22,500.
b. Some murine $a$ chains have a molecular weight of 52,000 (Grey 1969).
c. Leslie, Clem, and Rowe 1971.
d. The integer n associated with IgA can be 1, 2, 3, and higher.
e. Complement fixation refers to the classical pathway.

groupings termed subclasses on the basis of antigenic and structural properties. Four subclasses of human IgG immunoglobulin and two subclasses of IgA and IgM immunoglobulins have been observed. The heavy chains present in the IgG subclasses, for example, represent the products of closely linked genes encoding constant regions that arose recently from a common precursor gene (Gally and Edelman 1972) by a process of tandem duplication. I shall discuss the genetic relationships among mammalian immunoglobulin subclasses at greater length in Chapter 3.

## Heterogeneity of antibodies

The preceding discussion maintained that immunoglobulins are a related family of proteins that can differ in molecular size and other properties. Even within a single class of immunoglobulin, molecules isolated from a single normal individual possess a spectrum of amino acid sequences. This microheterogeneity was originally detected as heterogeneity in electrophoretic properties (Tiselius 1937) and in the as-

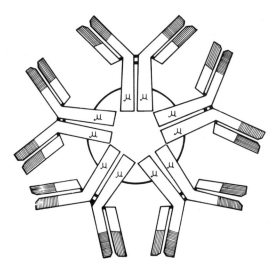

Fig. 10. Planar representation of polymerized (19S) IgM immunoglobulin. Variable regions are indicated by shading. Dark bars represent interchain disulfide bonds. In addition to light chains and $\mu$ chains, polymerized IgM molecules contain a third chain, the J chain, which is implicated in the formation of the pentameric molecule. The precise location of this chain is not known. This diagram does not give exact molecular dimensions.

sociation constant for equilibrium combination with antigen (Landsteiner and van der Scheer 1936). I will consider, first, the heterogeneity in association constant for binding of antibody to hapten. The equation $S + L \rightleftharpoons SL$ represents the binding of sites on a macromolecule (S) with small molecules or ligands (L). SL is the bound complex. At equilibrium, the association constant of this reaction can be written

$$K_A = \frac{(SL)}{(S)(L)}$$

which has the units liters/mole. This constant has been termed affinity, and it represents the binding energy of the union between hapten and antibody. In order to provide some intuitive feeling for measurements of the binding of antibody to antigen, I shall now outline the derivation of a binding equation originally developed by Scatchard (1949) which relates the number of moles of hapten molecules bound per mole of antibody to the ratio of bound hapten to free hapten. Operationally, the measurements to which the Scatchard plot is applied are made by techniques such as equilibrium dialysis in which the macromolecule is confined to one compartment whereas the ligand can diffuse freely throughout the system. Figure 11 provides a pictorial basis for this derivation which is a simplified version of Langmuir's (1916) original treatment of absorption of gases to solids.

Let n = the total number of sites on the macromolecules which are specific for hapten, h = amount of hapten bound, and c = concentration of free hapten. Then (n-h) = the number of free sites on the macromolecule. For the sake of convenience, express these quantities in moles/liter. The rate of binding is proportional to the number of sites available on the macromolecules and to the concentration of free hapten. This expression can be written

$$\text{rate}_b = k_b(n-h)c$$

where $k_b$ is the rate constant for binding. Similarly, the rate of release is proportional to the number of hapten molecules bound. This rate can be expressed as

$$\text{rate}_r = k_r h$$

where $k_r$ is the rate constant for release of hapten into solution.

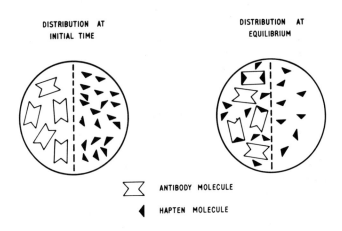

DISTRIBUTION  AT
INITIAL TIME

DISTRIBUTION  AT
EQUILIBRIUM

ANTIBODY  MOLECULE

HAPTEN  MOLECULE

Fig. 11. Schematic representation of haptens and antibody during an equilibrium dialysis experiment. The barrier (dotted line) is impermeable to antibody molecules, but hapten molecules can distribute freely across it.

At equilibrium,    $rate_b = rate_r$
Hence,    $k_b(n-h)c = k_r h$
Define    $K_A = k_b / k_r$
Now,    $h = K_A (n-h)c$

Rearrange to Scatchard from    $\dfrac{h}{c} = K_A n - K_A h$

An idealized plot of this equation is shown in Figure 12. The graph is a straight line with slope of $-K_A$ and x-intercept n; i.e., when $h/c = 0$, $h = n$. For antibodies of the IgG class, n = 2 and $K_A$ is the range of $10^5$ to $10^9$ liters/mole. The graph illustrated by line (A) in Figure 12 might apply for binding between a myeloma protein which is a homogenous immunoglobulin molecule and the hapten dinitrophenol (Eisen, Simms, and Potter 1968; Goetzl and Metzger 1970). In the case of antibodies produced by a normal animal, however, the pattern is quite different. The line representing this situation (B) is not straight in Figure 12 but exhibits considerable curvature because of the presence of antibodies of differing affinities in the population of purified antibodies. Mathematical techniques exist for straightening out such curves, and the magnitude of the factor required for this purpose gives an estimate of the degree of heterogeneity of binding affinities within a given antibody pool (Karush 1962).

Thus, antibody molecules to a given antigen represent a pool of molecules possessing slightly different combining sites. Heterogeneity of antibody molecules is also readily apparent if an immunoglobulin preparation from normal mammals is analyzed by electrophoretic or isoelectric focusing methods capable of resolving differences in electric charge. Figure 13A presents electrophoretic patterns obtained by subjecting normal human IgG immunoglobulin and a human myeloma protein to zone electrophoresis on cellulose acetate. The normal immunoglobulin migrates as a diffuse smear, whereas the myeloma protein gives a single sharp band indicative of limited charge heterogeneity. The discovery that proteins produced by men and mice suffering from multiple myeloma, a cancer of the lymphatic system, were homogeneous immunoglobulins (Waldenström 1952; Putnam 1957) facilitated attempts to elucidate the amino acid sequences of immunoglobulins. Charge heterogeneity is further reflected in Figure 13B, which depicts analysis of the polypeptide chains of reduced and alkylated normal and myeloma proteins by disc electrophoresis on polyacrylamide gels. Reduction and alkylation were performed to cleave disulfide bonds and stabilize the newly formed sulfhydryl groups, and the gel buffer contained 9M urea and acetic acid to disrupt hydrophobic interactions. Under these conditions, the light chains of normal immunoglobulin appear as a

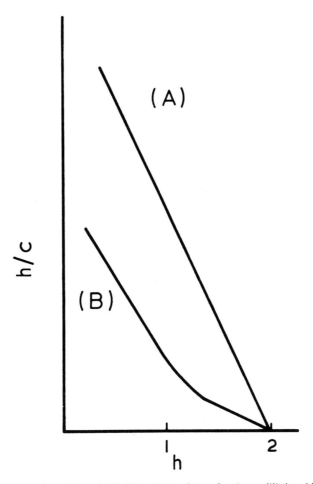

Fig. 12. Analysis by the method of Scatchard of data for the equilibrium binding of haptens to IgG antibodies. h is the number of moles of hapten
bound. C is the number of moles of free hapten. Operationally, a
number of cells such as those shown in Fig. 12 are set up with constant
amounts of antibody but the amount of hapten is varied. (A), an ideal
straight line which indicates uniformity of association constant. (B),
the usual result which shows pronounced curvature because of heterogeneity of antibodies with respect to association constant.

diffuse smear and the light chain of the myeloma protein gives a sharp
line. Although these conditions do not provide information on the charge
heterogeneity of heavy chains, they enable the resolution of heavy chains
of distinct classes. Origin 1 is normal human IgG immunoglobulin which
possesses the γ chain; origin 2 is a human IgM myeloma protein; and

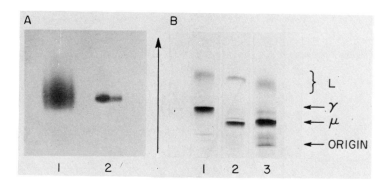

Fig. 13. Heterogeneity of immunoglobulins reflected in dispersity of electro-phoretic patterns. (A) Comparison of normal human IgG immuno-globulin (1) with a myeloma immunoglobulin (2) by electrophoresis on cellulose acetate membrane. Notice that the mobility of the normal immunoglobulin is very diffuse relative to that of homogeneous my-eloma protein. (B) Analysis of immunoglobulin polypeptide chains by polyacrylamide gel electrophoresis in acid urea. Under these conditions the light chains of normal human IgG immunoglobulin (1) migrate as a diffuse zone while that of an IgM myeloma protein (2) is a single sharp band. Normal goldfish IgM immunoglobulin (3) has a heavy chain similar to the human $\mu$ chain but has a diffuse envelope of light chains comparable to normal human IgG immunoglobulin. $\mu$, direc-tion of migration; L, $\gamma$, $\mu$, indicate the positions of light chains, $\gamma$ chains and $\mu$ chains respectively.

origin 3 contains normal IgM immunoglobulin of the goldfish. The $\gamma$ chains and $\mu$ chains are clearly separated by this technique which provided a quick, simple screening procedure to identify the types of heavy chains present in immunoglobulins of lower species. Charge heter-ogeneity of immunoglobulin chains is emphasized by Figure 14, which presents an idealized representation of isoelectric focusing patterns of normal (pooled) mouse IgG immunoglobulin and antibody produced by a single clone of antibody-forming cells. In this technique, proteins migrate through a gradient of pH, maintained by ampholines, until they find the region of their own isoelectric point. At this pH, their net charge becomes zero and their migration stops. Pooled IgG immunoglobulin exhibits a disperse pattern with the bulk of components distributed between pH 5 and pH 7. The monoclonal antibody exhibits considerable restriction in charge. More than one band is present (Williamson 1971) because of slight charge heterogeneity probably due to amidation of side chain carboxyl groups.

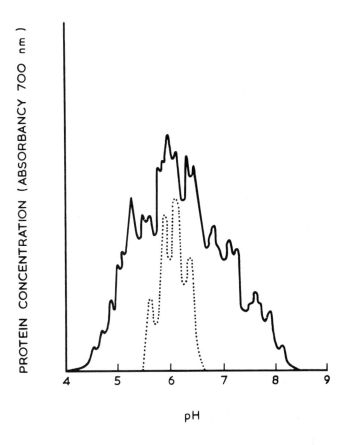

Fig. 14. Isoelectric focusing patterns of normal human IgG immunoglobulin and a homogeneous myeloma protein. This diagram represents the densitometric tracing at 700 nm of a stained pattern.

## Amino acid sequence analysis: Variable and constant regions

Hilschmann and Craig (1965) provided the first sequence data for immunoglobulins by elucidating the primary structure of light chains of human myeloma proteins. They used so-called Bence-Jones proteins which consist purely of homogeneous light chains (Edelman and Gally 1962), which are excreted in the urine of myeloma patients. These workers showed that light chains were composed of two distinct regions on the basis of amino acid sequence data. The κ light chains consist of

*27*

approximately 214 amino acid residues, with the amino-terminal 107 residues showing pronounced variability when κ chains from different myeloma patients were compared. The carboxy-terminal 107 residues, in contrast, were essentially invariant from patient to patient except for a leucine-valine interchange which occurred at residue number 191. This interchange was correlated with a genetic marker termed the *Inv allo-type* which was associated with the antigenic properties of κ light chains. The amino-terminal region has been designated the V, or variable, region and the invariant carboxy-terminal portion of the molecule is known as the C region. Allotypic distinctions result from relatively small amino acid interchanges in the constant regions of light and heavy chains. Such allotypic markers distribute in a Mendelian genetic manner within a breeding population.

Sequence analysis of human and murine immunoglobulins has established that approximately the first 110 amino acids from the amino terminus of both light chains and heavy chains are variable. The constant region in the light chain is also of this size. Edelman and Gall (1969) showed that the constant region of the myeloma protein Eu, which represented human IgG subclass IgG1, was comprised of three homologous regions each consisting of about 110 residues. This observation is illustrated in Figure 15. The $\gamma$1 chain is represented as a string of four discrete regions of similar size. The $NH_2$-terminal one is the V region. The remaining three are called $C_\gamma1$, $C_\gamma2$ and $C_\gamma3$. Although these regions are not identical, they show sufficient homology to each other to suggest that the gene specifying the total C region of the $\gamma$ chain probably arose by tandem duplication of a gene encoding a polypeptide of 110 amino acids. These regions constitute functionally distinct domains; for example $C_\gamma1$ combines with the $C_\kappa$ region of the light chain to form the Fab fragment and $C_\gamma2$ fixes complement. The internal homologous regions of the $\gamma$ chain show clear homology to the $C_\kappa$ region of the light chain. Although the V and C regions are homologous, the correlation is not as strong as that for the comparisons among C regions. Recent work by Putnam's group established that the $\mu$ chain consists of five domains (Putnam et al. 1972, 1973). The implications of these findings for the evolution of immunoglobulins will be discussed after the polypeptide chain structure of immunoglobulins of lower vertebrates has been analyzed.

The type of variation in amino acid sequence observed when variable regions of immunoglobulins are compared with each other is illustrated in Table 2, which summarizes data for a number of human κ chains. The problem of heterogeneity that must be faced in sequence analysis of normal immunoglobulin pools is evident in the results shown for the human pool. Ten of eighteen residues included here show interchanges

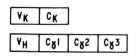

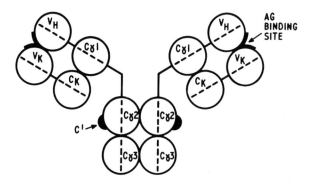

Fig. 15. Schematic diagrams of the IgG immunoglobulin molecule illustrating the domain (internal homology regions) arrangement within light chains and heavy chains. Upper figure, linear arrangement of homology regions within κ chain and γ chain. Lower figure, domains represented as compact structural regions linked by relatively flexible lengths of polypeptide chain. The positions of the antigen combining sites and the site for complement fixation (C′) are indicated.

of at least two amino acids. Individual myeloma light chains have unique sequences, and it appears that the normal pool can be considered an average of the various sequences of the individual κ chains. The sequences shown here represent the pattern typical of V-region diversity except for three regions of hypervariable sequence which do not fall within the residue positions listed. These areas occur from residues 24 to 34, 50 to 56, and 89 to 97 and are immediately COOH-terminal to the cysteines (cys 23 and cys 88), which form an interchain disulfide bridge in the V region. A diagram of amino acid sequence variability in human κ chains is given in Figure 16. Wu and Kabat (1970) performed a statistical study of the frequency of interchanges at each position, finding that certain regions show marked hypervariability. They predicted that such areas were probably involved in providing amino acids that determine the binding specificity of the antibodies. Their theoretical conclusions regarding the importance of hypervariable regions in combination with different antigens were confirmed in the experimental studies (Cebra et al. 1971; Goetzl and Metzger 1970; Haimovitch et al. 1972) which established that amino acid residues capable of binding affinity-labelling reagents mimicking haptens were confined to the three segments of

TABLE 2: N-terminal amino acid sequences of human $\kappa$ light chains.

| Protein | \multicolumn{9}{c}{Amino acid} | | | | | | | | |
|---|---|---|---|---|---|---|---|---|---|
|  | 1 | 2 | 3 | 4 | 5 | 6 | 7 | 8 | 9 |
| $V_\kappa I$ |  |  |  |  |  |  |  |  |  |
| ROY | Asp | Ile | Gln | Met | Thr | Gln | Ser | Pro | Ser |
| BJ | - | Val | - | - | - | - | - | - | - |
| CRA | - | - | - | - | - | - | - | - | - |
| HBJ 1 | - | - | Leu | - | - | - | - | - | Thr |
| PAP | - | - | - | - | - | - | - | - | - |
| LUX | - | - | - | Leu | - | - | - | - | Phe |
| EU | - | - | - | - | - | - | - | - | - |
| $V_\kappa II$ | Gly | Ile | Val | Leu | Thr | Gln | Ser | Pro | Gly |
| Nig | Lys | - | - | - | - | - | - | - | - |
| Rad | - | - | - | - | - | - | - | - | - |
| HBJ 5 | - | - | - | - | - | - | - | - | Asx |
| Win | Asp | - | - | - | - | - | - | - | Ala |
| Dob | - | - | Ile | Met | - | - | - | - | Ala |
| $V_\kappa III$ | Asp | Ile | Val | Met | Thr | Gln | Ser | Pro | Leu |
| Mil | - | - | - | Leu | - | - | - | - | - |
| Cum | Glu | - | - | - | - | - | Thr | - | - |
| Human | Asp | Ile | Glu | Met | Thr | Gln | Ser | Pro | Ser |
| Pool | Glu |  | Val | Leu |  |  |  |  | Gly |
|  |  |  |  |  |  |  |  |  | Ala |
|  |  |  |  |  |  |  |  |  | Val |
|  |  |  |  |  |  |  |  |  | Leu |
|  |  |  |  |  |  |  |  |  | Pro |

variable regions that possess hypervariable sequences (Givol et al. 1971; Press, Fleet, and Fisher 1971). This result obtained whether the studies were carried out using antibodies to a particular hapten, e.g., DNP, or myeloma proteins that exhibited high affinity binding capacity for the hapten. The experiments of Cebra and his colleagues (1971) will be discussed below.

Two salient features of V-region diversity have emerged from studies of the N-terminal sequences of immunoglobulin chains. First, the majority of amino acid interchanges represent single base changes in the nucleotide triplets coding for the particular residues. For example, isoleucine is coded by the triplet AUU whereas the codon for valine is GUU. The second point is that V regions can be factored into subgroups (see Table 2) which are probably encoded by separate germ.-line genes (Hood and Talmage 1970). The prototype sequences of $V_\kappa I$, $V_\kappa II$ and $V_\kappa III$, as given in the table, are variants that were observed. The $\lambda$ light

| residue position | | | | | | | | | |
|---|---|---|---|---|---|---|---|---|---|
| 10 | 11 | 12 | 13 | 14 | 15 | 16 | 17 | 18 | 19 |
| Ser | Leu | Ser | Ala | Ser | Val | Gly | Asp | Arg | Thr |
| - | - | - | - | - | - | - | - | - | - |
| - | - | - | - | - | Leu | Arg | - | - | - |
| - | - | - | - | - | - | - | ( ) | - | - |
| - | - | - | Val | - | - | - | - | - | - |
| - | - | - | - | - | - | - | - | - | - |
| Thr | - | - | - | - | - | - | - | - | - |
| Thr | Leu | Ser | Leu | Ser | Pro | Gly | Gly | Arg | Ala |
| - | - | - | - | - | - | - | - | - | - |
| - | - | - | - | - | - | - | Asp | - | - |
| - | - | - | - | - | - | - | - | ( ) | - |
| - | - | - | - | - | - | - | - | - | - |
| - | - | - | - | - | - | - | - | - | - |
| Ser | Leu | Pro | Val | Thr | Pro | Gly | Glu | Pro | Ala |
| - | - | - | - | - | - | - | - | - | - |
| - | - | - | - | - | - | - | - | - | - |
| Ser | Leu | Ser | Ala | Ser | Val | Gly | Asp | Arg | |
| Thr | Val | | Val | Thr | Pro | | Gly | | |
| | | | Leu | Val | Leu | | | | |

chains, which possess blocked $NH_2$-terminal groups consisting of pyrrolidone carboxylic acid, can be factored into five V subgroups (World Health Organization 1969b). In addition, the variable regions of heavy chains can be treated similarly. Present sequence data indicate that three types of variable regions exist; namely $V_\kappa$, $V_\lambda$ and $V_H$, each of which contain subgroups. $V_H$ is apparently shared by all heavy chain classes (Köhler et al. 1970; Wang et al. 1970; Levin et al. 1971).

The discovery that immunoglobulin molecules of a given class differed from one another in amino acid sequence within definite regions of the chains suggested that the differences associated with antigen binding specificity must reside within these variable regions. Cebra et al. (1971; Ray and Cebra 1972) have recently used affinity labels to provide evidence that the combining site of guinea pig antibodies to the dinitrophenol hapten (DNP) is formed from the variable regions of the light and heavy chains. Furthermore, they have obtained a partial sequence

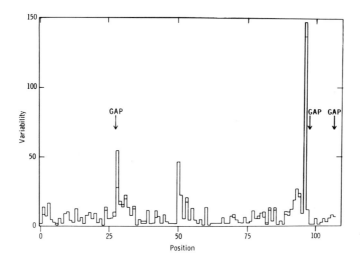

Fig. 16. Diagram illustrating the frequency of variation of amino acid sequence within human myeloma light chains plotted as a function of residue position. The graph plots variability versus residue position, where variability is defined as the number of different amino acids at a given position divided by the frequency of the most common amino acid observed in that position. A high frequency of amino acid interchanges occurred within three regions termed hypervariable regions. These corresponded to residues 24-34, 50-56 and 89-97. (From Wu and Kabat 1970).

of this portion of the guinea pig $\gamma_2$ heavy chain. Since the logic of their approach illustrates the underlying theme of much recent research, and their results suggest the arrangement of site region amino acids in antibodies and antigen-binding myeloma proteins of diverse mammalian species, I shall discuss the work of Cebra and his colleagues in some detail.

Basically, these workers reacted the pool of antibody elicited by injection of DNP with $^{14}$C-m-nitrobenzene diazonium tetrafluoroborate. This affinity label is chemically similar to DNP but is capable of forming a covalent bond to tyrosine. The molecule would thus mimic the hapten DNP and lodge within the antigen combining site of antibodies directed to this moiety, where it would become covalently linked to a nearby tyrosyl residue. This reaction is specific in that it can be inhibited by the presence of excess hapten, which competes for the active site. Moreover, normal immunoglobulin showed little propensity to take up the affinity label. The specifically labelled immunoglobulin was separated into light chains and $\gamma$ chains. The $\gamma$ chain was then cleaved by cyanogen bromide, which splits protein chains specifically at methionine residues (Gross and Witkop 1961). Such treatment usually produces a manageable number of fragments because the number of methionine residues in proteins is generally small. In this case, eight peptide fragments were obtained and arranged in order (Figure 17a). As Figure 17b illustrates, the affinity label was associated with tyrosyl residues in the V region. In general, two tryosyl residues were labeled, one at residue number 32, and the other in a region between residues number 98 and 120 which was so variable that it has not yet been specifically located. These studies and those of Goetzel and Metzger (1970) and Haimovich et al. (1972), who used myeloma proteins possessing high binding affinities for DNP, support the concept that V-region antibody sequence heterogeneity is correlated with antigen specificity. Both groups of workers found that a tyrosyl residue at position N-34 in the light chain of the protein MOPC 315 was specifically modified by affinity labeling reagents. Furthermore, Haimovich and his colleagues found that the heavy chain of this protein and that of MOPC 460 were each modified at lysyl residues at position 54. By comparison with Figure 16, it is evident that these labeled residues are confined to the hypervariable stretches within the variable regions.

Amino acid sequence data obtained for polypeptide chains of antibodies and myeloma proteins thus established that $\kappa$ chains, $\lambda$ chains, and heavy chains possess regions of invariant sequence and regions which show marked variability. The sequence variability was predominantly confined to approximately the N-terminal 110 residues of each chain, and particular sequences were found to be correlated with the recognition of distinct antigens. The mechanism by which sequence diversification can occur within an individual poses challenging questions that are integral to an understanding of immune function.

## Other Polypeptide chains

Polypeptide chains besides light and heavy chains have been observed to be associated with immunoglobulins. IgA molecules which occur in external secretions such as saliva or colostrum may contain a secretory component characterized by mass of 60,000 daltons and antigenically distinct from the light and $\alpha$ chains (O'Daly and Cebra 1971). Moreover, polymeric IgA immunoglobulins and 19S IgM immunoglobulins possess a J or joining chain having a molecular weight of approximately 20,000 (Halpern and Koshland 1970; Mestecky, Zikan, and Butler 1971). This chain is involved in determining the efficiency of polymerization of IgA or IgM monomer units into higher aggregates (Wilde and Koshland 1972; Chapuis and Koshland 1974) and probably

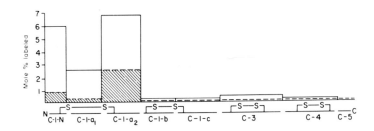

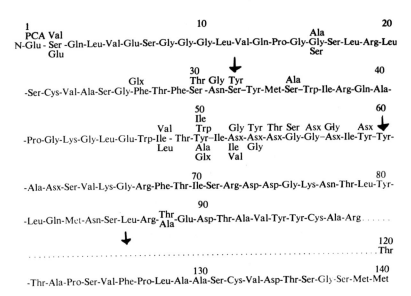

1                                        10                                    Ala                    20
PCA Val
N-Glu - Ser -Gln-Leu-Val-Glu-Ser-Gly-Gly-Gly-Leu-Val-Gln-Pro-Gly-Gly-Ser-Leu-Arg-Leu
Glu                                                                           Ser

                                        30      ↓
                    Glx                  Thr Gly Tyr            Ala                              40
-Ser-Cys-Val-Ala-Ser-Gly-Phe-Thr-Phe-Ser -Asn-Ser–Tyr-Met-Ser–Trp-Ile-Arg-Gln-Ala-

                                        50
                                        Ile                                                    60
                    Val        Trp      Gly Tyr Thr Ser Asx Gly      Asx ↓
-Pro-Gly-Lys-Gly-Leu-Glu-Trp-Ile - Thr-Tyr–Ile-Asx-Asx-Asx-Gly-Gly–Asx-Ile-Tyr-Tyr-
                    Leu        Ala      Ile  Gly
                               Glx      Val

                                        70                                                     80
-Ala-Asx-Ser-Val-Lys-Gly-Arg-Phe-Thr-Ile-Ser-Arg-Asp-Asp-Gly-Lys-Asn-Thr-Leu-Tyr-

                                        90
-Leu-Gln-Met-Asn-Ser-Leu-Arg-Thr/Ala-Glu-Asp-Thr-Ala-Val-Tyr-Tyr-Cys-Ala-Arg......
        ↓
                                                                                       120
..................................................................................Thr

                                        130                                           140
-Thr-Ala-Pro-Ser-Val-Phe-Pro-Leu-Ala-Ala-Ser-Cys-Val-Asp-Thr-Ser-Gly-Ser-Met-Met

Fig. 17. Location of the affinity label ¹⁴C-m-nitrobenzene diazonium tetro-
fluoroborate within the variable region of gamma heavy chains of
guinea pig antibodies to the dinitrophenyl hapten. (a) Moles of label
within ordered fragments obtained by digestion with cyanogen
bromide. S-S indicates the position of intrachain disulfide bonds. The
shaded areas indicate binding of the affinity reagent to γ chains of
immunoglobulins from unimmunized animals. The open bars indicate
binding of label to antibody γ chains. (b) Location of affinity labeled
tyrosyl residues within the sequence of the variable region of guinea
pig antibody γ chains. ↓ represents the location of a labeled residue.
. . . , region where the variability was so great that a unique sequence
could not be determined. (From Ray and Cebra 1972).

plays some role in determining the number of monomer units incorporated into the intact structure.

## *Structural properties and immunoglobulin evolution*

I have thus far emphasized the two major structural properties of immunoglobulins which will be relevant to studies of antibodies of lower vertebrates. The first property, the presence of light and heavy chains, allows the identification of distinct classes of immunoglobulins. This is an isotypic property because all normal human individuals contain IgG, IgM, IgA, IgD, and IgE immunoglobulins. In addition, these individuals contain subclasses within the different classes (see Chapter 3). The second salient property of immunoglobulins is the presence of V-region diversity in amino acid sequence, which is apparently correlated with antigen binding specificity. Another property, which is not unique, but has enabled the attainment of detailed knowledge of immunoglobulin genetics, is the existence of allotypes. Unlike class distinctions, allotypes are not present in all individuals of a species and represent predominantly the products of allelic genes for immunoglobulin constant regions. All of these physical or antigenic differences, of course, arise from distinctions in amino acid sequence. Class distinctions entail the greatest sequence divergence, e.g., human $\mu$ chains and $\gamma_1$ heavy chains differ in more than 60% of their constant region sequence (Putnam et al. 1972). Differences among subclasses are not so great, and antigenically distinct allotypes can differ at only one residue position in primary sequence.

The information available on immunoglobulins of man and mouse enables the formulation of questions which are directly answerable by chemical analysis of antibodies of lower vertebrates. These can be listed as follows: (1) Is the multichain structure characteristic of mammalian immunoglobulins a feature of antibodies produced by all vertebrate species? (2) Does one immunoglobulin class precede the others in phylogenetic emergence? This question is equivalent to inquiring which heavy chain type is present in primitive vertebrates. (3) Are antibodies of lower species heterogeneous in their amino acid compositions? If so, is the degree of heterogeneity comparable to that observed in mammals? Answers to these questions may help elucidate the genetic mechanisms governing the emergence of the distinct types of heavy chains, which determine particular immunoglobulin classes, and the origin of the diversity of variable regions.

The concepts presented here suggest that we consider the evolution of antibody structure under two categories. The first category is that posed in questions 1 and 2 above which deals with the problem of evolution of constant regions of the molecule. This probably occurred via a process similar to the emergence of other multichain proteins such as the hemoglobins. The second issue, that of the origin of V-region diversity, may entail problems at the level of somatic development as well as at the level of germ line evolution. Although analysis of immunoglobulins of lower vertebrates has not progressed as far as that of human and murine proteins, sufficient data have been accumulated to provide the major portion of the answer to each question.

The methodological approaches involved in the attempts to answer these structural questions comprise techniques used in protein chemistry and physical chemistry. In most cases, comparisons between immunoglobulins of diverse vertebrate species must be somewhat indirect, because detailed primary sequence has been obtained only from immunoglobulins of a few mammalian species. However, precise information regarding binding affinities for haptens and polypeptide chain structures exists for immunoglobulins of vertebrates representing all classes. Moreover, even in cases where primary sequence data are unavailable, statistical analysis of amino acid composition data provides a useful preliminary means of assessing relatedness among immunoglobulins of diverse species. The utility and possible limitations of such methods will be discussed within the context of specific comparisons to be made in the following pages.

# 3 Arrangement of Immunoglobulin Genes in Mammals

*Human γ chain subclasses*

Present knowledge of the number and arrangement of structural genes encoding immunoglobulin polypeptide chains is based largely upon studies of men, mice, and rabbits. The genetic information obtained for these species illustrates the nature of immunoglobulin subclasses and the manner in which the tandem arrangement of immunoglobulin cistrons might be implicated in the differentiation of antibody-forming cells and the generation of antibody diversity. A schematic diagram of the genetic region encoding human γ heavy chains is illustrated in Fig. 18. The cistrons for the IgG subclasses represent a closely linked array of genes, and the order shown is based upon linkage studies using allotypic markers (Litwin and Kunkel 1967; Natvig, Kunkel, and Litwin 1967). Although all such alloantigenic determinants must eventually be explained in terms of primary amino acid sequence, the correlation between sequence and antigenicity has been definitely established in only a few cases at present. Over twenty Gm, or γ chain, allotypes have been described in man. Certain markers are present only on individual subclasses; for example, Gm b (5) which is associated with the presence of phenylalanine residue at position 400 is found only in the γ3 chain. Other markers may be mapped onto distinct portions of the γ chain. This is illustrated by Gm z (17) and Gm a (1), which occur at positions 214 and 356-358 respectively. It is interesting that a number of allotypes are correlated with a single amino acid interchange, but others such as Gm a (1) and Gm y (22) are associated with multiple amino acid replacements. The

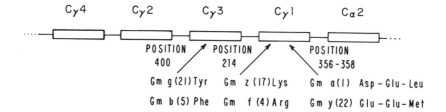

Fig. 18. Schematic diagram of the cistrons encoding the constant regions of heavy chains of the human IgG subclasses and the IgA$_2$ subclass. The order of the $\gamma$ cistrons is based upon linkage studies. The positions of certain allotypic markers and the amino acid sequences involved are illustrated for C$\gamma$3 and C$\gamma$1 genes.

Inv genetic markers on the human $\kappa$ chain correspond to a single amino acid interchange of valine and leucine residues at position 191.

## Arrangements of genes encoding immunoglobulins

It is difficult to draw precise conclusions from species to species because of subgroup and other variations within V-region sequences. However, the type of changes that occur from protein to protein appear to consist of standard interchanges caused primarily by single base changes in the genetic material. I shall consider the origin of V-region diversity in greater detail below but would point out at this time that a variety of data suggest that a given V-region gene must be in continuous juxtaposition with a C-region gene of the same chain type in order for synthesis of that immunoglobulin to occur. This requirement has prompted Gally and Edelman (1972) to define translocons consisting of a chromosomal segment which contains all of that structural information for a particular type of immunoglobulin polypeptide that is transmitted from one generation to the next. The cluster for $\kappa$ chains, for example, contains one C$_\kappa$ gene and probably three V region (V$_\kappa$I, V$_\kappa$II and V$_\kappa$III) gene arrays. Genetic evidence shows that the C$_\kappa$, C$\lambda$, and C regions of the heavy chains, e.g., C$_\mu$, C$_\gamma$, C$_\varepsilon$, etc., are not linked. The genes encoding the constant regions of the heavy chains, however, are closely linked. Figure 19 represents a schematic diagram of the chromosomal regions which encode human $\kappa$, $\lambda$, and heavy chains. The order of the cistrons specifying the constant regions of the $\gamma$ subgroups has been obtained by linkage studies (Litwin and Kunkel 1967; Natvig, Kunkel, and Litwin 1967). The rabbit heavy chain supergene would differ from this array because this species does not possess IgG subclasses. Although there exists no evidence for linkage between the V and C regions in man and the mouse because no V-region allotypic markers have been found, the rabbit heavy chains possess an allotype which may be located in the V region. This marker has been shown to be linked to allotypes derived from amino acid interchanges within the constant Fc region. Figure 19 illustrates the complexity of the genetic loci coding for immunoglobulins of mammals. I do not wish to imply that the immunoglobulin genetic systems of lower mammals are less complex than that of man, because the tentative arrangement of mouse loci would be extremely similar to the human model. Most likely, subclasses of other heavy chains exist. The detailed information on human and murine immunoglobulins was made possible largely because of the presence of homogeneous myeloma proteins which provide paradigms for chemical and immunological analyses. Such myeloma proteins have not been observed in rabbits or in the lower species to be considered below.

## Differential gene expression and clonal restrictions

The heavy chain and λ chain constant region cistrons are clustered in a linear array reminiscent of that proposed for the structural genes encoding hemoglobin $\beta$-type chains (Ingram 1963, Kabat 1972). It is worthwhile considering certain parallels between hemoglobins and immunoglobulins because both molecules normally occur as four chain units. Moreover, the genes encoding their chains appear to be aligned in a similar fashion and could be subject to analogous regulatory mechanisms. Normal hemoglobins are comprised of two alpha chains and two

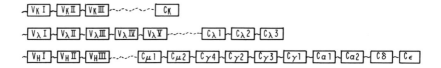

Fig. 19. Schematic diagram of one arrangement of the three linkage groups encoding immunoglobulin chains. The genes specifying the variable regions of light and heavy chains are drawn as if they form linear arrays which may lie on the same stretch of DNA as do their corresponding constant region genes. Other alternatives are possible. The linear set of v region genes may be located on an episome, for example. The model presupposes that a given v gene can, however, be translocated to the proper juxtaposition to the corresponding C region gene whether or not the two genes lie on the same chromosome.

non-alpha polypeptides. No covalent bonds occur between the subunits. The normal adult hemoglobin is hemoglobin A which has a structural formula $\alpha_2\beta_2$. Embryonic hemoglobin has a structure $\alpha_2\varepsilon_2$ and fetal hemoglobin $\alpha_2\gamma_2$. Another adult human hemoglobin contains a non-alpha chain termed $\delta$ which arose by a recent duplication of the gene encoding the $\beta$ chain (Ingram 1963). Figure 20 depicts the chromosomal regions containing the cistrons for the IgG subclasses. A parallel situation obtains for the cistrons encoding $\beta$, $\gamma$, $\delta$, and $\varepsilon$ chains of human hemoglobin (Kabat 1972). Different hemoglobins are expressed at various times during human ontogeny. Embryonic hemoglobin is the first to appear in the early embryo. It is subsequently replaced by fetal hemoglobin which is, in turn, superseded by adult hemoglobin A. All of these molecules contain $\alpha$ chains, but are distinguished by the presence of distinct $\beta$-type chains. The distinctive $\beta$-type chains are therefore formally similar to the immunoglobulin heavy chains, because the genes specifying them form a closely linked array which is not linked to cistrons encoding the common chains, $\alpha$ or light respectively. Kabat (1972) has proposed that the $\beta$-type chains are sequentially activated in development, thereby giving rise to the hemoglobins observed at various stages in the animal's life.

Immunoglobulins likewise show sequential appearance both in devel-

## CISTRONS ENCODING
## $\gamma$ CHAIN SUBCLASSES

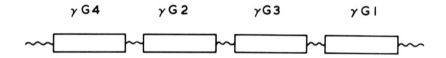

## CISTRONS ENCODING
## HEMOGLOBULIN NONALPHA CHAINS

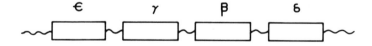

Fig. 20. Diagram illustrating the similarity in arrangement of genes encoding subclasses of the human $\gamma$ heavy chain and those specifying the nonalpha chains of hemoglobin.

oping mammals and as a consequence of immunization. The first immunoglobulin to appear in ontogeny of mammals is IgM, and it is generally the first to appear following primary antigenic stimulation. Subsequently, IgG antibodies usually comprise the major immunoglobulin synthesized. These phenomena are readily manifested at the level of circulating antibodies. In addition, a variety of results suggest that individual antibody forming cells of mammals can switch from producing IgM antibodies to IgG molecules (Nossal, Ada, and Austin 1965; Nossal, Warner, and Lewis 1971; Pernis, Forni, and Amante 1971). This event was originally considered to be highly unlikely because it would mean that one cell could synthesize antibodies of two different classes which possess the same combining specificity for antigen. However, recent studies of the V-region sequence of human heavy chains established that all of the heavy chains share a common pool of V regions. Thus, it would be possible for a cell to make IgM antibodies directed against a given antigen and then switch to IgG antibodies possessing the same combining site. For example, the original IgM antibody might be composed of a $\kappa$ chain ($V_\kappa{}^b C_\kappa$) and a $\mu$ chain ($V_H{}^b C_\mu$) both possessing variable regions of specificity b. If the gene encoding the $V_H$ region were translocated to become contiguous with a $C_\gamma$ gene, a $\gamma$ chain of the form $V_H{}^b C_\gamma$ possessing specificity to antigen b would be generated (see Table 3). I will consider genetic schemes for immunoglobulin evolution and diversification in Chapter 17. The translocation mechanism invoked here may prove to be an

TABLE 3: Model for expression of immunoglobulin genes in cell undergoing a switch from IgM synthesis to IgG synthesis.

| Stage | Genes read | Secretion products |
|---|---|---|
| IgM production | $V_\kappa^b C_\kappa$<br>$V_H^b C_\mu$ | IgM of specificity b |
| Intermediate stage (IgM + IgG) | $V_\kappa^b C_\kappa$<br>$V_H^b C_\mu$<br>$V_H^b C_\gamma$ | IgM; IgG of specificity b |
| IgG production | $V_\kappa^b C_\kappa$<br>$V_H^b C_\gamma$ | IgG of specificity b |

integral part of the regulation of immunoglobulin class expression and allotype expression, and could play a significant role in the diversification of the genes specifying V regions. Moreover, this event would cause a given lymphocyte to be restricted to a capacity to respond to one antigen, because it would form only that immunoglobulin for which it possessed a complete V-C ·cistron. The lymphocyte population would thus be clonally restricted in immunoglobulin expressed and, consequently, would be able to recognize only those antigens for which the particular V region exhibited binding specificity.

# 4 Putative Immune Phenomena in Invertebrates

A mechanism which confers induced resistance to infection would appear to give a selective advantage to all animals which are long lived and subject to attack by pathogens. For this reason, certain invertebrates as well as vertebrates might be expected to possess immune responses or analogous phenomena. This section will review the attempts to induce immunity in invertebrates. In addition, I will consider naturally occurring substances, particularly hemagglutinins, which bear certain similarities to antibodies inasmuch as they are capable of binding to erythrocytes or bacteria. These biological phenomena will be studied from the standpoint of their possible relevance as precursors of the immunological system as we know it in vertebrates.

## Diversity of invertebrate phyla

The task of generalizing and extrapolating from class to class and species to species is exceedingly difficult even within a single phylum such as that of the chordates, which includes all vertebrates. The problem is compounded in dealing with the invertebrates, a diverse collection of animal phyla which exhibit a myriad array of structural and developmental patterns. A knowledge of the phylogenetic relationships among invertebrate species is germane here because certain invertebrate phyla are related to the progenitors of vertebrates, whereas others are extremely distal to this evolutionary line.

Phylogenetic relationships among the invertebrates are illustrated in Figure 21. The stream of invertebrate evolution which led to vertebrates

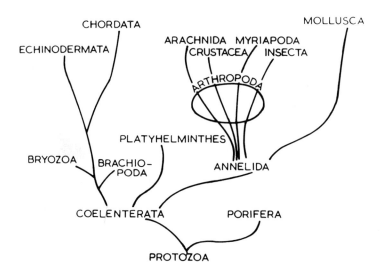

Fig. 21. General scheme for the phylogeny of invertebrates. The phylogenetic line leading to vertebrates includes Bryozoa, Brachiopoda, Echinodermata, and Chordata and is represented by the left branch of this diagram. The major invertebrate groups, the superphylum Arthropoda, and the phylum Mollusca represent the end stages of diverengences from the coelenterate group distinct from that which engendered the Chordata. (Redrawn from Romer 1966)

is the line represented by Bryozoa, Echinodermata, and protochordates. These taxa are classed as deuterostomates because in the embryogenesis of this group, the mouth and anus derive from separate pores of the blastula stage embryo. The vast majority of invertebrates are protostomates, where embryonic development entails the production of both orifices from a single blastopore by a process of splitting. This group includes all invertebrate phyla not included above. Modern speculation favors the hypothesis that the invertebrate echinoderms were ancestral to the protochordates (Romer 1962). Of the protochordates, the tunicates are considered the most likely ancestors of vertebrates (Berrill 1955). At this stage of our knowledge, generalizations regarding cellular and humoral defense mechanisms of invertebrates must be considered tentative, because too few species and phyla have been investigated to provide sufficient coverage of the diverse organisms included in this animal group. In addition, the stricture should be born in mind that most invertebrates are related to the stream of development that gave rise to vertebrates only in the very general sense that all living organisms must be related. Most probably induced mechanisms, where they exist, would be analogous to the immune phenomena of vertebrates rather than homologous.

As we will see in the course of this discussion, much biochemical and biological data have accumulated which enable us to substantiate this conclusion.

It is well documented, by work done as early as that by Mechnikoff in 1884, that invertebrates possess amoeboid cells which are capable of recognizing foreign material and reacting to it either by phagocytosis or encapsulation. A number of more recent studies including those of Salt (1963) and Cheng (1970) have concluded that these two responses are basic defense mechanisms of invertebrates when confronted with foreign pathogens. For example, bacteria and viruses are regularly phagocytized; but if a foreign material such as a miracidium (an infectious larval stage of a parasitic helminth) penetrates certain invertebrates, it would probably be encapsulated, since the parasite is a metazoan and is too large to be phagocytized and ingested by a single cell.

## Induced protective responses of insects

There have been a number of reports regarding induced responses in invertebrates dating back to classical studies of Noguchi (1903) and Cantacuzene (1923) which report the presence of inducible agglutins, lysins, and precipitins in a variety of invertebrate phyla as a result of challenge by red blood cells, bacteria, or serum from vertebrate species. These early immunization studies were reviewed by Cantacuzene in 1923 and Huff in 1940. I will not consider these early studies at any length because recent work has confirmed many of the early findings under more carefully controlled conditions and extended them in modern perspective (Bang 1967).

Insects, although generally short lived, are important because of their useful products, their destruction of crops, or their role as vectors of organisms pathogenic to man. For this reason insect pathology has been a subject of scientific investigation for over fifty years. It has been known since antiquity that insects are subject to a variety of diseases. Some species are particularly susceptible to infection by *Pseudomonas*, which is often lethal (Stephens 1959). In addition, a number of viral pathogens infect insects, e.g., a polyhedral virus, foul brood, which is lethal to honey bees. A number of studies have involved attempts to vaccinate insects using attenuated preparations of lethal bacteria (Stephens and Marshall 1962; Gingrich 1964).

The induced protective response which Gingrich observed in the large milkweed bug, *Oncopiltus fasciatus*, serves as a well studied example of an induced response in an insect. Gingrich (1964) showed that this bug, a representative of the Hemiptera, could produce a factor that was lytic for *Pseudomonas aeruginosa* after challenge with a single dose of a

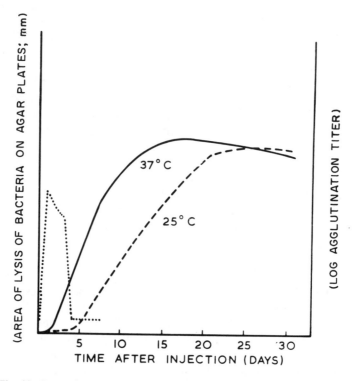

Fig. 22. Comparison of kinetics of appearance of the induced lysin of the milk-weed bug ( . . . .) with antibodies of the marine toad. Circulating anti-bodies in the toad appear more rapidly at 37°C (——) than at 25°C (---). Based upon data of Gingrich (1964) and Diener and Marchalonis (1970).

vaccine prepared by heat killing the bacteria. This vaccination also brings about the development of a short-lived resistance to infection by the living bacteria, which is correlated with the appearance of a lytic factor in the hemolymph of the insect. The time course of appearance of this lytic agent is shown in Figure 22. The factor is readily apparent within four hours of injection of the heat killed vaccine, and it reaches a peak by 24 hours. The amount of material diminished slowly between 24 and 72 hours then declined rapidly. There is a marked difference in kinetics of the lysin elaboration in this system and the appearance of antibodies in vertebrates. Two curves of the antibody production by toads (*Bufo marinus*) to *Salmonella adelaide* flagellin are also illustrated here. With animals maintained at 37°C, peak titer is attained at day 15 and a plateau exists for approximately 2 weeks. This situation is comparable to a typical response of mammals. The second toad curve provides a more apt comparison to the milkweed bug response because the animals were maintained at room temperature, as were the insects. The time course of appearance of antibodies is greatly protracted, with the maximum titer obtained at 21 days after injection of antigen.

Gingrich further observed that both protection by and induction of this bacteriolytic substance was non-specific. Injection of saline or merely puncture of the body of the insect by a needle was sufficient to protect the insect against further challenge with lethal doses of viable bacteria. Thus, in this case, the response cannot be considered a specific one to a definite challenge. It is more likely a general response to a variety of stimuli that perturb the physiological balance of the animal.

At this juncture I should mention that infection or trauma can induce the formation of the acute-phase protein termed C-reactive protein in mammals (Tillet and Francis 1930). This molecule binds to the C polysaccharide of *Pneumococcus* (MacLeod and Avery 1941). Since this protein was first observed in the serum of patients suffering from pneumonia, it was thought to be a type of antibody. In addition, it can be induced by any sort of extreme condition; chronic infections, burns, even physical injury. Initial analysis of its structure disclosed no obvious chemical similarity to immunoglobulins (Gotschlich and Edelman 1965), but comparison between its amino acid composition and that of antibodies suggests possible homology (Marchalonis and Weltman 1971). Subsequent studies may disclose some relationship between this phenomenon and the readily induced protective substances of invertebrates.

Gingrich (1964) found that induced resistance to infection in the milkweed bug could be transferred by transferring hemolymph from challenged animals to normal ones. In this case, the amount of protection which he obtained by using the hemolymph from the protected animals was much greater than that of the non-specific controls. He concluded that resistance can be transferred passively by transfer of serum as would occur if the serum contained a serum factor or antibody. Furthermore, Gingrich carried out assays to determine the biochemical characteristics of the lytic factor. He found that the lysin was extremely stable and could be subjected to extremes of heat and of pH without any loss of lytic activity. The material showed a much greater stability than would be expected if it were a protein. In addition, he fractionated the proteins in the serum in the attempt to determine directly if this factor was protein or associated with protein. His results indicated that the lysin was not a protein. The protein was precipitated from the immune hemolymph by trichloroacetic acid, a standard means for precipitating protein. The lytic factor remained in the supernatant after the protein had been removed. The possibility remains, however, that the lytic factor may be a small peptide. More recent studies by Seaman and Robert (1968) and Scott (1972) have shown that induced agglutinins of the cockroach are proteins.

Similar observations have been made with species representing other orders of insects. In particular, work has been carried out with the Lepidoptera (Stephens and Marshall 1962), and with Orthoptera (Seaman and Robert 1968; Scott 1972). These results provide a good example of the type of induced protection which can be observed in invertebrates. The properties of this response differ markedly from those that characterize induced immunity in vertebrates. The response is nonspecific. The kinetics of appearance of this protection are very rapid; surprisingly so, because these invertebrates are ectothermic and have to be maintained at relatively low temperatures in order to maintain proper homeostasis. Gingrich, for example, maintained his bugs at temperatures of 23°C or less. He found that higher temperatures were harmful to the insects. Figure 22 illustrates that temperature plays a major role in the kinetics of antibody production in the lower vertebrates. Within a temperature range which is physiological for the particular ectothermic species, circulating antibodies appear significantly faster at higher temperatures than at lower ones. The rapid appearance of protective factors in invertebrate hemolymph is puzzling if we wish to interpret it in terms of the immune response of vertebrates. A possible explanation for this discrepancy arises from the fact that the antibody response in vertebrates is dependent upon cell proliferation, whereas the protective response in insects is not (Hink 1970). This observation is yet another difference between the induced protective response of insects and the immune response of vertebrates.

## Induced humoral responses of crustaceans, coelenterates and molluscs

Induced protective responses have been found to occur in most invertebrate phyla. Evans and his colleagues (1968) have reported the induction of bacteriolytic factors in the lobster, a crustacean. There is no evidence that the factors responsible for this bacteriolysis bear any relationship to antibodies of the vertebrate immune system. They might consist of lytic or oxidative enzymes (Klebanoff 1967; Hogg and Jago 1970) released by cells in the hemolymph following aphysiological stimuli.

Other recent studies, particularly studies with coelenterates and with molluscs, have suggested that there may be specific induced factors in these types of invertebrates. For example, the sea anemone apparently responds to challenge of bovine serum albumin by producing a specific humoral factor that complexes with the serum albumin. It binds better to bovine serum albumin than it does to a related albumin, human serum albumin (Phillips 1966). Another factor was found by Michelson (1963) in the snail, *Australorbis glabratus*. If the infective miracidial stage of *Schistosoma*, the parasitic blood fluke, is injected into this type of snail, the snail produces some factor which is capable of immobilizing the miracidia. Evidence has been presented that this material is specific. However, its biochemical nature is problematic and further work is required to explain the phenomenon.

## Naturally-occurring hemagglutinins

In addition to the induced serum factors, there exist a number of natural hemagglutinins which have been studied in various invertebrate phyla. In all the cases where these have been investigated in some detail, they have been found to be proteins. These proteins, which bind to red blood cells of vertebrate species, are ubiquitous in distribution among invertebrate phyla. The classic study of Tyler and Scheer (1945) on the agglutinins of the spiny lobster encouraged further study into the nature of these proteins because they seem to show a number of general similarities to antibodies. The agglutinins were found to be electrophoretically slow moving proteins with the mobility of a $\beta$ globulin or possibly a fast $\gamma$ globulin. The electrophoretic mobility of these molecules is thus similar to that of immunoglobulins. In addition, the fact that they were large molecules makes them similar to immunoglobulins since immunoglobulins are among the largest of serum proteins.

*Limulus polyphemus*, the horseshoe crab, which is classed as a xiphosurid or a very primitive arachnoid, was also found to possess naturally-occurring hemagglutinins. The closest extant relatives of this living fossil are the scorpions, which are the oldest living animals found on land. *Limulus* has remained morphologically unchanged for over 300 million years, and its larval stage bears a very close similarity to the trilobites, which are long extinct. This unique animal has made great contributions to neurophysiology. Studies of Hartline and his colleagues (Hartline 1968) have elucidated the function of the compound eye in this animal. The simplicity of the eye and the nervous system of *Limulus* have allowed the complexity of lateral inhibition to be reduced to a system of mathematical equations. The lack of complexity and extreme age of this beast have made it a very useful animal for physiological and evolutionary studies.

Noguchi in 1903 found that this animal possessed in its hemolymph powerful agglutinins for vertebrate erythrocytes. Subsequent work by Elias Cohen (1968; Cohen, Rowe, and Wissler 1965) has confirmed this finding and furthermore has added some basic knowledge of the chemical properties of these agglutinins. Marchalonis and Edelman (1968a) isolated a protein from *Limulus* hemolymph which possessed hemagglutinating activity by a variety of techniques which had been previously used to isolate immunoglobulins from the serum of vertebrate species. This protein represented a homogeneous preparation and it was possible to compare its structure to that of vertebrate antibodies.

The hemagglutinin molecule as shown in Figure 23 is quite large. The intact protein species has a mass of 400,000 daltons and is comprised of 18 subunits. Each subunit has a molecular weight of 22,500. The representation of the intact molecule shown in Figure 23 is based upon hydrodynamic measurements and upon high resolution electron micrographs (Figure 24), which were taken by Fernandez-Moran (Fernandez-Moran, Marchalonis, and Edelman 1968). The molecule is roughly hexagonal in shape, has a hole in the center, and a diameter of approximately 100 Å. Its height is approximately 70 Å, so the molecule is a cylindrical toroid. The molecule bears little obvious similarity to the immunoglobulins that were discussed in Chapter 2. The obvious differences from immunoglobulins are: (1) The molecule contains only one type of subunit. The immunoglobulins contain two types, light chains and heavy chains. (2) The subunits are held together solely by non-covalent interactions. (3) No diffuse heterogeneity in the amino acid composition of these subunits was detected by electrophoresis or by analysis of peptides produced from this molecule by digestion with proteolytic enzymes. Another basic difference between this molecule and the immunoglobulins is the fact that the maintenance of the intact active structure required the presence of calcium ions. Calcium is necessary for the binding of the subunits together to give an active molecule. Subsequent studies by workers in other laboratories showed that hemagglutinins from invertebrate species were comprised of a number of apparently

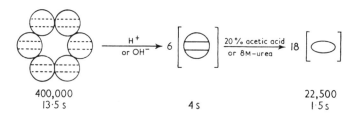

Fig. 23. Structural model based upon physicochemical measurements for the hemagglutinin molecule of the horseshoe crab, *Limulus polyphemus*. (From Marchalonis and Edelman 1968a).

Fig. 24. High resolution electron micrograph of the hemagglutinin of the horseshoe crab. Some of the ring shaped hemagglutinin molecules in the central area are distended, displaying their fine structure in various orientations (x 444,000). Asbestos filament insert (x 800,000) exhibits a lattice period of 7.3Å which serves as an accurate calibration standard. (From Fernandez-Moran, Marchalonis, and Edelman 1968).

identical subunits of molecular weight 15,000 to 20,000. Moreover, Ca++ was required to stabilize the active form of the molecules. Acton and his coworkers (1969) have found that oyster hemagglutinin is constructed in accordance with this basic pattern. Preliminary observations of hemagglutinins from another mollusc, the Murray mollusc from the Murray River in South Australia, have shown that this molecule likewise is constructed in accordance with this general scheme (Marchalonis and Jenkin, unpublished observation). Finstad et al. (1972) have recently confirmed the subunit structure of *Limulus* hemagglutinin. In addition they showed that an agglutinin of an echinoderm (*Asterias forbesii*) likewise is composed of subunits with molecular weights comparable to those characteristic of other species.

Although the chemical studies just cited add no support to the hypothesis that these naturally-occurring agglutinins may be precursors of immunoglobulins, final evidence must come from detailed knowledge of their amino acid sequences. It is possible that some homologies may exist, although on the basis of the data cited above it is unlikely. Finstad, Good, and Litman (1973) have determined the first 24 N-terminal residues of the *Limulus* hemagglutinin molecule and report that this sequence bears no obvious resemblance to the N-terminal sequences of immunoglobulin chains.

At this juncture it seems reasonable to ask what is the function of these agglutinins in the internal homeostasis of the animals in which they occur. Immunologists have long tried to make a case that these are protoantibodies, i.e., precursors of vertebrate antibodies. However, there is no evidence in favor of this hypothesis; but there is evidence, particularly that of McKay and Jenkin (1970), that these natural agglutinins may function as opsonins. Opsonins are factors that occur in the serum that facilitate phagocytosis of foreign materials. McKay and Jenkin (1970) have shown that the serum agglutinin of the lobster is necessary in order for the phagocytic cells to ingest certain materials. On this basis, the hemagglutinins of the Australian crayfish do resemble the immunoglobulins of vertebrates. One of the physiological roles of immunoglobulin is to serve as opsonin; however, other serum proteins such as C-reactive protein also facilitate phagocytosis of certain organisms (Kindmark 1971). The function of invertebrate agglutinin molecules and those of antibodies may be purely analogous, because we have no definite evidence that these substances are in any sense related to the immunoglobulins of vertebrates. Another possible function of these hemagglutinins is one that the immunologist would consider trivial, but may well be an important one among invertebrates. A study by Cohen, Roe, and Wissler (1965) has shown that the binding of *Limulus* hemagglutinin to erythrocytes is inhibited by the sugar N-acetyl-D-glucosamine. It is noteworthy that the binding is also inhibited by N-acetylneuraminic acid, a major charged sugar found on cell surfaces. Naturally-occurring hemagglutinins in vertebrates which are immunoglobulins are also inhibited by this acetylated sugar, and in fact, the acetylated sugar tested here occurs on the surfaces of erythrocytes (Jamieson and Greenwalt 1971). However, one must take into consideration that the bulk of the weight of the horseshoe crab is N-acetyl-D-glucosamine, but it is in the form of the polymer chitin (Foster and Webber 1960) which makes up the shell of the animal. So, in the case of the horseshoe crab, this molecule may be a transport molecule which carries these glucosamine units to and from the shell. In any case, these two illustrations represent a possible role that these protein hemagglutinins may play in the internal physiology of the invertebrates in which they are found.

## Graft rejection in annelids

Some interesting recent work has been reported on the ability of annelid worms to exhibit transplantation immunity. Studies by Cooper (1970) in California and Chateaureynaud-Duprat (1967) in France have independently claimed that pieces of skin grafted from one species of earthworm to another will be subsequently infiltrated by coelomocytes and rejected. Although the precise mechanism of this phenomenon remains to be ascertained, it is significant that such recognition and active rejection occurs. The reaction is most striking when grafts are made between different species of worms (xenografts). In addition, marked rejection may occur among races of the same species. Chateau-reynaud-Duprat, for example, found that subpopulations of *Eisenia foetida* from different areas of France rejected grafts from each other quite readily. Moreover, Cooper observed allograft rejections when bits of skin were transplanted between individuals of the same population of *E. foetida* (1970). The first-set grafts were rejected in a chronic fashion. If the grafted animals were given a second piece of skin from the same individual used in the original challenge, this second-set transplant was rejected at a faster rate than was the first. Bailey, Miller, and Cooper (1971) have been able to transfer the capacity to reject xenografts specifically by transfer of coelomocytes from sensitized animals, and on morphological grounds have described certain coelomocytes as lymphocytes and plasma cells, although the particular cell type responsible for the reaction is unknown.

## Immunity in invertebrates

Earlier I defined immunity to be an induced response that shows specificity toward the foreign material used as the inducing agent, and which involves the participation of cells of the lymphoid series. The subsequent discussion of antibodies and of antibody-forming cells in mammals established other criteria for the definition of adaptive immunity. The sequence of cellular events in mammals is relatively well known, and the organs which contribute to this process of antibody production, the lymphomyeloid complex, have been described in detail.

In this chapter, I have discussed induced responses of invertebrates to challenge of foreign material. In the case of the induced humoral factors, such as the induction of bacterial lysins and induced protection in insects, we observed: (1) that the specificity of the reaction was in no way comparable to that shown by antibodies; (2) that these substances were induced very rapidly; and (3) that these substances are probably produced without the presence of cell proliferation. A simple model to explain these observations might be the release of enzymes or possible other lytic contents from cells after the cells have been destroyed because of some sort of aphysiological stimulus.

Natural hemagglutinins may be a more promising subject of study for the comparative immunochemist since these are proteins which show a binding specificity for sugars very similar to that shown by antibodies to erythrocytes which are found in vertebrates. In addition, *Limulus* hemagglutinin and oyster hemagglutinin are composed of subunits which are about the same size as the light chain of vertebrate immunoglobulins. Although there are a number of striking differences between these molecules and antibodies, these agglutinins do merit further study, particularly at the level of amino acid sequence analysis.

Although annelid worms are not related to the stream of evolution which eventually gave rise to vertebrates (see Figure 21), transplantation experiments show that they do exhibit some ability to recognize specifically tissues from other individuals and species of worms and to destroy them. This capacity to reject grafts is transferred by coelomocytes. At present, the mechanisms of this process are not clear and the nature of recognition is unknown, but the situation with the annelid worms fits within the most general definition of immunity. Considerable work is required to establish that this phenomenon meets the stringent criteria defining adaptive immunity as it exists in vertebrates. The isolation, characterization, and comparison with vertebrate antibodies of the coelomocyte surface recognition unit are critical to establish homology with the immune response of vertebrates. At the present time, because of the lack of evolutionary relatedness between annelids and the chordate phylum, it is most reasonable to suppose that the observed graft rejection represents an example of the development of a defense mechanism analogous, rather than homologous, to the immune response of vertebrates.

## Generality of cell recognition capacities

It is well known that the ability to recognize distinct cell surface conformations is a property of embryonic tissues (Garber and Moscona

1972). This property is shared by very primitive animals such as sponges (Humphreys 1970; Margoliash et al. 1965), coelenterates (Campbell and Bibb 1970), and echinoderms (Reinish and Bang 1971). An example of this recognition follows. If sponges are pushed through a mesh, the individual cells formed will reaggregate to form new sponges. If sponges from two different species are pushed through a mesh and then the individual cell suspensions mixed, mixed sponges will not develop. Instead, the different types of cells segregate and form specific clusters (Humphreys 1970). They are obviously able to recognize surfaces and construct their own type of individual. Henkart and his coworkers (1973) have isolated and partially characterized a proteoglycan complex (47% amino acids; 49% sugars) responsible for species specific aggregation of *Microciona parthena*. The molecule is characterized by a strong net negative charge when analyzed by electrophoresis and an extremely large molecular weight ($2.1 \times 10^6$). The complex is comprised of smaller subunits of mass $2 \times 10^5$ daltons, apparently held together by noncovalent bonds (Cauldwell, Henkart, and Humphreys 1973). The intact structure displays a sunburst pattern with a central core of 800 Å surrounded by filaments 1,100 Å in length and 45 Å thick. Because no structural similarity to immunoglobulins is apparent, and therefore the likelihood of the recognition process being homologous to immune recognition is slight, the sponge recognition system probably represents a phenomenon analogous to immunological recognition.

The ability to recognize some sort of foreignness seems universal to all living forms. Even if these recognition and rejection mechanisms of invertebrates prove not to be directly related to corresponding immunological reactions of mammals, they deserve further critical experimental analysis because they may provide alternative solutions to problems of surveillance directed against tumor cells and infectious agents.

# 5  Origins of Vertebrates and Their Immune Capacity

## Phylogenetic considerations

In the previous chapter we considered the putative immune responses of protostomate invertebrates and their possible relationships to the antibody response of mammals. The present discussion deals with antibody formation in the lower vertebrates and its bearing upon the phylogenetic emergence of the immune system. Although the capability of lower vertebrates to produce antibodies was established over seventy years ago (Widal and Sicard 1897; Noguchi 1903), the technical approaches and depth of analysis available provided little detailed knowledge of the properties of the antibodies produced. In fact, basic issues such as the question of the degree of specificity characteristic of antibodies from ectothermic vertebrates was still open to dispute in relatively recent times (Sirotinin 1960; Everhart and Shefner 1966). For this reason, the studies to be described here will be presented in a phylogenetic, rather than a historical manner, and will rely heavily upon recent work employing carefully controlled experimental conditions. This chapter will provide general outlines of the immune reactions characteristic of the various vertebrate classes; later chapters will consider individual phenomena in detail.

Prior to considering immunological data, I shall give a brief summary of vertebrate phylogeny in order to provide a conceptual framework for understanding the emergence and interrelationships of the various classes of vertebrates. A detailed description of the paleontological emergence of vertebrate classes and species is given by Romer (1966) and Young

(1962). The exact details of vertebrate ancestry are lost in the distant past and, therefore, the origin of vertebrate species is open to speculation. At one time or another, most of the major phyla of invertebrates have been proposed as ancestors of the phylum Chordata, which includes the vertebrates. Even such unlikely candidates as *Limulus*, the horseshoe crab, were seriously considered for this role (Patten 1912). This proposal, although interesting and somewhat amusing, required inversion of the internal anatomy of this arachnid to fit the chordate mould and completely disregarded patterns of embryological development.

Berill (1955) makes a strong case supporting tunicates as the group of protochordates ancestral to vertebrates. Although the sessile, degenerate, adult forms of these species bear little resemblance to vertebrates, the free-living larval stage may well serve as a vertebrate prototype. In any case, these and other protochordates are small, soft-bodied animals which rarely become incorporated into the fossil record. Recognizable vertebrates have been found as fossils dating back to the Ordovician period. These forms were a type of agnathan or jawless fish known as ostracoderms (Figure 25). These small heavily armored creatures were more primitive than true fishes in a number of respects. In addition to the lack of jaws, they lacked paired fins and possessed only a single, dorsally-located nostril. These forms are long extinct, but

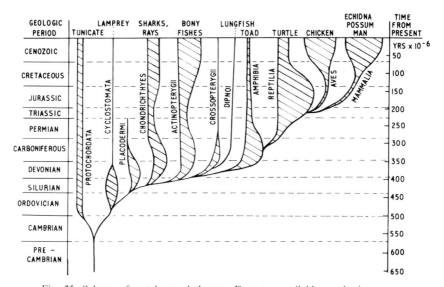

Fig. 25. Scheme of vertebrate phylogeny. Data are available on the immunological properties of the representative species indicated here. (Redrawn from Romer 1966).

modern cyclostomes, as exemplified by hagfish and lampreys, share many of their structural characteristics.

Jawed fishes arose nearly 400 million years ago during the Silurian period (Romer 1966). These forms, the placoderms, were true predators, being larger and much more agile than were the chunky bottom-feeding ostracoderms. Although these species are all extinct today, their descendants, the bony fishes (Osteichthyes) and cartilaginous fishes (Chondrichthyes), dominate modern oceans and lakes. These two major groups of fishes diverged from distinct placoderm lines during the late Silurian or early Devonian period. In a sense, the cartilaginous fishes represent the end of an evolutionary line. Although they presently exist in large numbers, morphologically they are very similar to their Devonian ancestors. From this standpoint, they have been classed as "living fossils." The bony fishes, however, possessed a progressive group which gave rise to higher vertebrates. The bony fishes are divided into two broad groups of species, the ray-finned fishes or Actinopterygii and the fleshy-finned fishes the Sarcopterygii. The vast majority of living fishes are ray finned, while there exist only a few species of fleshy-finned fishes. The fins of the Actinopterygii are composed only of a web of skin supported by horny rays, with flesh and bone being confined only to the base of the fin. The probability that these species could have evolved into forms possessing limbs capable of conquering the terrestrial environment was therefore small.

The sarcopterygian fishes were ancestral to amphibians and higher vertebrates. In the Sarcopterygii, bones homologous to the wrist and fingers actually extend into the fins, thereby providing the firm basis of support necessary for the development of walking as a means of escaping the aqueous environment. A particular group of fleshy-finned fishes, the crossopterygians or lobe fins, were directly ancestral to amphibians. Until very recently, this group was thought to be extinct. However, one species, the coelacanth, was discovered in the Indian Ocean in 1948. Another group of Sarcopterygii, the Dipnoi or lungfish, also exists today. These forms possess a number of morphological and embryological similarities to amphibians but are not directly in the line of emergence of higher vertebrates. In Romer's words (1967), they are "uncles of the tetrapods." Although the three extant lungfish species are by no means common and are restricted in distribution to limited sections of Australia, Africa, and South America, their immunological properties have been studied in some detail (Marchalonis 1969; Litman et al. 1971c). As I shall discuss below, this group of vertebrates may represent an important stage in the evolution of antibody structure.

Primitive amphibians diverged from the Crossopterygii during the Devonian period (Romer 1962, 1966). Modern amphibians consist of two major groups; namely, the Caudata, which are forms possessing tails, and the Salientia, which include frogs and toads. The Caudata can be further divided into two types which are urodeles, such as salamanders, and Apoda or legless amphibians. The amphibians represent a major adaptive step in vertebrate evolution because these forms were able to take advantage of the terrestrial environment, which had been denied to more primitive classes (Romer 1966). I shall present evidence that amphibians also represent a major transition stage in the evolution of the antibody-forming system. The fact that this group includes species that range from permanently larval states (neoteny) to forms which undergo radical morphological and biochemical metamorphoses, renders them critical to detailed studies of the ontogeny of immunity.

Figure 26 presents a schematic representation of the evolution of reptiles and their descendants. The stem reptiles, or cotylosaurs, which arose during the Carboniferous period, gave rise to a variety of forms, two of which, the thecodonts and the therapsids, were ancestral to birds and mammals, respectively.

The fact that birds and mammals diverged early in reptilian evolution warrants comment. The presence of primary immunological organs such as the bursa of Fabricius in birds does not guarantee that a homologue of this organ must exist in mammals. In this case, as in all evolutionary generalizations, possibilities of independent genetic events following divergence of two phylogenetic streams must be considered. If at all feasible, investigations should be pursued to determine if the property under study occurs in animals distal to the branch point of the two lines.

Reptilian stock diverged into essentially three major lines which are reflected in modern reptiles and nonreptilian forms sharing common ancestors among the cotylosaurs. The anapsids, which includes the turtles, and the therapsids, which were the progenitors of mammals, diverged from the cotylosaurs early in reptilian evolution (Romer 1966). Subsequently during the late Carboniferous or early Permian period, the Euryapsida and the Archosauria diverged from a common line of stem reptiles. The Euryapsida include familiar reptiles such as lizards and snakes. Archosauria comprised the group of "ruling reptiles" which included dinosaurs. All that remain of this once predominant group of animals are the crocodilians and birds, which bear marked structural similarity to their archosaurian ancestors (Walker 1972). Closely following the divergence of Euryapsida and Archosauria, the Rhynchocephalia emerged from the former line. One species of rhynchocephalian exists today, the tuatara (*Sphenodon punctatum*) which lives on isolated islands off the coast of New Zealand. This species, like the Australian lungfish, is a living fossil which has remained morphologically unchanged for over 200 million years (Simpson 1953; Romer 1966). It is possible that

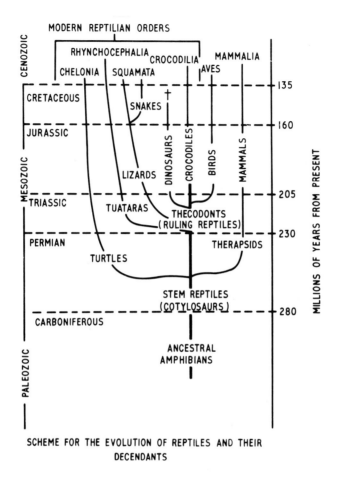

MODERN REPTILIAN ORDERS

SCHEME FOR THE EVOLUTION OF REPTILES AND THEIR
DECENDANTS

Fig. 26. Schematic representation of the evolution of modern reptiles and their
descendants. (Based upon Cohen 1971).

studies of such relict species will give insight into the structure of antibodies and the status of the lymphoid systems as they were early in
evolution, but evolutionary generalizations based upon living forms always contain an element of uncertainty and must be interpreted with
caution. All living vertebrates are modern, and we can only presume that
so-called primitive species or living fossils have retained the characteristics of their ancestors which diverged from other vertebrate lines in the
distant past.

## Progenitors of vertebrate immunity

With these evolutionary factors as a background, we can now outline
the phylogenetic emergence of the various aspects of vertebrate im-

munity. Table 4 presents a listing of general parameters of immunity that have been studied in representatives from each vertebrate class. Protochordates as represented by tunicates possess cells that have been characterized as lymphocytes both by light microscopy (Endean 1960) and electron microscopy (Overton 1966). Colonial tunicates exhibit a colony specificity in which members of one colony within a species do not fuse and form a common vascular system with members of other colonies (Oka and Watanabe 1960; Oka 1970; Tanaka 1973). This capacity is genetically determined by loci coding for colony specificity, so that colonies which share at least one of two alleles can fuse, whereas those which share none are incompatible (Oka 1970). Furthermore, recent evidence indicates that an active process of rejection occurs at the area of contact between the two colonies. This phenomenon requires further analysis to determine if the putative lymphocytes function in the recognition or destruction process and possess surface receptors homologous with immunoglobulins.

Oka (1970) pointed out that colonial ascidians such as *Botryllus primogenus* are hermaphrodites and that neither self-fertilization nor fertilization between individuals of the same colony occurs. He considered his observations of colony specificity in tunicates to be similar to the self-incompatibility which is well documented in plants (East and Mangelsdorf 1925). Burnet (1971) has analyzed these phenomena and believes that the genetics of recognition in colonial tunicates bears some similarities to the histocompatibility antigen systems present in mammals, although the degree of specificity is not as great. Histocompatibility antigens in mammals are encoded at the genetic level by multiple alleles, and the presence of even one different marker is usually sufficient to promote recognition and rejection in mammals (Hildemann 1972). Tunicate systems and typical vertebrate transplantation patterns will be compared in detail in Chapter 11. No data are available for the other protochordate groups, although Hildemann and Dix (1972) have recently reported that certain Australian echinoderms, which are in the deuterostomate lineage, can reject allografts and xenografts. The cellular responses of these pivotal species definitely deserve further investigation.

## *Parameters of immunity in primitive vertebrates*

The cyclostomes are considered the most primitive of living vertebrates, and studies of immunity in these species is one of the most challenging areas of comparative immunology. The hagfish, or *Myxinoidea*, was once considered the "negative hero" of immunity. Investigators were consistently unable to induce either antibody formation or cellular immunity in the Pacific hagfish (Papermaster et al. 1964; Good and Papermaster 1964). Their negative results were buttressed by obser-

TABLE 4: Immunological parameters exhibited by vertebrate classes.

| Class | Lymphocytes | Plasma cells | Thymus |
|---|---|---|---|
| *Tunicates*[a] | + (?)[b] | - | - |
| *Cyclostomes* | | | |
| Lamprey | + | - | Prim.[c] |
| Hagfish | + | - | N.F.[d] |
| *Elasmobranchs* | | | |
| Primitive | + | - | + |
| Advanced | + | + | + |
| *Holsteans* | | | |
| Bowfin | + | + | + |
| *Chondrosteans* | | | |
| Paddlefish | + | + | + |
| *Teleostei* | + | + | + |
| *Dipnoi* | + | + | + |
| *Amphibians* | | | |
| Urodeles | + | + | + |
| Anurans | + | + | + |
| *Reptiles* | + | + | + |
| *Aves* | + | + | + |
| *Mammals* | | | |
| Prototheria | + | + | + |
| Metatheria | + | + | + |
| Eutheria | + | + | + |

vations that hagfish possessed essentially no evidence of a lymphoid system (Good et al. 1966). This morphological observation may have resulted from the fact that primary lymphoid organs such as the thymus degenerate with age (Metcalf and Moore 1971) and immature hagfish have never been obtained for study. Larval lampreys, although they do not possess discrete lymphoid tissue, contain collections of lymphoid cells in the gill region (Good et al. 1966; Rowlands 1969). Thoenes and Hildemann (1970) maintained hagfish under conditions of temperature, feeding and seclusion simulating the creatures' natural environment and established that they possess a broad immunological competence. A natural precipitin titer to keyhole limpet hemocyanin was present in the serum, and this titer was specifically raised by immunization. Furthermore, hagfish were shown to synthesize antibodies to mammalian erythrocytes. Although antibody activity in the hagfish is associated with a high molecular weight protein that has been described as similar to IgM immunoglobulin (Thoenes and Hildemann 1970), no information is available regarding its polypeptide chain structure. In general, cyclostome antibodies are more difficult to isolate and characterize than those of

| Spleen | Lymph nodes | Bursa | Antibodies | Allograft rejection |
|---|---|---|---|---|
| - | - | - | ? | +(?) |
| Prim. | - | - | + | + |
| N.F. | - | - | + | + |
| + | - | - | + | + |
| + | - | - | + | + |
| + | - | - | + | + |
| + | - | - | + | + |
| + | - | - | + | + |
| + | - | - | + | + |
| + | - | - | + | + |
| + | +(?) | - | + | + |
| + | +(?) | +(?) | + | + |
| + | +(?) | + | + | + |
| + | +(?) | +(Eq.)[e] | + | + |
| + | + | +(Eq.) | + | + |
| + | + | +(Eq.) | + | + |

*Source:* Data reviewed in the following references: Good and Papermaster 1964; Grey 1969; Clem and Leslie 1969; Hildemann 1972; Marchalonis 1970; Marchalonis and Cone 1973b.

a. Tunicates represent a protochordate subphylum of the chordates. All other groups headings represent classes of vertebrates.

b. (?) indicates that some question exists regarding the exact homology of the lymphoid cells or structures under consideration, although such structures have been described.

c. Prim. = Primitive.

d. N.F. = not found.

e. Eq. = the presence of a lymphoid structure which serves an equivalent function is indicated.

placoderm-derived vertebrates for reasons which will become clear in the discussion of the other cyclostome branch, the lampreys.

Although initial studies indicated that lampreys mount feeble antibody responses to some antigens (Finstad and Good 1966), they have been found to respond to *Brucella* (Finstad and Good 1964), erythrocytes (Boffa et al. 1967; Pollara et al. 1970), and bacteriophage f2 (Marchalonis and Edelman 1968b). As shown in Figure 27, lamprey (*Petromyzon marinus*) antibodies to bacteriophage f2 were characterized by an electrophoretic mobility similar to that of elasmobranch anti-bodies and human IgM immunoglobulin. However, when antibody activity was assayed after fractionation of the antiserum on the basis of size, the phage neutralization was associated with a diffuse band of components with molecular weights ranging from some comparable to that of IgG immunoglobulin to components three to four fold heavier. The properties of these antibodies will be considered in detail in Chapter 6.

The basic phenomenon of vertebrate immunity exists at the level of cyclostomes. Not only is the capacity to form antibodies present, but cell-mediated immunity as manifested by graft rejection (Thoenes and Hildemann 1970) and the elaboration of a mixed lymphocyte reaction (Cooper 1971) occur in both hagfish and larval lampreys. Both of these responses are regarded as functions of thymus-derived lymphocytes of mammals (see Chapter 11 for further discussion). Therefore, we can conclude that cells possessing functions analogous to those of the B-cell lineage (antibody formation) and T-cell lineage (cell-mediated immunity) occur within the most primitive of extant vertebrates. However, at this time there exist no data which indicate that these functions are performed by separate cell lines. The possibility that one lymphocyte type which possesses characteristics attributable to both thymus-derived and bone marrow-derived lymphocytes of mice might occur in cyclostomes has not been excluded experimentally.

Chondrichthyes possess broad immunological competence which is reflected in their capacity to produce antibodies to a wide variety of antigens (Clem and Sigel 1963; Sigel and Clem 1966) and to reject skin allografts (Hildemann 1972). These observations obtain for both sharks and rays. Sharks bear a place of honor in studies of the phylogeny of immunity because a representative species, the smooth dogfish, was the first lower vertebrate to have its immunoglobulins studied in a detailed fashion (Marchalonis and Edelman 1965, 1966a). Other sharks, including the lemon shark (Clem and Small 1967), leopard shark (Suran and Papermaster 1967), and horned shark (Frommel et al. 1971), have now been investigated in some detail. Immunoglobulins of two species of ray have also been isolated and characterized in terms of polypeptide chain structure (Marchalonis and Schonfeld 1971; Johnston et al. 1971).

The most common representatives of bony fish or Osteichthyes are the teleosts, which are represented by familiar species such as goldfish and trout. The teleosts are the end stage of the line of ray-finned fishes or Actinopterygii. Recent evidence has established that representative ray-finned fishes possess the complete variety of antibody-forming cells and can carry out vigorous rejection of skin and scale allografts (Hildemann

1972; Cohen and Borysenko 1970). With respect to antibody formation, teleosts are vigorous producers of circulating antibodies (Ridgway, Hodgins, and Klontz 1966). Their antibodies consist chiefly of IgM molecules of intermediate size, and the binding affinity of teleost antibodies to hapten is comparable to that of higher vertebrates. Occasionally odd molecules are found in teleost serum that have been reported to possess antibodylike activity, but clearly are not immunoglobulins. Such molecules occur in other vertebrate classes, but have been studied in greatest detail in the teleosts. These immunological and quasi-immunological phenomena will be considered at length below.

Although lungfish, or Dipnoi, are not directly ancestral to amphibians and higher vertebrates, they are sufficiently close to have been termed "uncles of the tetrapods" (Romer 1966) and studies of their immunoglobulins may provide possible clues to the nature of immunoglobulins of the crossopterygians. Lungfish represent a transition from dependence upon an aquatic environment to the capacity to survive on land. Their immunoglobulin pattern shows a definite increase in complexity over the situation in lower fishes, because two distinct classes of immunoglobulins are present in the serum. The Australian lungfish (*Neoceratodus forsteri*) is the most primitive of the three lungfish existing today. It bears a striking anatomical similarity to the prototype form, *Ceratodus*, which lived during the Devonian period. Its immunoglobulins (Marchalonis 1969) and those of the African lungfish, *Protopterus aethiopicus*, (Litman et al. 1971c) have been studied in some detail. The results are very similar for both species, which suggests that the gene duplication responsible for the emergence of two distinct heavy polypeptide chains probably occurred before the ancestors of these two species became geographically isolated over 200 million years ago.

Antibody production in anuran amphibians resembles that of mammals in general cellular aspects and in types of antibodies produced (Marchalonis 1971b). Although the present discussion must consider primarily studies of anurans because of the relative lack of investigation of antibody formation in urodeles and apodans, I shall review the information available on these amphibian groups. The paucity of immunological data on amphibians is surprising because these species have long been favorite vertebrate experimental subjects of developmental biologists. The entire sequence of embryological development has been elucidated for many species (Shumway 1940; Taylor and Kollros 1946; Nieuwkoop and Faber 1967), and a variety of surgical procedures for cloning, transfer of cell nuclei (Briggs and King 1952; Fischberg, Gurdon, and Elsdale 1958; Volpe and McKinnell 1966), and formation of chimeras have been developed (Volpe 1971). Amphibians occupy a crucial branch point in the phylogenetic emergence of antibodies, and detailed knowledge of their immune reactions might help elucidate some basic immunological mechanisms.

Although much of the early literature provided examples suggesting a lack of immune responses in anurans (Goodner 1926; Bisset 1948), recent studies have shown that these species often rival mammals in the vigor of their immune reactions. Frogs and toads respond to a variety of antigens including erythrocytes, bacteria, viruses, and heterologous serum proteins (Trnka and Franek 1960; Alcock 1965; Marchalonis and Edelman 1966b; Lykakis 1969; Marchalonis 1971b), and even to adult bullfrog hemoglobins (Maniatis, Steiner, and Ingram 1969). In subsequent chapters, I will present data which justify the conclusion that antibody production in anuran amphibians resembles that of mammals in general cellular aspects and in types of antibodies produced. Antigen is localized in toad jugular bodies by a process similar to follicular localization in rat lymph nodes. A series of lymphoid cells, originating with lymphocytes and terminating with plasma cells, arise by division from stimulated precursor cells and synthesize antibodies (Diener and Marchalonis 1970). All anurans studied possess distinct types of immunoglobulins that correspond to the IgM and IgG classes of man in size and polypeptide chain structure. These species represent the lowest groups of vertebrates that contain antibodies distinct from the IgM class that are of a similar size to IgG molecules of mammals. Certain identification of the anuran $\gamma$-type heavy chain as homologous to the $\gamma$-chain of mammals requires detailed amino acid sequence information.

The presence of multiple immunoglobulin classes in urodeles requires further investigation. To date, only high molecular weight IgM has been found in the serum of urodeles (Marchalonis and Cohen 1973; Fougereau, Houdayer, and Dorson 1972).

The amphibians provide experimental opportunities to approach four basic immunological questions which are relatively inaccessible in mammals. (1) Temperature plays a major role in the time of appearance of antibody forming cells and antibodies in ectothermic species. The exact mechanistic details of this phenomenon have not been determined, but present evidence suggests that alteration of temperature may be a functional means of disengaging antigen recognition from antibody synthesis. (2) Different types of antigen induce qualitatively dissimilar antibody responses. Flagella antigens elicit a pure IgM response that apparently lacks the capacity for immunological memory and tolerance. The mechanisms underlying this schism in antibody production raise analogies to thymus-dependent and independent responses of mammals and obviously merit future detailed investigation. (3) Two pieces of evidence suggest that cell collaboration may be involved in the immune responses of anurans to erythrocytes (Chapters 11 and 14; Cone and Marchalonis 1972) and of urodeles to hapten carrier systems (Ruben et al. 1973). These results indicate that not only did cellular and humoral immunity arise early in vertebrate phylogeny, but the mechanisms of cell collaboration were present at the level of primitive amphibians. (4) The fact that anurans possess a free living larval stage that shows a progressive development of immunological competence renders them an excellent system for studying the ontogenetic emergence of immunoglobulin synthesis. The larvae can provide information on the activation of genes specifying immunoglobulins and the regulation of cellular synthesis of these proteins. This particular area represents a juncture at which the developmental properties of the anurans aid the immunologist in elucidating basic mechanisms of antibody production, while the application of specialized immunological techniques enables the developmental biologist to delve further into the mechanisms of differentiation.

## Antibody production in reptiles and birds

Reptiles are the most neglected vertebrate class in terms of studies of their immune responses. However, sufficient data are available to establish that they possess multiple classes of immunoglobulins and can exhibit the entire range of cellular and humoral immune responses (Cohen 1971). Various species of reptiles have demonstrated the capacity to form antibodies to proteins (Grey 1966; Evans et al. 1966; Lerch, Huggins, and Bartel 1967), bacterial antigens (Maung 1963; Marchalonis, Ealy, and Diener 1969), erythrocytes (Kassin and Pevnitskii 1969), and haptens (Benedict and Pollard 1972). The capacity of reptiles to demonstrate clear-cut immunological memory will be discussed below.

It is amply documented that birds, especially the domestic chicken, are vigorous antibody producers (Dreesman et al. 1965) and express a variety of cell mediated immune reactions comparable to those exhibited by mammals (Hildemann 1972).

The pattern of immune responses and immunoglobulin types in reptiles and birds is consistent with the trend first observed in amphibians. The reptiles and birds which have been studied possess serum IgM immunoglobulin and, in addition, at least one other class defined by the presence of a heavy chain distinct from the $\mu$ chain. Despite the fact that birds rival mammals in vigor and extent of immune reactions, they do not possess a lymphatic system as complex as that of mammals, and reptiles possess even less organized lymphatic tissue than do birds. The chicken, however, has a primary lymphoid organ associated with the cloaca which generates the cells which are competent to produce antibodies. This B-lymphocyte system is distinct from the thymus dependent cellular immune system. The dichotomy of immune responses in the chicken correlates very well with the distinction between cellular immunity and humoral immunity in the mouse and man (Warner 1972). These systems may represent examples of convergent evolution in which two divergent species evolved similar means of solving the same immunological problem, and will be considered in detail in Chapter 11.

## Phylogenetic aspects of antibody formation in mammals

The ancestors of mammals diverged from the group of reptiles termed therapsids approximately 200 million years ago. Three subclasses of mammals exist today. (1) Monotremes, or protherian mammals, retain a number of reptilian characteristics. These egg laying mammals are represented by only two extant species, the spiny anteater or echidna, and the duck-billed platypus. Both species represent relict populations which are limited in distribution to Australia. (2) Marsupials, the metatherian or pouched mammals, are viviparous, as are higher mammals, but cannot nourish their offspring *in utero*. Young marsupials are born at developmental stages comparable to early embryonic development in higher mammals and leave the uterus and migrate to the pouch, where they become attached to the mother's teat. Marsupials are generally restricted to Australia, where the early isolation of this large land mass prevented the migration of more advanced mammalian species, but the common opossum, *Didelphis virginiana*, represents a marsupial which occurs widely throughout North America. (3) Eutherian mammals possess a placenta, which serves as a means of nourishing and removing waste products from the developing embryo.

Paleontologists are convinced that the progenitors of all three mammalian subclasses were therapsid reptiles (Romer 1966), but it is not clear that all three groups are derived from the same species of therapsids. In fact, Romer has proposed that the monotremes evolved separately from existing marsupials and placental mammals.

The processes of immunoglobulin evolution that were operative in premammalian species continued to function within mammals. Although amphibians, reptiles, and birds possess low molecular weight immunoglobulins representing immunoglobulin classes distinct from the IgM class, molecules clearly similar to the IgG class first appeared within the primitive mammals (see Chapter 8). The protherian echidna and the metatherian bush-tailed opossum both possess 7S immunoglobulins containing heavy chains similar to the $\gamma$ chain. In addition, the partial N-terminal sequence of the echidna $\gamma$ chain indicates that a substantial fraction of these molecules have a sequence comparable to that of the $V_H III$ subgroup of human heavy chain variable regions (Atwell, Hunt, and Marchalonis, unpublished observations). Since the genes encoding distinct heavy chains probably arose via gene duplication, the event which generated the gene specifying γ chains must have occurred prior to the phylogenetic divergence of protheria, metatheria and eutheria from the therapsid line of reptiles. Other immunoglobulin classes such as IgA and IgE are also widely distributed among the mammals. Moreover, readily detectable antigenic similarities occur among the immunoglobulins of eutherian mammalian species, although the strength of cross-reactivity is correlated generally with phylogenetic relatedness.

Duplication events comparable to those which generated distinct heavy-chain classes also operated during the evolution of mammalian orders and families. The IgG subclasses of man, for example, are encoded by four closely linked autosomal genes which arose during the emergence of primates. Subgroups of various heavy chains have also emerged during speciation within other mammalian orders. Furthermore, the genes specifying variable regions have also undergone evolutionary variation within the mammals. Different V-region subgroups, for example, predominate in different species.

All aspects of the immune system are fully operative in all mammals studied. The trend of immunoglobulin evolution in mammals is toward increase in the number of immunoglobulin classes, demonstrating that evolutionary variation is an ongoing process.

## *General properties of antibodies of vertebrate species*

The general physicochemical properties of antibodies of vertebrate species provide a basis of comparison among vertebrate species, and data regarding the molecular weights of antibodies aid one in deciding if immunological memory exists within certain lower species. Two of the major properties by which proteins can be resolved are their electric charge and molecular weight. Since antibodies from representatives of all vertebrate classes have been isolated by techniques which make use of these properties, it is possible to present a general comparison of vertebrate immunoglobulins. Figure 27 shows a comparison among antibodies of diverse vertebrate species by zone electrophoresis on starch blocks. The block can be cut into slices and the protein eluted into physiological medium so that antibody activity can be localized in discrete fractions (Kunkel 1954). Although the distribution patterns of bulk proteins differed to a marked degree in the species depicted, antibody activity was always found, under the electrophoretic conditions applied here, in the more slowly migrating components (isoelectric points of pI's ranging from approximately 5.0 to 7.5), which were originally termed γ and β globulins (Marchalonis and Edelman 1965;

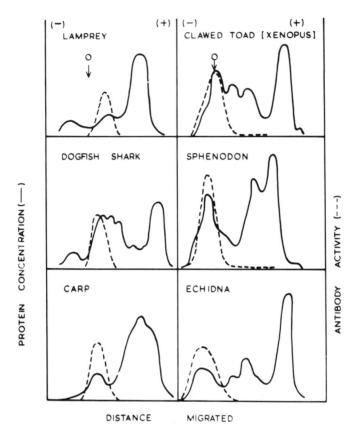

Fig. 27. Zone electrophoresis on starch blocks of antisera from representatives of six vertebrate classes. Protein concentration is given by the solid lines; antibody activity by the dotted lines. The samples were loaded on the block at the origin (marked ↓) and separated in an electric field. The pattern obtained for the echidna is typical of mammalian sera; the large anodally migrating peak is albumin and the slowest peak is γ globulin. Direction of migration is left to right. Electrophoresis was carried out at pH8.6 in barbital buffer (ionic strength $\Gamma/2 = 0.1$).

Clem and Sigel 1966; Marchalonis and Cone 1973b). Antibodies in the lamprey, the dogfish shark, and the carp occur in fractions migrating as fast γ or β components. This mobility is comparable to that of the IgM immunoglobulins of mammals. Antibody activity in the echidna, a monotreme or egg-laying mammal, resembles that of higher mammals in occupying the most cathodal fractions (Atwell, Marchalonis, and Ealey 1973). This mobility corresponds to that of human IgG immunoglobulin. Antibodies of the sphenodon (Marchalonis, Ealey, and Diener 1969), the most primitive of living reptiles, and the clawed toad (Marchalonis, Allen, and Saarni 1970), a primitive anuran amphibian, migrate slightly faster than do the γ globulins of the echidna. Proteins which migrated as γ globulins were present in the lamprey and shark but were not antibodies (Marchalonis and Edelman 1965, 1968b; Clem and Small 1967). These proteins were shown by studies involving the binding of radioactive iron to be transferrins. Transferrins of the lamprey (Webster and Pollara 1969) and lemon shark (Clem and Small 1967) have been described in some detail. Since not all proteins that resemble immunoglobulins in electrophoretic mobility are actually immunoglobulins, we must interpret early searches for the presence of antibodies in lower species that were based upon electrophoresis of serum alone (Engle et al. 1958; Good and Papermaster 1964) with great care.

Antisera of lower vertebrates have also been fractionated by ultracentrifugation on linear gradients of sucrose, a technique which allows estimation of the molecular size of proteins (Martin and Ames 1961). Mammalian antibodies can be resolved into two major components on the basis of size; the IgM molecules, which are characterized by a sedimentation coefficient of approximately 19S and the IgG proteins, having a sedimentation coefficient of 6.7S. Antibody activity usually appears earliest after immunization in the IgM component (Uhr 1964). As time after challenge increases, the vast bulk of antibodies are IgG molecules. This situation obtains clearly in mammals (Uhr 1964) and in anuran amphibians (Uhr, Finkelstein, and Franklin 1962; Marchalonis and Edelman 1966b) immunized with bacteriophage antigen. The presence of high and low molecular weight antibodies in other species represented in Figure 28 are not as clear-cut and require individual attention. The sphenodon possesses both IgM immunoglobulins and a distinct class of 7S immunoglobulin; however, the antigen used in the study (Marchalonis, Ealey, and Diener 1969) illustrated here was *Salmonella adelaide* flagella, which induced the formation of only IgM antibodies. The low molecular weight immunoglobulins present in the serum were isolated for chemical studies which will be described below. The carp, a teleost fish, possessed only IgM immunoglobulin in its serum (Marchalonis 1971a). However, instead of molecules characterized by the formula $(L_2\mu_2)_5$, these proteins consisted of tetramers of the form $(L_2\mu_2)_4$ (Shelton and Smith 1970). The existence of IgM molecules of intermediate size is an interesting occurrence in teleosts and suggests that factors regulating the degree of polymerization of IgM subunits are not constant from species to species. The dogfish shark possessed 19S antibodies to hemocyanin of the horseshoe crab, although it also contained 7S immunoglobulin in its serum (Marchalonis and Edelman 1965, 1966b).

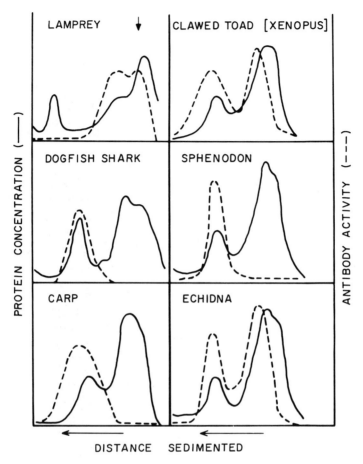

Fig. 28. Analysis of antisera from representatives of six vertebrate classes by ultracentrifugation on linear gradients of sucrose. Direction of sedimentation is from right to left. Protein concentration (———); antibody activity (---). In the echidna pattern, which typifies mammalian patterns, two peaks of antibody activity occur. The first peak is high molecular weight IgM (19S) antibodies, the second corresponds to IgG (7S) antibodies.

The fact that only 19S antibody (IgM pentamers) was found probably reflects the fact that the animals were bled at one month post immunization, whereas Clem and Small (1967) detected antibody activity in the 7S fraction obtained from animals bled at periods greater than six months after injection.

As I shall discuss below, shark 19S and 7S immunoglobulins contain the same light chains and heavy chains. Therefore, they represent different states of aggregation of molecules within the same class.

The size distribution of antibody activity in lamprey serum is unique because antibody activity is spread over a diffuse area ranging from molecules characterized by sedimentation coefficients of approximately 7S to 16S (Marchalonis and Edelman 1968b). These results indicated that lamprey antibodies to bacteriophage f2 possessed both light chains and heavy chains, but these polypeptide chains were not covalently linked via disulfide bonds (Marchalonis and Edelman 1968b). In this context, I should emphasize that hemoglobins (Braunitzer 1966) and thyroglobulins (Aloj, Salvatore, and Roche 1967) of lampreys show a much greater tendency to dissociate than do their homologues in higher vertebrates. Evidence that the lamprey antibody to bacteriophage f2 is recognizable immunoglobulin will be given in the Chapter 6, which presents a more detailed analysis of the polypeptide chain structure of IgM-like antibodies of lower vertebrates.

## *Synopsis*

It is now apparent that representatives of all classes of existing vertebrates contain cells of the lymphoid series, can form circulating antibodies in responses to a variety of foreign antigens, and exhibit cellular recognition and destruction reactions such as allograft rejection. Lymphoid tissues of cyclostomes are relatively disorganized, but all higher vertebrate classes possess definite lymphoid structures corresponding to the thymus and spleen of mammals. Certain lower vertebrates also possess lymphoid aggregates in areas which are not lymphoid in mammals, such as the anterior portion of the kidney of teleost fishes. The minimal requirements for antibody responses are the presence of circulating lymphocytes and some aggregates of lymphocytes. Direct evidence for collaboration between T lymphocytes and B lymphocytes does not yet exist below the phylogenetic level of avians, although indirect approaches suggest that this phenomenon is general among vertebrates. Analyses of lymphocyte structure and function in lower forms have only been initiated at this time. Careful investigation of the properties and functions of lymphocyte-like cells of protochordates is crucial to the problem of delineating the origins of vertebrate immunity.

# 6 Emergence of IgM Immunoglobulins

Data have been reviewed earlier that established that representatives of all classes of true vertebrates can form circulating antibodies in response to challenge with foreign antigen. These antibodies possessed generally similar electrophoretic properties which confined them to the so-called $\beta$ and $\gamma$ globulin fractions of serum (isoelectric points ranging from approximately pH 5.0 to pH 7.5). Furthermore, all of the species studied formed antibodies characterized by sedimentation coefficients sufficiently great to suggest that the mass of the molecules was considerably larger than that of IgG immunoglobulin, the predominant antibody class of mammals. In addition, some animals produced lower molecular weight antibodies that were generally comparable in size to IgG molecules. In this chapter, I shall present a more detailed analysis of the structural properties of immunoglobulins of the more primitive vertebrate classes such as cyclostomes, Chondrichthyes and Osteichthyes. I will raise the hypotheses that immunoglobulins of these animals bear a strong resemblance to molecules of the IgM class of mammals, that IgM immunoglobulins occur within vertebrates of all classes, and that the evolution of these molecules was a markedly conservative process.

## Immunoglobulins of cyclostomes

At this time antibodies of hagfish have not been obtained in a form suitable for detailed chemical analysis. It has been possible, however, through electrophoresis, ion-exchange chromatography, and gel filtration to isolate a single component of lamprey serum that possesses antibody activity to bacteriophage f2 (Marchalonis and Edelman 1968b). At high protein concentrations, this protein exhibits a sedimentation coefficient of 6.6S in the ultracentrifuge. Rabbit antibodies directed to this purified protein give a reaction of identity when tested against the higher molecular weight (14S) antibody (Marchalonis and Edelman 1968b). Figure 29 is a graph of molecular weight versus concentration of the lamprey 6.6S antibody as determined by measurements performed in the analytical ultracentrifuge. A strong concentration dependence was observed, with a molecular weight of approximately 150,000 at relatively high concentrations. These results indicate that the lamprey molecule dissociates and aggregates by a mechanism utilizing non-covalent interactions. A purified polypeptide chain component, the light chain, of the same protein is shown for comparison. The mass of this molecular species, determined in a dissociating solvent consisting of 20% acetic acid, was invariant as the concentration of protein increased.

The above data establish that lamprey antibody can be isolated, although it possesses an unexpected tendency to dissociate. In order to assess further the degree of similarity to well-characterized antibodies of mammals, it was necessary to determine whether this molecule consisted of light and heavy polypeptide chains. Unfortunately sufficient 14S antibody could not be obtained for detailed studies, possibly because the molecule dissociates during purification. However, the 6.6S antibody was obtained in amounts suitable for partial chemical analysis. Since the two antibodies are antigenically identical, the higher molecular weight protein is probably an aggregate of the 6.6S form. The lamprey antibody was analyzed by gel electrophoresis in acid urea and it was found that: (a) it contained polypeptide subunits resembling light chains and $\mu$ heavy chains of mammals in gel penetration; (b) the light chains gave a diffuse pattern similar to that exhibited by light chains of normal human immunoglobulin under these conditions; and (c) these polypeptide chains could be separated merely by exposing the intact molecule to dissociating solvents. The dissociation behavior was confirmed by studies involving gel filtration and analytical ultracentrifugation, which also corroborated the molecular weights of the polypeptide chains.

The capacity of lamprey antibody to dissociate into polypeptide chains without cleavage of interchain disulfide bonds was, at the time, a unique finding among immunoglobulins. However, independent studies of certain IgA myeloma proteins of man (Grey et al. 1968; Jerry, Kunkel, and Grey 1970) and mouse (Abel and Grey 1968; Warner and Marchalonis 1972) established that these proteins lacked disulfide bonds covalently linking the light chains to the heavy chains (Grey et al. 1968). A situation such as this lack of disulfide bonds strikes the investigator as aberrant if it is first discovered in a primitive species. However, this

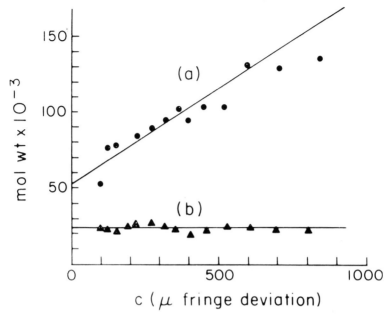

Fig. 29. Graph of variation of molecular weight as a function of concentration of lamprey 7S immunoglobulin ( •—• ) and purified light chain (Δ—Δ). Measurements were made by equilibrium ultracentrifugation under conditions developed by Yphantis (1964). The molecular weight determination of the light chain was carried out under dissociating conditions and is independent of concentration. Determination of the molecular weight of the intact molecule was carried out in a physiological buffer and shows a strong concentration dependence, indicating dissociation at low concentrations and aggregation at high protein concentration.

result, like many to follow, had its counterparts in more acceptable mammalian systems. On the basis of present data it is not possible to decide whether the lamprey immunoglobulin represents a situation in which the H-L interchain disulfide bond had not yet emerged or had been present but was lost. Such a change probably requires only the gain or loss of one-half of a cysteine amino acid residue, probably through a single nucleotide change in the DNA codons of the amino acid involved.

Another key point of the lamprey results is that the light chain seems to show a heterogeneity in electrophoretic mobility comparable to that of normal human light chain, which represents a pool of molecules differing in variable region sequence. This observation is consistent with the fact that lampreys can respond to a variety of antigens. The exact degree of this heterogeneity remains to be determined, which can only be done by using more sensitive techniques.

Lamprey antibodies were first studied by Boffa et al. (1967) who characterized them as electrophoretic γ globulins with sedimentation coefficients of 6.6S and 19S. Litman et al. (1970) more recently isolated lamprey proteins possessing agglutinating activity to human erythrocyte "O" antigens. These molecules showed a tendency to dissociate spontaneously into subunits, but no evidence of light polypeptide chains was found. The antibody consisted of four non-covalently-linked units of approximately 90,000 daltons which are larger than μ chains (M. W. 70,000). In addition, these workers subjected the purified protein to analysis by circular dichroism and found its spectrum differed markedly from that of immunoglobulin (Litman et al. 1970). Although it is possible that this molecule is a polymer of μ-type heavy chains similar to those of the antibody to bacteriophage f2 described previously, it may also be a totally distinct protein. In line with this proposal, the antibody to "O" antigen migrates electrophoretically as an α globulin which is distinctly faster than that of the bacteriophage antibody, although a lack of light chains might alter the electrophoretic mobility of the aggregate. The argument that this is a μ-chain polymer cannot be disregarded, particularly since Choi and Good (1971) have observed a similar molecule in bursectomized chickens. In any case, further chemical studies are required to establish the relationship of this molecule to those of other vertebrates. The similarities of these proteins to certain nonimmunoglobulin proteins that possess binding specificity for carbohydrates will be discussed below.

Another protein molecule present in lamprey serum deserves attention. Unimmunized lampreys possess high titer agglutination activity to horse erythrocytes that is associated with an extremely large protein having a sedimentation coefficient of 48S. This molecule is composed of low molecular weight subunits that require cleavage of disulfide bonds to separate them (Marchalonis and Edelman 1968b). This protein is similar to invertebrate natural hemagglutinins in two respects. First, agglutinating activity is potentiated by the presence of $Ca^{++}$ ions. In the second place, the amino acid composition of this protein resembles that of the agglutinin of the horseshoe crab more closely than it does those of vertebrate antibodies. Since cyclostomes occupy a phylogenetic position bridging that of the prevertebrates and true vertebrates, it is possible that they share the biochemical properties of each. The natural hemagglutinin molecule may provide direct evidence of this.

Present data establish that both lampreys and hagfish can form antibodies to a variety of antigens. Lamprey antibodies to bacteriophage f2 resemble those of higher vertebrates in polypeptide chain structure and diversity. The basic unit of antibody structure probably consists of 2 light chains and 2 $\mu$-type heavy chains. In addition, higher polymers of this $L_2\mu_2$ unit occur and possess antibody activity. Amino acid composition data have been obtained for lamprey polypeptide chains, and these will be compared with those of other heavy chains later in this chapter in order to assess the relatedness among these polypeptides. Lamprey immunoglobulins are characterized by a tendency to dissociate, which probably results from a lack of L-H interchain disulfide bonds. This observation is in accordance with reports that other proteins of lampreys such as hemoglobins (Braunitzer 1966) and thyroglobulins (Aloj, Salvatore, and Roche 1967) possess a greater tendency to dissociate than do their homologues in higher species. The attainment of detailed chemical information about lamprey antibodies is critical to the development of schemes of immunoglobulin evolution. Such data may prove extremely difficult to obtain, however, because the amount of immunoglobulin present in the serum of these cyclostomes is approximately one-fiftieth that found in mammalian serum (Marchalonis and Edelman 1968b).

## Immunoglobulins of Chondrichthyes

In Chapter 5, I presented data obtained for the smooth dogfish, *Mustelus canis*. This species possesses immunoglobulins with an electrophoretic mobility comparable to that of human IgM immunoglobulin, which can be resolved into two components on the basis of molecular weight. Clem and Small (1967) and Sigel and Clem (1966) observed antibody activity to influenza virus, bovine serum albumin, and the dinitrophenyl hapten occurring in both components during long-term immunization studies. The antigenic relationship of the 17S and 7S immunoglobulins of the dogfish (Marchalonis and Edelman 1965) and other sharks (Clem and Small 1967) resembles that described above for the lamprey. The components are antigenically identical when tested against one another using rabbit antiserum to either protein. This observation indicates that both immunoglobulins belong to the same class.

To confirm that both molecular species in the dogfish shark represent the same immunoglobulin class, it was necessary to isolate and to compare the heavy polypeptide chains of both immunoglobulins. The chemical properties of the heavy chain also provided an indication of the mammalian class to which the elasmobranch proteins can be considered homologous. With these goals in view, the behavior of purified dogfish 17S and 7S immunoglobulins were analyzed by gel filtration under dissociating conditions (Marchalonis and Edelman 1965). This approach allowed separation of polypeptide chains on the basis of molecular size.

As illustrated in Figure 30, intact molecules of both immunoglobulins were too large to be within the fractionation range of Sephadex G-100 equilibrated with 6M urea, 1M propionic acid, and emerged at the void volume of the column. This result contrasts with the situation observed for lamprey immunoglobulins which dissociated into light and $\mu$-like chains upon contact with solvents disrupting noncovalent interactions. It was necessary to cleave disulfide bonds in order to resolve the shark immunoglobulins into polypeptide chains. Both dogfish immunoglobulins were composed of polypeptides emerging at elution volumes comparable to those characteristic of human light chains and $\mu$ chains under these conditions. The isolated light chains possessed a molecular weight of approximately 20,000 and the heavy chains of approximately 70,000. These values corresponded closely to those reported for mammalian light chains and $\mu$ heavy chains.

The identity of the corresponding polypeptide chains of the two immunoglobulin components was corroborated by polyacrylamide gel electrophoresis in acid urea, antigenic analysis, mapping of tryptic peptides, and amino acid compositions. These properties, moreover, strengthened the conclusion that the shark antibodies resembled those of the IgM class of mammals. From the relative yields of heavy chains (70%) and light chains (30%) obtained by gel filtration and their molecular weights, (980,000 for the 17S molecule and 180,000 for the 7S protein), it was concluded that equal numbers of light and heavy chains were present in the molecules (Marchalonis and Edelman 1966a). The mass relationships were consistent with the conclusions that the 7S protein consists of 2 light chains and two heavy chains, and the 17S molecule exists as a pentamer of this unit. By analogy with the IgM immunoglobulins of higher species (World Health Organization 1964) these structures were presented $L_2\mu_2$ and $(L_2\mu_2)_5$. Similar structures were proposed for immunoglobulins of the lemon shark (Clem and Small 1967) and horned shark (Frommel et al. 1971). In addition to light chains and $\mu$-type heavy chains, the high molecular weight IgM immunoglobulins of the leopard shark possessed a small amount of acidic polypeptide of molecular weight 20,000, which was covalently bound to the molecule by disulfide bonds. Klaus et al. (1971) claimed that this polypeptide is analogous to the J chain of mammalian IgA and IgM immunoglobulin.

The tentative identification of elasmobranch immunoglobulins as IgM type gained further support from their relatively high carbohydrate contents (Marchalonis and Edelman 1966a; Clem and Small 1967) and from physical studies employing electron microscopy (Feinstein and Munn 1969; Parkhouse, Askonas, and Dourmashkin 1970) and circular dichroic analysis (Litman et al. 1971d). Feinstein and Munn (1969) have taken electron micrographs of dogfish shark 17S immunoglobulins

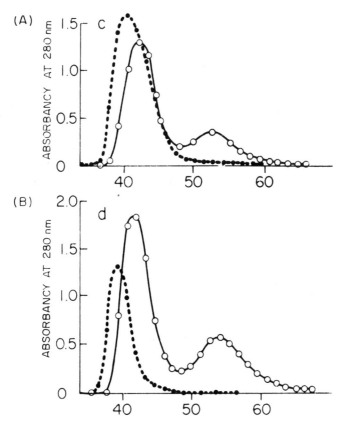

Fig. 30. Separation of light and heavy polypeptide chains of smooth dogfish immunoglobulin by gel filtration on Sephadex G-100 under dissociating conditions. (A) o——o, 17S immunoglobulin reduced and alkylated in the presence of 8M urea; ● --- ●, 17S immunoglobulin dissolved in 8M urea, but not reduced. (B) o——o, 7S immunoglobulin reduced and alkylated in the presence of 8M urea; ● --- ●, 7S immunoglobulin dissolved in 8M urea but not reduced. The dissociating solvent used for gel filtration was 6M urea, 1M propionic acid. Notice correspondence in elution positions of light chains of 17S and 7S immunoglobulin, and also for heavy chains of both immunoglobulins. Intact immunoglobulins are too large to be resolved by the gel; these emerge at the void volume of the column. (From Marchalonis and Edelman 1965).

showing that these proteins possess the five branched cyclic symmetry and dimensions characteristic of mammalian IgM immunoglobulins (Feinstein and Munn 1969; Parkhouse, Askonas, and Dourmashkin 1970). Litman et al. (1971d) analyzed circular dichroic spectra of immunoglobulins from horned sharks and leopard sharks and seven other lower vertebrate species. This technique allows the estimation of the degree of $\alpha$ or $\beta$ helix characteristic of the hydrogen-bonded secondary structure of proteins. Immunoglobulins of all species except the lamprey possessed very similar spectra suggesting that their secondary structure consisted primarily of pleated sheets, and very little evidence for $\alpha$ helix was obtained. Moreover, IgG and IgM of mammals were distinguishable by this technique, and Litman and his coworkers concluded that the immunoglobulins of sharks, paddlefish, and bowfin are extremely similar to mammalian IgM immunoglobulins in secondary structures. The lamprey immunoglobulin studied here differed from the other proteins in exhibiting a high degree of $\alpha$ helix. This molecule was the antibody to human "O" erythrocytes (Litman et al. 1970) which differed markedly from the antibody to bacteriophage f2 considered by Edelman and Marchalonis (1968b). The problem of homology among immunoglobulin heavy chains of vertebrate species will be discussed further at the close of this chapter.

Although 19S IgM molecules are known to dissociate into 7S subunits upon mild reduction (Deutsch and Morton 1957), the presence of large amounts of naturally-occurring 7S IgM was a striking find when it was initially observed in 1965. However, various disease states in man, such as ataxia telangiectasia (Stobo and Tomasi 1967), are accompanied by the presence of abnormally large amounts of low molecular weight IgM immunoglobulin. In addition, trypanosome infections in mammals induce 7S IgM antibodies (Frommel et al. 1970). Recent studies, furthermore, have established that the major class of immunoglobulin found on the surfaces of T lymphocytes (Marchalonis and Cone 1973a; Moroz and Hahn 1973; Boylston and Mowbray 1973; Ladoulis et al. 1974), B lymphocytes (Vitetta et al. 1971; Marchalonis, Cone, and Atwell 1972) and cells from a variety of lymphoid tumors (Klein and Eskeland 1971; Marchalonis and Cone 1973a; Marchalonis, Atwell, and Haustein 1974) is the 7S form of IgM immunoglobulin. Clem and his colleagues (Clem and Leslie 1969; Fidler, Clem, and Small 1969) have attempted to delineate the physiological roles and interrelationships of shark 17S and 7S IgM antibodies. They found that (a) the 7S molecule is neither a precursor nor a degradation product of the 17S molecule, and (b) the 17S molecules are localized to the intravascular spaces of the body whereas the 7S species distributes both extravascularly and intravascularly. Although these two proteins represent the same immunoglobulin class, Clem and Leslie (1969) conclude that the two molecular forms of shark IgM functionally mimic IgM and IgG humoral immunoglobulins of mammals.

The discussion thus far has considered that shark immunoglobulins are composed of polypeptide chains similar to those of mammalian IgM antibody but has not approached the question of heterogeneity within these chains. Polyacrylamide gel electrophoresis of shark light chains established that these proteins resemble those of mammals in electrophoretic heterogeneity (Marchalonis and Edelman 1965, 1966a; Clem and Small 1967; Frommel et al. 1971). This result suggests that shark immunoglobulins contain variable regions comparable to those described in mammalian immunoglobulins. Limited sequence work confined to the amino-terminal position of light and heavy chains of leopard sharks provided direct evidence for this conjecture (Klaus, Nitecki, and Goodman 1971). I shall return to this point in further detail in Chapter 9. Major technical difficulties of two types are encountered in sequence studies of shark immunoglobulins. One obstacle stems from the heterogeneity of the variable region, which presents a challenge because plasma cell tumors secreting homogeneous immunoglobulins have not been observed in these species. To circumvent this problem, Clem and Leslie (1971) injected sharks with pneumococcal polysaccharide antigens which induce antibodies characterized by restricted heterogeneity in rabbit (Krause 1970). This approach has proved feasible in that the sharks also made relatively homogeneous antibodies as assessed by comparison with normal immunoglobulin polypeptide chains by polyacrylamide gel electrophoresis. N-terminal sequence data of shark antibodies are given in Chapter 9. The second difficulty arises because a proportion of shark light chains and heavy chains possess blocked aminoterminal residues which are not amenable to standard techniques of amino acid sequence analysis. Such blocked light-chain molecules have been taken to resemble $\lambda$ light chains of mammals and the conclusion has been drawn that $\kappa$ and $\lambda$ light chains were present at the phylogenetic level of elasmobranchs (Suran and Papermaster 1967; Hood et al. 1967). I shall return to this problem and that of the phylogenetic emergence of variable region subgroups below.

The choice of shark immunoglobulins as the first lower vertebrate antibodies to be studied in detail proved fortunate, because the physical similarities to mammalian IgM molecules were so pronounced that identification was relatively straightforward. Present evidence based upon two species of rays indicates that these elasmobranchs also possess IgM immunoglobulin, but the molecular state is distinct from that of sharks. Figure 31 illustrates comparisons by polyacrylamide gel electrophoresis in acid urea of polypeptide chains of immunoglobulins of man and the northern stingray (*Dasyatis centroura*). The stingray immunoglobulin possesses light chains which give a diffuse band comparable to that of normal human light chains. The ray heavy chain resembles the human $\mu$ chain in penetration of the gel. The similarity of the ray chains to

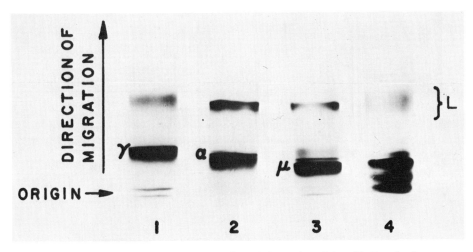

Fig. 31. Comparison by polyacrylamide gel electrophoresis in acid urea of polypeptide chains of stingray (*D. centroura*) and human immunoglobulins. The proteins were reduced and alkylated in the presence of 9M
urea. Origins: 1, human IgG immunoglobulin; 2, human IgA immunoglobulin (myeloma protein); 3, human IgM immunoglobulin
(myeloma protein); 4, stingray immunoglobulin. L, position of human
light chains. $\gamma$, $\alpha$, and $\mu$ refer to the position of human $\gamma$, $\alpha$, and $\mu$
heavy chains, respectively. (From Marchalonis and Schonfeld 1970).

human light and $\mu$ chains was further illustrated by measurements of
molecular weights (22,000 and 72,000 respectively) and amino acid compositions. The molecular weight of the intact immunoglobulin molecule,
however, corresponded to neither that of classical 19S IgM or its 7S subunit. Figure 32 compares this immunoglobulin to the human IgM and
IgG proteins by gel filtration on Sepharose 6B. It is clearly distinct in
molecular size from the two human proteins. The molecular weight
obtained from its elution position on a calibrated column of the Sepharose gel was approximately 360,000. Thus, IgM immunoglobulin from
*D. centroura* probably consists of two units of the form $L_2\mu_2$. No 7S IgM
was observed in the serum, although a small amount of 19S IgM was
present. Johnston et al. (1971) have isolated immunoglobulin from the
southern ray *D. americana*, finding that the major component was a pentamer of units, each consisting of two light chains and two heavy chains.
It is interesting that IgM polymers of different sizes exist in different
species of ray. The pentamer in *D. americana* is similar to the IgM polymers of sharks and man, whereas the dimeric form occurring in *D.
centroura* represents a polymerized IgM immunoglobulin which has not
been found in other species. The factors which uniquely determine the

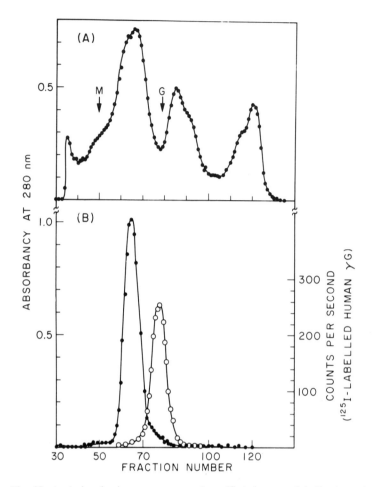

Fig. 32. Analysis of stingray serum and purified immunoglobulin by gel filtration on Sepharose 6B. (A) Whole stingray serum. ●——●, absorbance at 280nm; M, elution volume of human IgM immunoglobulin (M.W. 950,000); G, volume at which human IgG (M.W. 150,000) elutes from the column. (B) Purified stingray immunoglobulin. ●——●, absorbance at 280 nm; o——o, counts per minute of [125]I-labeled human IgG which was added as an internal standard. (From Marchalonis and Schonfeld 1970).

tendency of IgM subunits (mass 180,000 daltons) to form cyclic polymers of discrete size remain to be determined, but they are most likely a function of the properties of the J chain (Kownatzki 1973). The finding of an IgM polymer of a size intermediate between the usual pentamer and monomer forms in *D. centroura* provides a direct introduction to the immunoglobulins of Osteichthyes or bony fishes.

## IgM immunoglobulins of bony fishes

Bony fishes form antibodies to viral (Uhr, Finkelstein, and Franklin 1962; Sigel and Clem 1965, 1966), cellular (Chiller et al. 1969; Chiller, Hodgins, and Weiser 1969), and protein (Hodgins, Weiser, and Ridgway 1967; Everhart and Shefner 1966; Avtalion 1969; Trump 1970; Marchalonis 1971a) antigens. In the case of protein antigens, teleost species such as the goldfish readily produce precipitating antibodies which resemble those of rabbits in specificity for the particular protein (Everhart and Shefner 1966; Marchalonis 1971a). Moreover, teleosts readily synthesize antibodies to haptens such as DNP (Smith and Potter 1969; Clem and Small 1970; Frenzel and Ambrosius 1971; Ambrosius and Fiebig 1972). Immunoglobulins from representatives of the three major groups of bony fishes have recently been isolated and partially characterized. Table 5 presents data on the molecular weights of the intact molecules and their constituent polypeptide chains. The sedimentation coefficients of immunoglobulins of all species were considerably lower than that of mammalian IgM macroglobulin. In the cases cited here, electron microscopic analysis and measurements of molecular weights based upon equilibrium ultracentrifugation confirm this difference and indicate that high molecular weight antibodies consist of tetramers of the form $(L_2\mu_2)_4$. I should point out that sedimentation coefficients usually do not provide exact information regarding molecular size because these are measurements of the rate at which a macromolecule sediments in a unit centrifugal field. This rate is dependent upon the shape of the molecule as well as its molecular weight. Moreover, such hydrodynamic measurements are valid only when extrapolated to a protein concentration of zero under standard conditions of temperature and viscosity. Since most workers have not carried out their experiments under stringent conditions, a large amount of variability exists in reported values for the sedimentation coefficients of immunoglobulins. This situation is pronounced in the case of IgM macromolecules, whose sedimentation coefficients show a strong concentration dependence (Miller and Metzger 1965). This variation with concentration is illustrated in Figure 33. The tetrameric conformations of IgM immunoglobulins of the gar (Acton et al. 1971a), paddlefish (Acton et al. 1971b), and carp (Shelton and Smith 1970) were determined directly by electron microscopy, a technique which obviates the above difficulty in interpretation.

Another major point regarding immunoglobulins of bony fishes is that generally only one species of immunoglobulin, the intermediate IgM polymer, exists in the serum (Shelton and Smith 1970; Acton et al. 1971c; Marchalonis 1971a). Immunoglobulins characterized by molecular weights lower than that of this protein have been isolated from the

TABLE 5: Immunoglobulins of bony fishes.

| Species | Sedimentation coefficient | Intact |
|---|---|---|
| Paddlefish (*Polydon spathula*) Chondrostean[1, 2] | 19S (14.2S) | 870,000 660,000 |
| Gar (*Lepisoteus platyrincus*) Holostean[3] | 14S | 610,000 |
| Bowfin (*Amia calva*) Holostean[4] | 13.6S 6.3S | (700,000) |
| Grouper (*Epinephelus itaria*) Teleost[5] | 16S 6.4S | 700,000 120,000 |
| Carp (*Cyprinis Carpio*) Teleost[6] | 16S | (740,000) |
| Goldfish (*Carassius auratus*) Teleost[7] | 16S | |
| Plaice (*Pleuronectes platessa*) Teleost[8] | 12S | |

bowfin (Litman et al. 1971b) and the grouper (Clem 1971). Uhr and his colleagues (1962) provided preliminary evidence for the existence of low molecular weight immunoglobulin in the goldfish; however, more detailed studies have recently established that this species possesses only 16S IgM immunoglobulin (Trump 1970; Marchalonis 1971a). Low molecular weight immunoglobulins of the gar and bowfin possess light chains characterized by a mass typical of that of mammalian light chains. The heavy chains, however, possess a molecular weight considerably lower than that of the $\mu$ chain. Because the 6.3S and 13.6S immunoglobulins of the bowfin were antigenically identical, Litman et al. (1971b) proposed that the low molecular weight heavy chain of the bowfin represents the $\mu$ chain which has had a polypeptide segment of about 110 amino acid residues deleted from it. Clem (1971) found that the low molecular weight heavy chain of the grouper shared some but not all of the antigenic properties of the $\mu$ chain. The chain differed considerably in size from the $\mu$ chain, having a mass approximately 30,000 daltons less than that of the $\mu$ chain. Moreover, the low molecular weight heavy chain lacked carbohydrate, whereas the $\mu$ chain was rich in this moiety. Both Litman and colleagues (1971b) and Clem (1971) consider the low molecular weight heavy chains to represent fragments of the $\mu$ chain. Two pos-

| Molecular weights | | Structural formula |
|---|---|---|
| Light chain | Heavy chain | |
| 23,500 | 75,300 | $(L_2\mu_2)_n$ |
| | (58,000) | n probably 4 |
| 22,000 | 70,000 | $(L_2\mu_2)_4$ |
| 24,000 | 70,000 | $(L_2\mu)_4$ |
| 24,000 | 52,000 | $L_2\mu_2^F$ |
| 22,000 | 70,000 | $(L_2\mu_2)_4$ |
| 22,000 | 40,000 | $(L_2\mu_2^F)$ |
| 24,000 | 71,400 | $(L_2\mu_2)_4$ |
| 23,300 | 72,600 | $(L_2\mu_2)_n$ |
| | | n probably 4 |
| 22,000 | 70,000 | $(L_2\mu_2)_n$ |

*Sources:* (1) Pollara et al. 1968; (2) Acton et al. 1971b; (3) Bradshaw, Clem and Sigel 1969, 1971; (4) Litman et al. 1971a; (5) Clem 1971; (6) Shelton and Smith 1970; (7) Marchalonis 1971a; (8) Fletcher and Grant 1969.

$\mu^F$ indicates that the heavy chain with a molecular weight of 40,000 appears to be a fragment of the $\mu$ chain. Possibly it arose through a deletion of one homology region of the Fc portion of the $\mu$ chain.

sibilities can be proposed to account for the presence of these low molecular weight immunoglobulin subunits. Either the molecule is a proteolytic degradation product of $\mu$ chain or it is the naturally synthesized product of a gene which resulted from duplication of the $\mu$-chain gene within the recent ancestors of the species considered. If the latter alternative proves to be correct, the duplicated gene underwent a deletion which accounts for the lower molecular weight of its product than that of the $\mu$-chain gene. The implications of mass deletions of approximately 12,000 daltons and multiples of this unit will be discussed below in the context of the evolution of immunoglobulins by addition or deletion of gene segments corresponding to individual domains of the molecules. Since the low molecular weight immunoglobulins bind antigen (Clem 1971; Litman et al. 1971b), the domains must have been deleted from the Fc regions of the molecules.

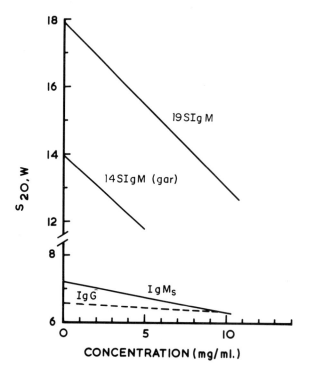

Fig. 33. Graph of the variation of sedimentation coefficient as a function of
protein concentration for human 19S IgM, human IgM subunit
(IgM$_S$), human IgG immunoglobulin, and gar 14S IgM immunoglobulin. Sedimentation coefficients are not hydrodynamically valid unless
extrapolated to zero concentration. (Based upon Miller and Metzger
1965, and Acton et al. 1971a).

## Tentative evidence for IgM subclasses in elasmobranchs and teleosts

The preceding discussion of low molecular weight immunoglobulins
or immunoglobulin fragments in certain osteichthyian fishes raises the
possibility that genes encoding heavy chains might have duplicated
during the evolution of the separate vertebrate classes. A series of such
events clearly occurred during the evolution of the primates, generating
a series of tandemly arranged genes specifying the γ-chain subclasses
(see Chapter 3). It is conceivable that genetic events of this nature might
have been part of the process of immunoglobulin evolution in any
extant vertebrate class or family, but definitive evidence on this point
is lacking for immunoglobulins of lower species.

Two studies of the antigenic relationships among immunoglobulins present in lower species bear upon the issue of the origin of subclasses. Gitlin, Perricelli and Gitlin (1973) investigated physical and antigenic properties of 19S and 7S antibodies of sixteen species of sharks. They found that each species possessed at least two but not more than four types of immunoglobulin which were distinguished by the antigenic properties of their heavy chains. The mako shark (*Isurus oxyrinchus*), an ancient galeoid shark, possesses two antigenically distinguishable immunoglobulins each of which exist in both 19S and 7S forms. Some galeoid sharks possess three antigenically distinct immunoglobulins while the most complicated situation is that of a squaloid shark, the spiny dogfish (*Squalus acanthias*). Four distinct immunoglobulins were detected in this species; three existed both as 19S and 7S molecules and another was found only in 19S form. Since the molecular species appear to coexist normally in the serum of all sharks, the distinct heavy chains are isotypic molecules that are probably encoded by separate germ line genes. Because the properties of heavy chains of all species of sharks and rays are extremely similar and resemble the $\mu$ chain of mammals, the distinct immunoglobulins of the sharks can be considered subclasses of the IgM type. The ancestors of sharks might have possessed a gene specifying a $\mu$-type chain and, during the course of evolution, the gene could duplicate to form a tandem array of heavy chain genes. Each gene would then exist as a separate genetic unit, and mutation would result in the emergence of antigenic differences. This sequence of events parallels the phylogenetic development of IgG subclasses in mammalian evolution (see Chapters 3 and 15).

An example of antigenically distinct immunoglobulins was also observed in teleost fishes. Trump (1970) carried out investigations of goldfish antibodies to bovine serum albumin, finding that goldfish possess only immunoglobulin characterized by a sedimentation coefficient of 16S. However, within these 16S molecules, he could distinguish two populations of antibodies on the basis of electrophoretic mobility and antigenic properties. Trump discounted the possibility that the antigenic difference represented allelic variation, because rabbit antiserum to goldfish IgM was used in the experiments and it seemed doubtful that this reagent could detect minor antigenic variations in such a distant species. Such reagents tend to detect major differences such as antigenic determinants characteristic of immunoglobulin of the species or determinants defining major immunoglobulin classes within the species. Relatively small differences within immunoglobulin classes, e.g., allotypic differences, require special reagents such as rheumatoid factors from individuals of the same species for their detection (Grubb 1970). Although the possibility of distinct light-chain types cannot be excluded at this time, it is unlikely that antigenically distinct light chains are responsible for Trump's results because all immunoglobulin classes of higher vertebrates contain both $\kappa$ and $\lambda$ light chains. Thus, the cause of the electrophoretic and antigenic differences between the two IgM subpopulations probably resides in the heavy chain. As in the shark example, it is conceivable that the ancestral $\mu$-chain gene of osteichthyian fishes underwent a duplication within the evolution of the goldfish, and two IgM subclasses resulted. The existence of such subclasses among Osteichthyes indicates that the mechanism of the generation of immunoglobulin classes and subclasses by gene duplication occurs within all classes of vertebrates.

## IgM immunoglobulins of Dipnoi

The Dipnoi possess high molecular weight serum immunoglobulins characterized by a sedimentation coefficient of 19.0S (Marchalonis 1969; Litman et al. 1971a). In addition, a higher molecular weight component, which is probably a dimer of this unit, is present at lower levels. Such higher aggregates are frequently observed for mammalian IgM immunoglobulins. The IgM-like immunoglobulin is composed of light chains (M. W. 22,000) and heavy chains which resemble the $\mu$ chain of man in electrophoretic mobility and mass (M. W. 70,000). Although the molecular weight of the intact IgM molecule has not been precisely determined, it is likely that the structure is similar to the pentameric IgM molecules of elasmobranchs and mammals. Lungfish, in addition, possess a low molecular weight immunoglobulin that represents a class distinct from the IgM molecules and constitutes the major serum immunoglobulin in these species. The structural properties of this immunoglobulin will be considered in Chapter 8.

## IgM immunoglobulins of the tetrapods

The preceding data establish the fact that all representatives of the lower vertebrates studied possess immune macroglobulins composed of polypeptide chains which resemble light chains and $\mu$ chains of man in molecular weights and electrophoretic properties. Moreover, except for the case of the cyclostomes, these polypeptides are linked covalently via disulfide bonds in a manner corresponding to that observed in human IgM polymers. Vertebrate species more advanced than fishes also possess immune macroglobulins in their serum. In some cases, e.g., the turtle (Grey 1966), such proteins comprise a relatively large fraction of total serum proteins. Table 6 summarizes data obtained for immune macroglobulins of amphibian, reptilian, avian and lower mammalian species. In the cases where physical measurements were carried out, the intact molecules possessed molecular weights of approximately 900,000, a value which was consistent with a pentameric structure. This point was established directly by electron microscopic visualization in the case of turtle IgM molecules (Acton et al. 1972). An interesting structure was obtained for *Xenopus* IgM, however, which exists as a cyclic polymer containing six units (Parkhouse, Askonas, and Dourmashkin 1970). High molecular weight IgM immunoglobulins of other anurans have masses consistent with a pentameric arrangement (Atwell, J. L. unpublished observations; Parkhouse, R. M. E. personal communication). Recent studies of the reassembly of human IgM immunoglobulins show that the J chain plays an important role in controlling the assembly of IgM subunits (Kownatzki 1973). Polymers assembled from reduced subunits in the absence of J chain formed large aggregates, whereas those formed in the presence of J chain existed chiefly as pentamers. The observations with immune macroglobulins of lower species showed that the numbers of units incorporated into the intact structure is a species specific factor, e.g., 5 for sharks, 4 for teleosts, and 6 for *Xenopus*. This probably reflects the symmetry of the J chain produced in the different species.

With one exception, all of the representative tetrapods included in Table 6 possess classes of immunoglobulin which are formally distinct from the IgM molecules described here. These immunoglobulins, which possess heavy chains that are readily distinguishable from $\mu$ chains, will be discussed in Chapter 8. The exceptional case, that of the urodele amphibian, *Necturus*, bears upon the evolutionary level of the origin of $\gamma$-like heavy chains and warrants further consideration. Other urodeles also appear to possess only high molecular weight IgM molecules (Fougereau, Houdayer, and Dorson 1972; Ambrosius et al. 1970). We estimated the molecular weight of the intact *Necturus* IgM immunoglobulin to be approximately 950,000, a value comparable to that of polymerized IgM molecules of higher species. Upon reduction and alkylation followed by analysis by polacrylamide gel electrophoresis, this molecule was resolved into heavy chains similar to $\mu$ chains and light chains comparable to those of other vertebrates (Figure 34). The *Necturus* light chains exhibited an interesting feature that was previously observed for light chains of the bullfrog (Marchalonis and Edelman 1966b). The chains migrate as a number of discrete bands when subjected to gel electrophoresis in acid urea. This observation suggests that the charge distribution pattern differs from that characteristic of mammalian light chains. Furthermore, it raises the possibility that light chains of amphibian species may show a decreased heterogeneity rela-

TABLE 6: Properties of IgM immunoglobulins of tetrapod vertebrates.

| Species | Vertebrate class | Intact |
|---|---|---|
| Mudpuppy (*Necturus maculosus*)[1] | Urodele amphibian | 950,000 |
| Bullfrog (*Rana catesbeiana*)[2] | Anuran amphibian | N.D. |
| Marine toad (*Bufo marinus*)[3] | Anuran amphibian | 880,000 |
| *Xenopus laevis*[4, 5, 6] | Anuran amphibian | N.D. N.D. |
| Sleepy lizard (*Tiliqua rugosa*)[7] | Reptile | 953,000 |
| Turtle (*Pseudymus seripta*)[8] | Reptile | 850,000 |
| Chicken (*Gallus domesticus*)[9] | Avian | 890,000 |
| Echidna (*Tachyglossus aculeatus*)[10] | Monotreme mammal | 950,000 |

tive to that of other species. If this possibility is verified, amphibian antibodies would prove extremely suitable for amino acid sequence analysis.

If the mudpuppy *Necturus* is truly representative of all urodeles, the finding that only IgM-like immunoglobulins are present raises provocative questions bearing upon the evolution of immunoglobulin structure and function. Baldwin and Cohen (1970) have shown that urodeles are capable of producing enhancing antibodies which prolong the survival of allografts. In mammals, enhancing antibodies have been shown to be IgM immunoglobulins, and IgM antibodies are generally considered to be cytotoxic for transplanted cells (Voisin 1971). The allograft studies in urodeles then imply that either some trace quantities of another immunoglobulin class is present, or that high molecular weight IgM antibodies can function as enhancing antibodies. No evidence in support of the former alternative was observed by Marchalonis and Cohen (1973). Precedents exist for the latter hypothesis because secretory immunoglobulins in fish, which possess only IgM immunoglobulins, are IgM immunoglobulins (Fletcher and Grant 1969; Bradshaw, Richards, and Sigel 1971). Secretory immunoglobulins in mammals, however, belong to the IgA class.

The pronounced similarity between the $\mu$ chains of lower species and those of man is illustrated in Figure 30, above, which compares the

| Molecular weights | | | |
| --- | --- | --- | --- |
| L-chain | $\mu$-chain | Molecular formula | %CHO |
| 22,000 | 70,000 | $(L_2\mu_2)_5$ | N.D.[a] |
| 22,000 | 72,000 | $(L_2\mu_2)_5$? | 10 |
| 22,500 | 67,000 | $(L_2\mu_2)_5$ | 7.6 |
| 22,000 | 70,000 | | N.D. |
| 26,700 | 74,500 | $(L_2\mu_2)_6$ | N.D. |
| 22,400 | 77,000 | $(L_2\mu_2)_5$ | N.D. |
| 22,500 | 67,000 | $(L_2\mu_2)_5$ | 6.7 |
| 22,000 | 70,000 | $(L_2\mu_2)_5$ | (Hexose) 2.6 |
| 22,500 | 69,000 | $(L_2\mu_2)_5$ | (Hexose) 6.4 |

*Sources:* (1) Marchalonis and Cohen 1973; (2) Marchalonis and Edelman 1966b; (3) Acton et al. 1972a; (4) Marchalonis, Allen, and Saarni 1970; (5) Haji-Azimi 1971; (6) Parkhouse, Askonas, and Dourmashkin 1970; (7) Wetherall 1969; (8) Acton et al. 1972c; (9) Leslie and Clem 1969; (10) Atwell, Marchalonis, and Ealey 1973.

a. N.D. = not determined.

polypeptide chains of stingray IgM with those of human IgM immunoglobulin using polyacrylamide gel electrophoresis in acid urea. This technique resolves proteins on the basis of both charge and size. Polyacrylamide gel electrophoresis of reduced, alkylated immunoglobulin chains can also be performed in the presence of the anionic detergent, sodium dodecylsulfate (SDS). Separation in this system is predominantly based upon molecular weight as long as the proteins do not contain appreciable amounts of carbohydrate, which decreases the amount of SDS bound per mass unit (Glossman and Neville 1971; Segrest et al. 1971). Accepting this constraint on the use of gel electrophoresis in SDS as a means for the quantitative determination of heavy-chain mass, it is nevertheless possible to obtain excellent resolution of immunoglobulin chains by carrying out electrophoresis in detergent-containing buffers. Figure 35 presents a comparison of reduced alkylated immune macroglobulins isolated from diverse vertebrate species with standard IgM, IgG and IgA immunoglobulins of man. The $\mu$-like heavy chains of all vertebrate species possess extremely similar mobilities under these con-

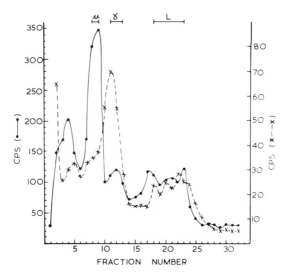

Fig. 34. Analysis of polypeptide chains of IgM immunoglobulin of *Necturus* (•——•) and IgG-like immunoglobulin of the marine toad (x--x) by polyacrylamide gel electrophoresis in acid urea. The proteins were labeled with ¹²⁵I-iodide and reduced and alkylated in the presence of 8M urea. $\mu$, $\gamma$, and L refer to the position of human $\mu$ chain, $\gamma$ chain, and L chain respectively, as resolved under these conditions. (From Marchalonis and Cohen 1973).

ditions, and this mobility is readily distinguishable from those of the $\alpha$ and $\gamma$ chains. Light chains of all the vertebrate species studied also exhibit extremely similar mobilities, a result which is consistent with the fact that they all are characterized by a mass of approximately 22,000 daltons.

## Amino acid composition analysis: An approach to the phylogeny of immunoglobulins

Although amino acid sequence data are not available for many lower species, precise amino acid compositions have been obtained for light and heavy chains of immunoglobulins of species ranging from lampreys to man. A number of statistical approaches that enable some estimation of protein homology when their amino acid compositions are known have recently been devised (Metzger et al. 1968; Marchalonis and Weltman 1971; Harris and Teller 1973). These methods are not as precise and sensitive for quantifying homologies among proteins as methods based upon the nucleic acid codons associated with the amino acid sequence (Fitch 1966; Fitch and Margoliash 1967), but they have

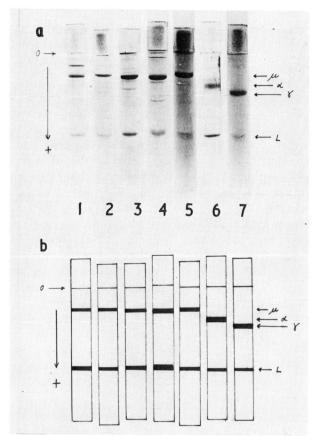

Fig. 35. a. Polyacrylamide gel electrophoresis in discontinuous sodium dodecylsulfate-containing buffers of reduced high molecular weight immunoglobulins from diverse vertebrate species. From left to right: 1. Stingray IgM, 2. *Necturus* IgM, 3. Toad IgM, 4. Echidna IgM, 5. Human IgM, 6. Mouse IgA, 7. Human IgG. O. indicates the origin of the running gel and → + shows the direction of migration of the protein bands. b. Tracing of each gel shown in a; gel tracings are in similar order to photographs above.

proven to be useful first approximations in cases where sequence data were unavailable. I shall describe briefly the operation of one of these parameters and its application to the phylogeny of vertebrate immunoglobulins. I will use the S$\Delta$Q measure (the sum of squared differences) to support the identification of the heavy chain of immunoglobulins of fishes as similar to $\mu$ chain and to provide evidence that this polypeptide is characterized by a relatively conservative evolutionary history.

When amino acid sequences of proteins are unknown, the statistical parameter SΔQ can serve as a first approximation of the degree of relatedness among proteins (Marchalonis and Weltman 1971). This parameter is defined as follows:

$$S\Delta Q = \Sigma_j (X_{ij} - X_{kj})^2$$

where i and k identify the particular proteins compared, and $X_j$ is the content (residues per 100 residues of carbohydrate-free protein) of a given amino acid of type j. In comparisons within families of proteins such as the cytochrome c proteins and the hemoglobins, where extensive sequence data are available, SΔQ was shown to correlate significantly with differences in amino acid sequence. This is illustrated in Figure 36, which plots scatter diagrams of SΔQ versus differences in amino acid sequence (AAD) for proteins of the hemoglobin-myoglobin family (Figure 36a) and cytochrome c proteins (Figure 36b) of vertebrate species. A linear correlation existed in both cases.

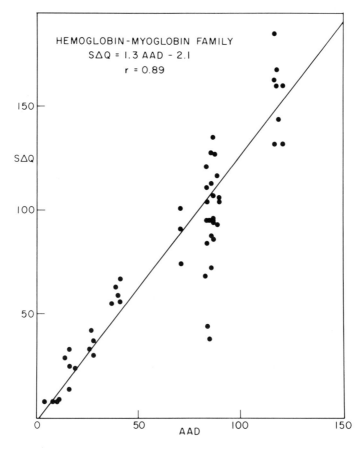

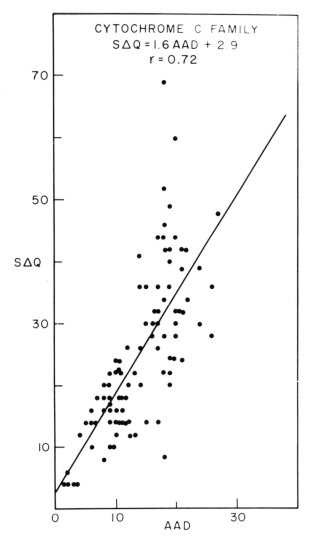

Fig. 36. Graphs (scatter diagrams) of the statistical parameter $S\Delta Q$ based upon amino acid composition versus the number of residue differences in amino acid sequence (AAD). A. The hemoglobin-myloglobin family of proteins; B. the cytochrome c family of proteins. Highly significant linear correlations were obtained in both cases. (From Marchalonis and Weltman 1971).

In order to obtain an estimate of the frequency of occurrence of $S\Delta Q$ values within groups of proteins known to be related and those considered distinct (unrelated in an evolutionary sense), cumulative frequency diagrams of $S\Delta Q$ were constructed for unrelated proteins, hemoglobins, cytochromes and immunoglobulins. Figure 37 presents cumulative

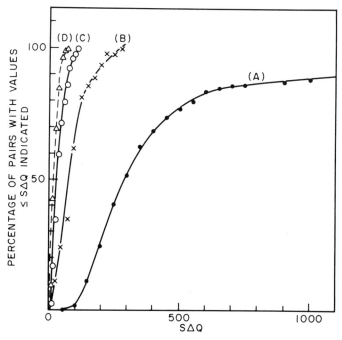

Fig. 37. Cumulative frequency diagrams of the SΔQ parameter for (A) unrelated proteins (820 comparison pairs), (B) hemoglobins (from 16
vertebrate species), (C) immunoglobulins (from 8 vertebrate species),
and (D) cytochrome c proteins (from 14 vetebrate species). Comparisons within the latter three protein families were made only among
vertebrate species. (From Marchalonis and Weltman 1971).

frequency diagrams for comparisons among: (A) unrelated   proteins
(820 comparison pairs); (B) hemoglobins from 16 vertebrate species;
(C) immunoglobulins from 8 vertebrate species, and (D) cytochrome
c proteins of 14 vertebrate species. Clearly, the unrelated proteins
(curve A) showed a much greater dispersity in SΔQ values than did the
groups of proteins known to be homologous. The median in curve A
was 300 SΔQ units, and 12 percent of the pairs of proteins differed by
over 1000 SΔQ units. Only 2 percent of unrelated pairs possessed
SΔQ values within 100 SΔQ units of each other, and none of those
tested concurred within 50 SΔQ units of each other. These observations
provide an empirical way of determining if two proteins are related. In
contrast, the median values for hemoglobins, cytochromes and immunoglobulins were 80, 20 and 30 respectively. The cytochrome c family
showed the least deviation in SΔQ value, which was consistent with the
hypothesis that these proteins have had a relatively conservative evolutionary history (Nolan and Margoliash 1968), with only three accepted

point mutations per 100 residues for each hundred million years (Dayhoff 1969; Kimura and Ohta 1971). The hemoglobin family has evolved at a moderate rate, characterized by the incorporation of eleven to thirteen accepted point mutations per 100 residues per hundred million years (Dayhoff 1969; Kimura and Ohta 1971). Thus, this approach to the estimation of protein homology provides a quantitative means of determining relatedness among unknown proteins and, when applied to well-characterized proteins, gives results in accordance with accepted models.

The immunoglobulin family (curve C in Figure 36) appears less disperse than does the hemoglobin family and similar to the cytochrome c proteins in the distribution of S$\Delta$Q values. This phenomenon is illustrated further in Tables 7, 8 and 9. Table 7 cites values of S$\Delta$Q computed for cytochrome c proteins of six vertebrate classes. The maximum difference value observed was 48 S$\Delta$Q units. The relatively low values observed were in accordance with amino acid sequence data, which show that the maximum difference between the sequences of proteins compared in Table 7 was twenty-seven out of 104 positions (Dayhoff 1969). The hemoglobin chains represent a more disperse collection of polypeptides (Table 8). $\alpha$ chains are clearly related to $\alpha$ chains of other species, and $\beta$ chains to $\beta$ and $\gamma$ and $\delta$ polypeptides of other species. The marked difference between $\alpha$ chains and $\beta$ chains is consistent with sequence data indicating that these polypeptides diverged from each other early in vertebrate evolution (Ingram 1963; Braunitzer 1966). Furthermore, the $\alpha$ chain of the carp, a teleost fish, is quite distinct from that of mammals. The $\alpha$ chains of man and the carp differ by seventy-one out of 141 residues in primary sequence (Dayhoff 1969). Even more striking differences are found when the $\alpha$ and $\beta$ chains of higher species are compared with the hemoglobins of tuna (a teleost fish) and the lamprey (a cyclostome). Table 8 also illustrates the sensitivity of the S$\Delta$Q approach to detect similarity among proteins. The human $\beta$ chain differs from the human $\delta$ chain by only eight S$\Delta$Q units, a value consistent with evolutionary schemes suggesting that the $\delta$ chain arose by recent duplication of the gene coding for the $\beta$ chain (Ingram 1963; Braunitzer 1966).

Table 9 presents data obtained by comparing immunoglobulin heavy chains from various primitive vertebrate species with those of man. The lower vertebrates considered here represent species that differ from man in possessing only a single class of immunoglobulins. Although the intact antibody molecules within these species may exist in various molecular forms, these forms contain the same heavy chain that resembles the human $\mu$ chain in molecular weight, carbohydrate content, and electrophoretic mobility. Sea lampreys (*Petromyzon marinus*) represent cyclostomes, the most primitive of living vertebrates. The smooth dog-

TABLE 7: SΔQ comparisons among cytochrome C proteins of vertebrate species.

| | Man | Rabbit | Rattlesnake | Tuna | Dogfish | Lamprey |
|---|---|---|---|---|---|---|
| Man | 0 | | | | | |
| Rabbit | 16 | 0 | | | | |
| Rattlesnake | 26 | 22 | 0 | | | |
| Tuna | 42 | 26 | 28 | 0 | | |
| Dogfish | 38 | 36 | 36 | 32 | 0 | |
| Lamprey | 44 | 44 | 48 | 36 | 14 | 0 |

*Source:* Data from Dayhoff 1969.

TABLE 8: S $\Delta$ Q comparisons among hemoglobins of vertebrate species.

| | Human $\alpha$ | Rhesus $\alpha$ | Mouse $\alpha$ | Carp $\alpha$ | Human $\beta$ | Rhesus $\beta$ | Rabbit $\beta$ | Human $\delta$ | Human $\gamma$ | Tuna | Lamprey |
|---|---|---|---|---|---|---|---|---|---|---|---|
| Human $\alpha$ | 0 | | | | | | | | | | |
| Rhesus $\alpha$ | 8 | 0 | | | | | | | | | |
| Mouse $\alpha$ | 25 | 14 | 0 | | | | | | | | |
| Carp $\alpha$ | 101 | 91 | 74 | 0 | | | | | | | |
| Human $\beta$ | 95 | 84 | 111 | 127 | 0 | | | | | | |
| Rhesus $\beta$ | 121 | 104 | 128 | 135 | 8 | 0 | | | | | |
| Rabbit $\beta$ | 84 | 72 | 86 | 106 | 29 | 33 | 0 | | | | |
| Human $\delta$ | 95 | 86 | 107 | 104 | 8 | 9 | 24 | 0 | | | |
| Human $\gamma$ | 117 | 93 | 95 | 96 | 63 | 55 | 59 | 56 | 0 | | |
| Tuna | 93 | 105 | 93 | 51 | 110 | 119 | 103 | 83 | 77 | 0 | |
| Lamprey | 169 | 183 | 167 | 102 | 249 | 278 | 191 | 223 | 196 | 151 | 0 |

Source: Data from Dayhoff 1969.

101

TABLE 9: S∆Q comparisons among immunoglobulin heavy polypeptide chains of vertebrate species.

| | Lamprey μ | Catfish μ | Gar μ | Paddle-fish μ | Dogfish μ | String ray μ | Human μ | Human α | Human γ |
|---|---|---|---|---|---|---|---|---|---|
| Lamprey μ | 0 | | | | | | | | |
| Catfish μ | 31 | 0 | | | | | | | |
| Gar μ | 70 | 26 | 0 | | | | | | |
| Paddlefish μ | 41 | 17 | 21 | 0 | | | | | |
| Dogfish μ | 46 | 12 | 20 | 14 | 0 | | | | |
| Stingray μ | 40 | 17 | 27 | 27 | 13 | 0 | | | |
| Human μ | 27 | 14 | 18 | 19 | 15 | 14 | 0 | | |
| Human α | 57 | 44 | 37 | 55 | 48 | 45 | 26 | 0 | |
| Human γ | 64 | 26 | 25 | 36 | 29 | 36 | 21 | 48 | 0 |

fish (*Mustelus canis*) and the stingray (*Dasyatis centroura*) are elasmobranchs. The catfish (*Ictalurus punctatus*), the gar (*Lepisosteus osseus*) and the paddlefish (*Polyodon spathula*) represent the Teleostei, Holostei, and Chondrostei, respectively. The progenitors of all these species diverged from the evolutionary line which was ancestral to mammals more than 200 million years ago (Romer 1966).

The results in Table 9 support the identification of immunoglobulins of primitive vertebrates as similar to mammalian IgM molecules on the basis of physiochemical properties of the heavy chain. Furthermore, the degree of similarity among all of the heavy chains of these divergent species is remarkable. This relatedness is emphasized by comparisons between lamprey hemoglobin and the $\alpha$ and $\beta$ chains of human hemoglobin. The lamprey protein differs from these polypeptides by 169 and 249 S$\Delta$Q units respectively, whereas the heavy chain of lamprey immunoglobulin (antibody to bacteriophage f2) differs from the human macroglobulin heavy chain ($\mu$ chain) by only twenty-seven units. The low S$\Delta$Q values observed for comparisons among all of the heavy chains are comparable to those obtained for the cytochrome c group of proteins.

Similar analyses were performed among light chains of vertebrate species ranging from lamprey to man. The results are depicted in the cumulative frequency diagram shown in Figure 38, which compares the distribution of S$\Delta$Q values for light chains and heavy chains. The distribution of S$\Delta$Q values for light chains shows more dispersity than that of heavy chains. This is probably because of the early emergence of distinct $\kappa$ and $\lambda$-type light chains which differ markedly from each other in amino acid sequence in both the variable and constant portions of the molecule (See Chapter 7). This suggestion gains support from S$\Delta$Q comparisons among seven human myeloma $\kappa$ chains and seven human $\lambda$ chains. The $\kappa$ chains differed from each other by an average of 18 S$\Delta$Q units, the $\lambda$ chains from one another by about 19 S$\Delta$Q. The $\kappa$ chains differed from the $\lambda$ chains, however, by 47 S$\Delta$Q units, thereby suggesting an early evolutionary divergence from the ancestral light chain. This difference between $\kappa$ and $\lambda$ light chains is closer to that found between $\alpha$ and $\beta$ chains of mammalian hemoglobins (approx. 82 S$\Delta$Q units), which diverged from their common ancestor 400 million years ago (Zuckerlandl and Pauling 1967). The light chain data, nevertheless, support the contention that immunoglobulin evolution was a relatively conservative process. Moreover, S$\Delta$Q comparisons among light and heavy chains of diverse vertebrate species indicate the homologous origins of these peptides, which differ from each other by an average of 52 S$\Delta$Q units over all species considered. As Metzger et al (1968) first pointed out, light chains and heavy chains of human immunoglobulins differ from each other less in amino acid composition than do the $\alpha$ and $\beta$ chains of human hemoglobin.

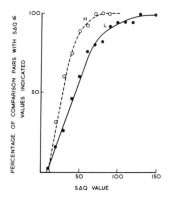

Fig. 38. Cumulative frequency diagrams of the S$\Delta$Q parameter for heavy (H) and light (L) polypeptide chains of immunoglobulins of vertebrate species ranging from lamprey to man.

## Phylogenetic distribution of IgM immunoglobulins

A common theme emerges from studies of immune macroglobulins present in the serum of representatives of all vertebrate classes. Such molecules are composed of light chains and heavy chains similar to the $\mu$ chain of mammalian IgM immunoglobulin. Although these species are considered to possess immunoglobulins homologous to the IgM class, the state of polymerization of the molecules ranges from monomers of the form $L_2\mu_2$ to covalently linked hexamers of this unit. The extent of polymeric association is reliable within a species or class but may differ from class to class. The characteristic form of polymerized IgM in man and shark is a pentamer, for example, whereas it is a hexamer in *Xenopus laevis* and a tetramer in bony fishes. In addition, substantial amounts of the monomeric unit occur in the serum of sharks. Another variant on the basic IgM structural motif is the situation in lampreys, where disulfide bonds are lacking and a variety of aggregates possess antibody activity to bacteriophage. The characteristically distinct states of covalent association within the various species and classes suggests that some factor, in addition to the light and heavy chains, directs the establishment of the spatial relationships within IgM polymers. Such a factor is probably the J chain, or joining chain, which occurs in mammalian IgM and IgA polymers and has been found in elasmobranch pentamers.

Analysis of immunoglobulins of pre-tetrapod vertebrates establishes that the basic patterns of immunoglobulin structure and, therefore, its genetic basis had emerged and become stabilized early in the evolution of vertebrates. Furthermore, the general structural features of the primitive immunoglobulin in phylogeny, namely IgM, have been conserved in all vertebrate species.

# 7 Emergence of Light Chains

The data of previous chapters establish that light polypeptide chains, as well as $\mu$-like heavy chains, are ubiquitously distributed among vertebrate species. The light chains of all species studied possess a mass of approximately 22,000 daltons and exhibit heterogeneity in electrophoretic mobility. Light chains, like $\mu$ chains, thus represent early developments in the phylogeny of vertebrate immunity and it is of interest to determine whether the light chains of primitive vertebrates fall into the $\kappa$ and $\lambda$ types described for higher vertebrates. Two presently available pieces of information bear upon this question: the first derives from limited amino acid sequence data for light chains of lower vertebrates, the second from comparisons of amino acid sequences of human and murine myeloma light chains. In the case of light chains, aminoterminal sequence data provide information on the issue of whether a chain is $\kappa$ or $\lambda$ as well as the problem of the phylogenetic distribution of variable region diversity. Amino-terminal sequences of heavy chains, in contrast, do not allow distinction among heavy chain classes because all classes share a common pool of variable region sequences.

## Comparisons of light chain amino-terminal sequences

The amino acid sequence data given in Table 10 provide information regarding phylogenetic patterns of variation in V regions of light chains. Strong homologies exist between the $\kappa$ chain of man and the light chains of the paddlefish, the leopard shark, and the African lungfish. Since $\lambda$ chains generally possess blocked N-terminal groups (pyrrolidone car-

TABLE 10: N-terminal amino acid sequence of vertebrate light chains.

| Species | Residue number | | | | | |
|---|---|---|---|---|---|---|
| | 1 | 2 | 3 | 4 | 5 | 6 |
| Lamprey[1] | Asp | - | - | - | - | - |
| Leopard[2] shark | Asp | Ile | Val | Leu | Thr | Glx |
| | - | Pro | Ile | Met | - | - |
| | - | - | - | Val | - | - |
| Paddlefish[3] | Asp | Ile | Val | Ile | Thr | - |
| | - | - | - | Leu | - | - |
| Lungfish[4] | Asp | - | - | Leu | Thr | Glx |
| Chicken[5] | - | - | Ala | Leu | Thr | Gln |
| κ (Human)[6] | Asp | Ile | Val | Met | Thr | Gln |
| | Glu | Val | Gln | Leu | - | - |
| | - | Met | Leu | - | - | - |
| λ (Human)[6] | PCA | Ser | Val | Leu | Thr | Gln |
| | | Tyr | Ala | | Ala | |
| | | | Glu | | | |

*Sources:* (1) Marchalonis and Edelman 1968b; (2) Klaus et al. 1971; (3) Pollara et al. 1968; (4) Litman et al. 1971c; (5) Kubo, Rosenblum, and Benedict 1971; (6) Hood and Talmage 1970.

boxylic acid, or cyclic glutamine) most studies in lower vertebrates have concentrated on κ-like chains which have free aspartic acid or glutamic acid in this position and are, consequently, easier to analyze. Klaus, Nitecki, and Goodman (1971) have pointed out, however, that the first six amino acid residues of κ and λ chains are very similar and that the data presently available do not allow a distinction between κ and λ type to be made for shark or paddlefish immunoglobulins. Thus, the light chains of primitive vertebrates are obviously homologous to those of mammals, but present sequence data do not indicate the class to which the former belong. More extensive data have been obtained for light chains of purified chicken antibodies to the dinitrophenyl hapten by Kubo, Rosenblum and Benedict (1971) who propose this unblocked light chain is more similar to human λ chain than it is to κ chain. These data are tabulated (Table 11) in a comparison of the first eighteen amino-terminal residues of the variable region subgroups of human κ and λ chains and chicken light chains. The chicken light chain preparation is clearly homologous to both κ and λ chain V regions, but the similarity to λ is slightly more pronounced. This result emphasizes the difficulties involved in establishing definite relationships among immunoglobulin chains of lower species and those of man on the basis of limited amino acid sequence data. This point warrants consideration because the presence of blocked amino-terminal residues on light chains has been taken as evidence that the chains in question were λ chains (Hood et al. 1967).

Table 12 is based primarily upon the data of Hood et al. (1967) and illustrates the distribution of κ and λ chains among mammalian and avian species. Saluk, Drauss, and Clem (1970) have demonstrated the presence of antigenically distinct light chains in the alligator and tentatively identified them as homologues of mammalian κ and λ chains. It is interesting that the ratio of κ chain to λ chain varies in different species. About 95% of mouse light chains are κ, whereas virtually all horse light chains represent the λ class. Cohn (1972) has interpreted the relative fractions of κ and λ chains in the various species as indicators of the ratio of structural genes encoding the respective light chains which are present within the particular species.

Sequence data indicate that V regions of murine light chains can be classified into subgroups analogous to those of man. However, it is not certain that the mouse V-region subgroups are exactly homologous to those of human chains. Svasti and Milstein (1972) have compared the V-region sequences of two mouse κ myeloma light chains, MOPC 21 and MOPC 41, with the basic sequences of human $V_\kappa I$ $V_\kappa II$ and $V_\kappa III$ (Table 13). The comparison suggests that the two mouse $V_\kappa$ sequences do not fall into any of the human basic subgroup patterns. Moreover, the two mouse sequences differ as much from each other as they do from the human sequences. This observation suggests that the ancestral V genes which gave rise to these proteins existed prior to the separation of human and mouse evolutionary lines. $V_H$ regions of mammals show a clearer relationship to human $V_H$ subgroups than that observed for the comparison between the $V_\kappa$ sequences of mouse and man. Kehoe, Hurvitz, and Capra (1972) and Kehoe and Capra (1972) report that heavy chains of certain canine and feline myeloma proteins could be unequivocally assigned to the $V_H III$ subgroup of human $V_H$ regions. Furthermore, these V-region sequences showed 90% identity in γ comparison across the species, although certain residues were specific for the particular species investigated.

The common ancestry of κ and λ light chains was recently emphasized by work of Stanton et al. (1974), which established that both chains possess a relatively conserved pentapeptide. The peptide-containing residues 68 through 73 of chicken λ chain has the sequence Arg-Phe-Ser-Gly-Ser-Lys. The sequences of the corresponding peptides in λ chains of duck, turkey, and various mammals including sheep, mouse, and man were either identical or very similar to this sequence. The

TABLE 11: Comparison of amino terminal sequences of chicken light chains and the

| Chain | Residue | | | | | | | |
|---|---|---|---|---|---|---|---|---|
| | 1 | 2 | 3 | 4 | 5 | 6 | 7 | 8 |
| $V_\kappa$ I | Asp | Ile | Gln | Met | Thr[a] | Gln | Ser | Pro |
| $V_\kappa$ II | Glu | Ile | Val | Leu | Thr | Gln | Ser | Pro |
| $V_\kappa$ III | Asp | Ile | Val | Met | Thr | Gln | Ser | Pro |
| Chicken L | - | - | Ala[b] | Leu | Thr | Glu | - | Pro |
| $V_\lambda$ I | PCA | Ser | Val | Leu | Thr | Glu | Pro | Pro |
| $V_\lambda$ II | PCA | Ser | Ala | Leu | Thr | Gln | Pro | Ala |
| $V_\lambda$ III | - | Tyr | Val | Leu | Thr | Gln | Pro | Pro |
| $V_\lambda$ IV | PCA | Ser | Ala | Leu | Thr | Gln | Pro | Pro |
| $V_\lambda$ V | - | Ser | Glu | Leu | Thr | Gln | Pro | Pro |

λ chain of the human myeloma protein Bo, for example, was identical in residues 68-73 whereas the λ chain of myeloma protein Kern differed only in having a serine residue at position 73. This peptide was common also to κ chains, even in species as primitive as the nurse shark. Residues 68-72 were identical to the λ-region sequence, but residue 73 consisted of Gly rather than Lys. Rabbit, man, and mouse κ chains also contained corresponding sequences either identical or very similar to the prototype. The function of this highly conserved region is unknown, but it has been preserved for over 400 million years and must fulfill some key structural role in all light chains. The authors suggested that antisera produced against the pentapeptide Arg-Phe-Ser-Gly-Ser might prove a useful reagent in visualizing cell-associated protoimmunoglobulin receptors of invertebrates and possible primitive T-cell receptors (see Chapter 17).

## Constant region sequences of light chains: The emergence of κ and λ chains

Constant region sequences of human and murine light chains are tabulated in Table 14, which lists residues located near the beginning of the C region and those located at the carboxyl terminals of the molecules. A pattern arises which shows that the κ chain of man is similar to the κ chain of mouse, and that a parallel picture obtains for the respective λ chains. Moreover, the number of residues common to all five pro-

V-region prototypes of human κ and λ chains.

| number | | | | | | | | | |
|---|---|---|---|---|---|---|---|---|---|
| 9 | 10 | 11 | 12 | 13 | 14 | 15 | 16 | 17 | 18 |
| Ser | Ser | Leu | S̄er | Āla | Ser | Val | Ḡly | Asp | Arg |
| Gly | Thr | Leu | S̄er | Leu | Ser | Pro | Ḡly | Glu | Arg |
| Leu | Ser | Leu | Pro | Val | Thr | Pro | Ḡly | Ḡlu | Pro |
| Ala | S̲e̲r̲ | Val | S̲e̲r̲ | Āla | Glu | Leu | Ḡly̲ | Ḡlu | Thr |
| - | S̲e̲r̲ | V̲a̲l̲ | S̲e̲r̲ | Gly | Ala | Pro | Ḡly̲ | Gln | Arg |
| - | S̲e̲r̲ | V̲a̲l̲ | S̲e̲r̲ | Gly | Ser | Pro | Ḡly̲ | Gln | Ser |
| - | S̲e̲r̲ | V̲a̲l̲ | S̲e̲r̲ | V̲a̲l̲ | Ser | Pro | Ḡly̲ | Gln | Thr |
| - | S̲e̲r̲ | Ala | S̲e̲r̲ | Gly | Ser | Pro | Ḡly̲ | Gln | Ser |
| - | Ala | V̲a̲l̲ | S̲e̲r̲ | V̲a̲l̲ | Ala | Leu | Ḡly̲ | Gln | T̲h̲r̲ |

*Sources:* Human data taken from W.H.O. *Bull.* 41: 975 (1969). Chicken data taken from *Kubo* et al. 1971.

a. Line above indicates identity with $V_\kappa$.

b. Line below indicates identity with $V_\lambda$.

teins (34%) bears witness to the common evolutionary origin of κ and λ chains. Overall, κ and λ chains of man are approximately 40% identical in amino acid sequence. We can use this index of similarity to make a simple estimation of the time at which the genes encoding κ and λ chain diverged. To do this, assume that the rate of evolution of immunoglobulins approximates that of hemoglobin chains. The $\alpha$ and $\beta$ chains of human hemoglobin are 45% identical and are thought to have diverged from one another approximately 400 million years ago during the Ordovician period. The rate of change in amino acid sequence is equal to 13 accepted point mutations per 100 residues per 100 million years (Zuckerkandl and Pauling 1965), which is close to 13 amino acid residues per 100 residues per 100 million years, because the great majority of amino acid replacements result from single base changes. If immunoglobulin light chains evolved at a rate similar to that of hemoglobin $\alpha$ and $\beta$ chains, κ and λ chains have a calculated divergence time of 460 million years, a value which suggests that these chains should be present at the phylogenetic levels of cartilaginous fishes. This computation indicates that divergence of this ancestral light chain gene into the prototype κ and λ genes probably occurred early in vertebrate evolution. Calculations of this nature are not exact and should not be taken

TABLE 12: Distribution of $\varkappa$ and $\lambda$ light chains among vertebrate species.

| Species | Class | Kappa chains | Lambda chains |
|---------|-------|--------------|---------------|
| Aligator[a] | Reptile | + | + |
| Chicken[b] | Avian | ? | + |
| Turkey | Avian | + (50%)[c] | + (50%) |
| Rabbit | Mammal (Lagomorph) | + (90%) | + (10%) |
| Mouse | Mammal (Rodent) | + (95%) | + ( 5%) |
| Guinea pig | Mammal (Rodent) | + (75%) | + (25%) |
| Baboon | Mammal (Primate) | + (75%) | + (25%) |
| Man | Mammal (Primate) | + (60%) | + (40%) |
| Dog | Mammal (Carnivore) | + (10%) | + (90%) |
| Mink | Mammal (Carnivore) | + (<5%) | + (>95%) |
| Pig | Mammal (Artiodactyl) | + (50%) | + (50%) |
| Cow | Mammal (Artiodactyl) | + (10%) | + (90%) |
| Horse | Mammal (Perissodactyl) | + (<5%) | + (>95%) |

*Source:* Unless stated otherwise, data are taken from Hood et al. 1967.

a. Saluk, Drauss, and Clem 1970.

b. Kubo, Rosenblum, and Benedict 1971.

c. Figures in parenthesis refer to the percentage of total light chain within the species contributed by either $\varkappa$ or $\lambda$ chain.

TABLE 13: Differences (expressed as minimum number of base changes/100 residues compared) between human and mouse myeloma light chain C and V regions.

| | MOPC 21 | MOPC 41 | MOPC 70 | Human $V_\varkappa$ I | Human $V_\varkappa$ II |
|---|---------|---------|---------|--------------|---------------|
| MOPC 41 | 55 | | | | |
| MOPC 70 | 53 | 54 | | | |
| Human $V_\varkappa$ I | 42-43 | 37-35 | 45-44 | | |
| Human $V_\varkappa$ II | 45 | 53 | 46 | 41-46 | |
| Human $V_\varkappa$ III | 46 | 47 | 44 | 37-40 | 32 |

*Source:* Svasti and Milstein 1972.

TABLE 14: Comparisons of constant region sequences of human and mouse light chains.

| Chain | Residues | | | | | | | | |
|---|---|---|---|---|---|---|---|---|---|
| | 119 | 120 | 121 | 122 | 123 | 124 | 125 | 126 | 127 |
| Human ϰ | Thr | Val | Ala | Ala | Pro | Ser | Val | Phe | Ile |
| Mouse ϰ | Ala | Asx | Ala | Ala | Pro | Thr | Val | Ser | Ile |
| Human λ | Pro | Lys | Ala | Ala | Pro | Ser | Val | Thr | Leu |
| Mouse λ 1 | Pro | Lys | Ser | Ser | Pro | Ser | Val | Thr | Leu |
| Mouse λ 2 | Pro | Lys | Ser | Thr | Pro | Thr | Leu | Thr | Val |

| | 128 | 129 | 130 | 131 | 132 | 133 | 134 | 135 | 136 |
|---|---|---|---|---|---|---|---|---|---|
| Human ϰ | Phe | Pro | Pro | Ser | Asp | Glu | Gln | Leu | Lys |
| Mouse ϰ | Phe | Pro | Pro | Ser | Ser | Glu | Gln | Leu | Thr |
| Human λ | Phe | Pro | Pro | Ser | Ser | Glx | Glx | Leu | Glu |
| Mouse λ 1 | Phe | Pro | Pro | Ser | Ser | Glu | Glu | Leu | Thr |
| Mouse λ 2 | Phe | Pro | Pro | Ser | Ser | Glx | Glx | Leu | Lys |

| | 137 | 138 | 139 | 140 | 141 | 142 | 143 | 144 | 145 |
|---|---|---|---|---|---|---|---|---|---|
| Human ϰ | Ser | Gly | Thr | Ala | Ser | Val | Val | Cys | Leu |
| Mouse ϰ | Gly | Gly | Ser | Ala | Ser | Val | Val | Cys | Phe |
| Human λ | Ala | Asn | Lys | Ala | Thr | Leu | Val | Cys | Leu |
| Mouse λ 1 | Glu | Asn | Lys | Ala | Thr | Leu | Val | Cys | Thr |
| Mouse λ 2 | Glu | Asn | Lys | Ala | Thr | Leu | Val | Cys | Leu |

| | 199 | 200 | 201 | 202 | 203 | 204 | 205 | 206 | |
|---|---|---|---|---|---|---|---|---|---|
| Human ϰ | Ala | Cys | Glu | Val | Thr | His | Glu | Gly | |
| Mouse ϰ | Thr | Cys | Glx | Ala | Thr | His | Lys | Thr | |
| Human λ | Ser | Cys | Glx | Val | Thr | His | Glu | Gly | |
| Mouse λ 1 | Ser | Cys | Glu | Val | Thr | His | Glx | Gly | |
| Mouse λ 2 | Thr | Cys | Glx | Val | Thr | Asx | Glx | Gly | |

| | 207 | 208 | 209 | 210 | 211 | 212 | | 213 | 214 |
|---|---|---|---|---|---|---|---|---|---|
| Human ϰ | Leu | Ser | Ser | Pro | Val | Thr | . . . | Lys | Ser |
| Mouse ϰ | Ser | Thr | Ser | Pro | Ile | Val | . . . | Lys | Ser |
| Human λ | Ser | Thr | . . . | . . . | . . . | Val | Glu | Lys | Thr |
| Mouse λ 1 | His | Thr | . . . | . . . | . . . | Val | Glu | Lys | Ser |
| Mouse λ 2 | His | Thr | . . . | . . . | . . . | Val | Glx | Lys | Ser |

| | 215 | 216 | 217 | 218 | 219 | 220 | 221 |
|---|---|---|---|---|---|---|---|
| Human ϰ | Phe | Asn | Arg | Gly | Glu | Cys | |
| Mouse ϰ | Phe | Asn | Arg | Asn | Glu | Cys | |
| Human λ | Val | Ala | Pro | Thr | Glu | Cys | Ser |
| Mouse λ 1 | Leu | Ser | Arg | Ala | Glu | Cys | Ser |
| Mouse λ 2 | Leu | Ser | Pro | Ala | Glu | Cys | Leu |

*Source:* Gally and Edelman 1972.
Numbering according to human ϰ .

seriously except as a general indication of whether molecules diverged early or late in vertebrate evolution.

Some interesting aspects of the evolution of $\kappa$ and $\lambda$ chains within mammalian species arise from direct inspection of the data listed in Table 14. Mouse $\kappa$ chain is more closely related to human $\kappa$ chain than it is to human $\lambda$ chain, and vice versa for $\lambda$ chains. However, human $\kappa$ chain and mouse $\kappa$ chain are 63% identical, while human $\lambda$ and mouse $\lambda$ are about 80% identical. These conclusions are emphasized in Table 15, which cites differences between the entire constant regions of human and murine $\kappa$ and $\lambda$ chains. Svasti and Milstein (1972) comment that the observation that human and mouse $C_\lambda$ regions differ by 37 base changes per 100 residues whereas human and mouse $C_\kappa$ chains differ by 52 such changes, suggests that $\kappa$ chains may evolve at a faster rate than do $\lambda$ chains. A detailed treatment of computer approaches to the determination of relationships among immunoglobulin chains of eutherian mammals is given in Smith (1973).

## Synopsis

Immunoglobulins composed of equal numbers of light and heavy polypeptide chains are present in representatives of all vertebrate classes. The limited amino-terminal sequence data presently available are insufficient to decide whether the light chains of elasmobranchs, chondrosteans, and dipnoïd fish resemble $\kappa$ or $\lambda$ chains of higher mammals. More extensive data suggest that chickens possess $\lambda$-type light chains. Extrapolation of amino acid sequence differences among human and murine $\kappa$ and $\lambda$ chains implies that these chains diverged from a common ancestor approximately 400 million years ago. Further studies of light chains of lower species are required in order to determine whether the gene duplication that generated the genes encoding $\kappa$- and $\lambda$-type light chains actually represents an ancient event in vertebrate evolution.

TABLE 15: Human and mouse $C_L$ -region differences, expressed as the minimum number of base changes/100 residues compared.

|                | Human $C_x$ | Mouse $C_x$ | Human $C_\lambda$ |
|----------------|-------------|-------------|-------------------|
| Mouse $C_x$    | 52          |             |                   |
| Human $C_\lambda$ | 87       | 82          |                   |
| Mouse $C_\lambda$ | 90       | 90          | 37                |

*Source:* Data of Svasti and Milstein 1972.

# 8 Distribution of Immunoglobulins Distinct from the IgM Class

## General nature of non-IgM immunoglobulins

In Chapter 6 evidence was presented supporting the conclusion that IgM-like immunoglobulins are present in all vertebrate classes. This type of immunoglobulin appears to comprise the predominant, if not the only, antibody class of lower vertebrates. Beginning with the phylogenetic level of dipnoid fishes and continuing through amphibians and higher vertebrates, immunoglobulin classes appear which differ unequivocally from the IgM type. The definitive characteristic of these new immunoglobulin molecules derives from the fact that their heavy chains are distinct antigenically and physically from the $\mu$ chain found in the same species. In accordance with the structural principles of mammalian immunoglobulin classes, antigenically and structurally similar light chains are associated with the different heavy-chain types. It now becomes pertinent to inquire whether these non-IgM immunoglobulins, which are usually of low molecular weight (150,000 to 200,000), bear any direct similarity to the well-characterized immunoglobulin classes of man. Furthermore, we must consider functional interrelationships between the IgM and non-IgM classes and possible selective pressures which might have resulted in the emergence of the heavy chains specifying these new immunoglobulin classes.

## The IgN immunoglobulin of the Dipnoi

Sedimentation analyses obtained for purified serum immunoglobulins of the Australian lungfish (Marchalonis 1969) indicate that two types of these proteins are present. The first is a high molecular weight immunoglobulin fraction which contains proteins characterized by sedimentation coefficients of 19.0S and 23S, with the latter probably constituting a dimer of the former. Such higher aggregates are frequently observed for mammalian IgM immunoglobulins. The second immunoglobulin of the lungfish possesses a sedimentation coefficient of 5.9S when extrapolated to zero concentration. This value is definitely lower than that of the usual lower molecular weight antibodies of mammalian species. The intact 5.9S protein has a molecular weight of 120,000, rather than the 150,000 characteristic of IgG immunoglobulins. The lungfish IgM and 5.9S immunoglobulins are antigenically related through the presence of common light chains, but possess antigenically distinct heavy chains. These properties are illustrated in Figure 39, which presents the electrophoretic patterns obtained when reduced alkylated immunoglobulins of man, rat, and lungfish are compared by starch gel electrophoresis in acid urea. The usefulness of gel electrophoretic techniques in preliminary identification of heavy polypeptide chains is readily apparent. The $\mu$ chains of man, rat, and the lungfish possess identical mobilities. The $\gamma$ chains of man and rat penetrate the gel to a comparable degree and this position is quite distinct from that of the $\mu$ chain. The heavy chain of the lungfish 5.9S immunoglobulin is distinct from both $\mu$ chain and $\gamma$ chain of mammals. The fact that it enters the gel more deeply than does the $\gamma$ chain suggests that it possesses a lower molecular weight than that of the $\gamma$ heavy chain. Direct measurement of this parameter confirms this conjecture. The molecular weight of the lungfish heavy chain is 38,000; whereas that of the macroglobulin is 70,000, the characteristic value for the $\mu$ chain (Marchalonis 1969). All light chains possess comparable mobilities and molecular weights. Light chains of both lungfish immunoglobulins give evidence of diffuse heterogeneity. At the time the low molecular weight heavy chain of the lungfish was discovered, it was considered a unique find and termed the $\nu$ chain. Other studies to be discussed below suggest that similar molecules occur in higher vertebrates. Litman et al. (1971a) reported that the immunoglobulins of the African lungfish are essentially similar to those of its Australian cousin and, moreover, obtained information on the sequence of amino acid residues at the amino-terminus of the light chain (Table 10). Such data are critical to a detailed assessment of similarity or differences among vertebrate immunoglobulins.

## IgG-like immunoglobulins of amphibians

A general feature of the antibody-forming response of mammals to a number of antigens, such as bacteriophage or proteins, is that the anti-

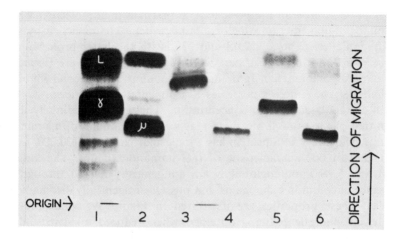

Fig. 39. Comparison by starch gel electrophoresis in acid urea of reduced alkylated immunoglobulins of man, Australian lungfish, and rat. Origins: (1) normal human IgG immunoglobulin (L, light chain; γ, gamma chain); (2) human IgM myeloma protein (μ, mu chain); (3) lungfish 5.9S immunoglobulin; (4) lungfish 19S immunoglobulin; (5) rat normal IgG immunoglobulin; (6) rat normal IgM immunoglobulin. Notice the correspondence of the heavy chains of IgM immunoglobulins of all three species. The γ chains of man and the rat have identical mobilities; the heavy chain of lungfish 5.9S immunoglobulin penetrates the gel to a greater degree than do the gamma chains. (From Marchalonis 1969).

bodies which appear soon after immunization tend to consist mostly of IgM molecules whereas IgG antibodies predominate later (Uhr 1964). Although 7S antibodies do appear in sharks, these are low molecular weight IgM units (Marchalonis and Edelman 1965, 1966a; Clem and Small 1967) and relatively prolonged immunization is necessary for them to be produced (Sigel and Clem 1966). Anuran amphibians, in contrast, exhibit a readily discernible transition from IgM antibodies to low molecular weight (7S) non-IgM antibodies following immunization.

Uhr, Finkelstein, and Franklin (1962) first reported that an anuran amphibian, the bullfrog, formed both 19S and 7S antibodies in response to immunization with bacteriophage Φ X174. However, studies of the polypeptide subunit structure of these immunoglobulins were required in order to determine their class relationships. Marchalonis and Edelman (1966b) confirmed the presence of 19S and 7S antibodies using bacteriophage f2 as antigen (Figure 40). Bullfrogs were given a single injection of the antigen, and sera obtained at various times after immunization were fractionated by ultracentrifugation on linear gradients of

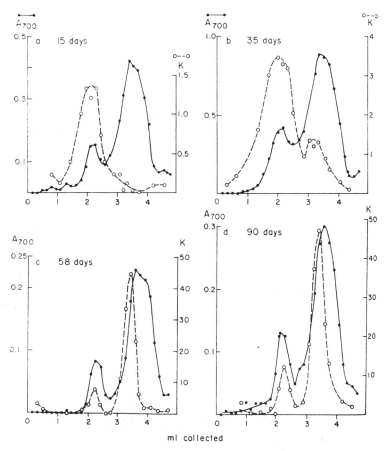

Fig. 40. Sequence of appearance of immunoglobulin in immunized bullfrogs as shown by ultracentrifugation on linear gradients of sucrose. $A_{700}$ (●——●), protein concentration measured by absorbancy of the Folin reaction at 700 nm; K (o---o), bacteriophage neutralization coefficient expressed in ml × $mg^{-1}$ × $min^{-1}$. Sedimentation proceeded from right to left. (From Marchalonis and Edelman 1966b).

sucrose to resolve proteins of different molecular weights. Early antibody activity was limited to the 19S molecules. Low molecular weight antibodies first appeared at 35 days after immunization. By day 58, over 90% of the antibody activity was localized in the 7S proteins.

An interesting feature of the kinetics of appearance of 19S and 7S antibodies to bacteriophage f2 in the bullfrog is illustrated in Figure 40. Although the bulk of the antibody activity is in the 7S immunoglobulin at the later times, the amount of 19S antibody is still rising. This observation contrasts with the usually accepted model in mammalian systems where 7S (IgG antibody) exerts a negative feedback effect that first shuts off IgM production and eventually brings about a cessation of the IgG response (Uhr and Möller 1968). The response of the frog, therefore, bears a superficial similarity to that of mammals; but a difference that may shed light on the mechanisms of feedback inhibition by antibody is also present.

Another feature of these diagrams that warrants discussion is the slowness of the response relative to that of *B. marinus* shown in Figures 21 and 56. The bullfrogs studied here were maintained at 22°C, whereas the toads were kept at an ambient temperature of 37°C. Widal and Sicard (1897) first observed the critical importance of temperature on antibody formation in ectothermic species. This effect will be considered in more detail below.

The 19S and 7S immunoglobulins of the bullfrog show some resemblance to the IgM and IgG classes, respectively, of mammals in polypeptide chain structure. They are antigenically related through the presence of shared light chains and distinguishable by the fact that the former possess $\mu$ chains and the latter possess antigenically distinct $\gamma$-type heavy chains. Figure 41 depicts the electrophoresis of immunoglobulins of man and bullfrog in acid urea. The proteins were reduced and

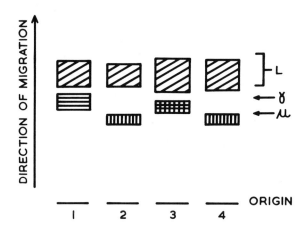

Fig. 41. Comparison of polypeptide chains of reduced alkylated immunoglobulins of man and bullfrog by starch gel electrophoresis in acid urea. Origins: (1) human normal IgG immunoglobulin; (2) human normal IgM immunoglobulin; (3) bullfrog 6.7S immunoglobulin; (4) bullfrog IgM immunoglobulin. L, position of light chains; $\gamma$, position of human $\gamma$ chains; $\mu$, position of human $\mu$ chain. Notice that the mobility of the heavy chain of the bullfrog 6.7S protein is slightly retarded relative to the human $\gamma$ chain. (Based upon Marchalonis and Edelman 1966b).

alkylated to cleave interchain disulfide bonds, and electrophoresis was performed in the presence of urea and formic acid in order to disrupt hydrophobic interactions and facilitate the separation of light chains from heavy chains. The frog proteins possess light chains which exhibited a mobility and diffuse heterogeneity similar to that of light chains from normal IgG and IgM immunoglobulin. The heavy chain of the frog 19S protein penetrated the acrylamide gel to an extent comparable to that of the human $\mu$ chain; that of the anuran 7S protein is clearly distinct from the $\mu$ chain but not quite identical in gel penetration to the $\gamma$ chain. Heavy chains of all low molecular weight anuran amphibians studied showed this property.

In addition to *B. marinus* and *R. catesbeiana*, which represent advanced species of anuran amphibians, the immunoglobulins of the clawed toad *Xenopus laevis* were also studied. On the basis of morphological and karyotypic considerations, this species is considered more primitive than either of the above species. *Xenopus* responded to injection of *Limulus* hemocyanin by producing 19S and 7S antibodies which constituted distinct classes showing some resemblance to IgM and IgG in size and polypeptide chain structure.

Although it is clear that the heavy chains of the low molecular weight immunoglobulins of the toad (Diener and Marchalonis 1970; Acton et al. 1970, 1972a) and *Xenopus* (Marchalonis, Allen, and Saarni 1970; Hadji-Azimi 1971) are distinct from that of the IgM immunoglobulin, there are some uncertainties regarding the molecular weights of these chains. Marchalonis and Edelman (1966b) and Acton et al. (1972a) reported molecular weights of 53,000 for the 7S Ig heavy chain of both the bullfrog and the marine toad, respectively. Hadji-Azimi (1971) found a molecular weight of 65,000 for the *Xenopus* chain. The last value is considerably higher than that expected for a $\gamma$-type chain. Hadji-Azimi measured molecular weights by gel electrophoresis in the detergent, sodium dodecylsulfate. Use of this method is formally valid only if the polypeptides analyzed contain no carbohydrate moieties. If no carbohydrate is present, polypeptides bind an equal amount of detergent per mass unit, and migration into the gel is a function only of size of the protein and it is independent of charge. Heavy chains, however, are rich in carbohydrate, and hence may exhibit a mobility that does not strictly reflect size. The molecular weight of the bullfrog 7S Ig heavy chain has been measured by equilibrium ultracentrifugation (Marchalonis and Edelman 1966b) and that of *Xenopus* by a gel electrophoretic procedure which obviates the problem of differing amounts of carbohydrate (Parish and Marchalonis 1970). In both cases, the values obtained were similar to those of human $\gamma$ chain; i.e., approximately 50,000. Atwell and Marchalonis (1975) recently analyzed the heavy chain of

B. *marinus* 7S immunoglobulin by gel filtration, polyacrylamide gel electrophoresis in sodium dodecylsulfate, and polyacrylamide gel electrophoresis in gels of varying concentrations. In all cases, the estimated molecular weight was approximately 58,000, a value significantly higher than that of mammalian $\gamma$ chain. Moreover, the intact molecular weight determined by gel filtration was 170,000. The toad IgG-like molecule eluted ahead of the human IgG molecule and possessed a higher molecular weight. These results emphasize the requirement for detailed information of primary structure as an absolute criterion in the determination of relatedness among proteins. The present data clearly establish that anuran amphibians possess at least two distinct classes of immunoglobulins that bear certain resemblances to the IgM and IgG classes of mammals.

An additional factor that requires that care be taken in the assignment of direct homology between the $\gamma$-like chain of advanced amphibians and the true $\gamma$ chain of mammals arises from studies of immunoglobulins of urodele amphibians. Although Cohen and his colleagues (Baldwin and Cohen 1970, 1971; Cohen 1969, 1971) have made a thorough investigation of allograft rejection in urodeles, our knowledge of antibody formation and immunoglobulins in this group of primitive amphibians is sparse at present. The ancestors of the urodeles diverged early from those of the anurans. The clawed toad *Xenopus* is among the primitive anurans which had ancestors located close to the branch point of the two groups. The third group of amphibians, the Apoda, are not included here, since no studies of these relatively rare species have been reported. Three groups of investigators have reported that axolotls (*Ambystoma mexicanum*) can form antibodies to bacteriophage. Ching and Wedgewood (1967) observed both high molecular weight and low molecular weight antibodies to the bacteriophage $\Phi X 174$. In contrast, Fougereau and his colleagues (Fougereau and Houdayer 1968; Fougereau, Houdayer, and Dorson 1972) and Ambrosius et al. (1970) found only high molecular weight antibodies to other phages. In studies of another urodele, Marchalonis and Cohen (1973) found that unimmunized *Necturus* possess only 19S IgM immunoglobulins (see Chapter 6). No 7S immunoglobulin was detectable. A similar result was obtained in studies by Fougereau, Houdayer, and Dorson (1972), who characterized the immunoglobulin fraction of axolotl antiserum by gel filtration. If urodeles lack the genes for the $\gamma$-like chain of anurans, either they were lost in urodele evolution, or these genes emerged within the phylogenetic development of the anurans. This evolutionary argument supports the physicochemical differences cited above, which undermine the concept of direct homology between the anuran $\gamma$-like chains and mammalian $\gamma$ chains.

The possibility exists that amphibian immunoglobulins might prove useful in comparative amino acid sequence studies. Hadji-Azimi (1969) found that *Xenopus* afflicted with a certain lymphoid tumor possess large amounts of serum immunoglobulin. At this stage it is unknown whether this immunoglobulin represents a homogeneous product similar to that secreted by myeloma tumor cells of man and mouse, or merely excess secretion of normal heterogeneous immunoglobulin. *Xenopus* is subject to a variety of lymphoid tumors, and its response to these malignancies has been investigated in some detail (Ruben 1970).

## Immunoglobulins of reptiles and birds

Too few species of reptiles and birds have been investigated in detail to enable one to make confident generalizations regarding the types of immunoglobulins present and their exact relationships to those of other vertebrate species. Physicochemical studies of immunoglobulins of three chelonian species have been carried out in some detail. Leslie and Clem (1972), Chartrand et al. (1971) and Benedict and Pollard (1972) have isolated and partially characterized high molecular weight and low molecular weight immunoglobulins of *Pseudemys seripta, Chelydra serpentina* (snapping turtle), and *Chelonia mydas* (sea turtle), respectively. The high molecular weight immunoglobulin of the three species is comparable to human IgM macroglobulin in size, polypeptide chain structure, and carbohydrate content (see Chapter 6). The major low molecular weight antibody of the three species possessed a mass of approximately 120,000. This was the only low molecular weight immunoglobulin in *C. serpentina*. The protein was composed of two light chains (M. W. 22,500) and two heavy chains characterized by a molecular weight of about 38,000. In this respect, the serum immunoglobulins of the turtle resemble those of Dipnoi rather than those of amphibians. As illustrated in Figure 26, the ancestors of the chelonians diverged early from those of other reptiles. A second low molecular weight class of *P. seripta* immunoglobulin was characterized by a sedimentation coefficient of 7.5S and a molecular weight of 180,000. This molecule contained two light chains (M. W. 22,500) and two heavy chains (M. W. 67,500). Although the analyses of immunoglobulins of other reptilian species have not been pursued to the extent of those performed on turtle immunoglobulins, data are available that indicate that the chelonian pattern may not obtain for tuataras (Marchalonis, Ealey, and Diener 1969), and lizards (Wetherall 1969). It may, however, be quite similar to that found in certain birds.

The tuatara possesses a 19S immunoglobulin and at least one class of 17S immunoglobulin (Marchalonis, Ealey, and Diener 1969). Immunological studies and analyses by polyacrylamide gel electrophoresis indicate that these molecules possess identical light chains and distinct heavy chains. The heavy chain of the 7S immunoglobulin penetrates the gel to a degree similar to that shown by the $\gamma$-like heavy chain of the bullfrog. It does not migrate as if its molecular weight were 38,000. The posssible existence of such a low molecular weight species has not been excluded, however. Wetherall (1969) has obtained more detailed results for 19S and 7S immunoglobulins of the Australian sleepy lizard (*Tiliqua rugosa*). The 19S immunoglobulin was characterized by a mass of 950,000 daltons and the 7S protein by a mass of 158,000 daltons. Both immunoglobulins had light chains of mass 22,000. The heavy chain of the high molecular weight immunoglobulin was comparable in mass to the human $\mu$ chain, whereas that of the 7S immunoglobulin was 51,000 daltons. This value is comparable to that of the mammalian $\gamma$ chain. Thus it is clear that reptiles possess IgM immunoglobulins and at least two distinct low molecular weight immunoglobulins. Provisionally I shall designate the 19S immunoglobulin as IgM, the 7.5S immunoglobulin as IgG (Y) by analogy with the chicken, and the 5.7S protein as IgN by analogy with the lungfish.

The types and distribution of immunoglobulins in birds extend the pattern described for reptiles. The two species that have been studied in greatest detail are ducks and chickens. The chicken possesses at least three classes of immunoglobulin, which have been tentatively identified as IgM, IgG and IgA (Ivanyi and Cerny 1965; Dreesman et al. 1965; Szenberg, Lind, and Clarke 1965; Lebacq-Verheyden, Vaerman, and Heremans 1972). The former two classes are serum antibodies, and the IgG class predominates. The IgA-like molecules are found in low levels in serum but occur in large quantities in bile (Lebacq-Verheyden, Vaerman, and Heremans 1972). The designation of the 7.1S immunoglobulin as IgG has been questioned by Leslie and Clem (1969). Although the protein gives an arc similar to that of rabbit IgG when analyzed by immunoelectrophoresis, its heavy chain possesses a molecular weight of about 65,000, a value significantly higher than that of the $\gamma$ chain. For this reason, these workers have proposed that the intact molecule be termed IgY to distinguish it from the true IgG of higher mammals. In any case, the chicken possesses two major classes of serum immunoglobulin; one resembling IgM and the other resembling IgG to some degree. The duck represents a very interesting situation. Its immunoglobulin pattern is almost a compromise between that of the chicken and that of the turtle. Three serum immunoglobulin classes are present. The major one resembles the 5.7S immunoglobulin of the turtle (Grey 1966; Zimmerman, Shalatin, and Grey 1971). The other two resemble the 19S IgM class and the 7.1S class of the chicken or the turtle *P. seripta*. The

7S immunoglobulins of the chicken and duck are antigenically related (Zimmerman, Shalatin, and Grey 1971), a fact which strengthens the conclusion that they are homologous proteins. Although further data are required to provide positive identification of the 5.7S immunoglobulins as similar to a particular class of immunoglobulins of other vertebrates, I propose to tentatively designate these as IgN after the 5.7S immunoglobulin of lungfish.

The presence of IgN-type antibodies is, thus, widely distributed among vertebrate species. Such molecules occur in Dipnoi (Marchalonis 1969; Litman et al. 1971), reptiles (Chartrand et al. 1971; Benedict and Pollard 1972; Leslie and Clem 1972), and birds (Zimmerman, Shalatin, and Grey 1971). The IgN-like molecule constitutes the major serum immunoglobulin class in ducks (Grey 1967) but has not been found in the serum of chickens. The presence of this component as a minor immunoglobulin in the chicken is suggested by a study of Choi and Good (1971) in which surface immunoglobulin of bursa lymphocytes contained a heavy chain characterized by a mass of approximately 40,000 daltons. These workers reported that this polypeptide was antigenically related to the $\mu$ chain. Furthermore, two recent studies suggest that marsupials (Thomas et al. 1972) and rabbits (Steward et al. 1969) may possess antibodies with properties generally similar to those described for the IgN class. Although more information is required to enable identification of the 7.1S immunoglobulins in terms of mammalian immunoglobulins, the IgM class of birds appears comparable to that of man. Mehta, Reichlin and Tomasi (1972) claim that $\mu$ chains of these proteins resemble those of man in antigenic properties and electrophoretic mobility. This observation is consistent with the hypothesis developed in Chapter 6 that IgM immunoglobulins emerged early in vertebrate evolution and have had a rather conservative phylogenetic history.

## Non-IgM antibodies of lower mammals

In Chapter 2 the immunoglobulins of man were used as an example of the existence of multiple immunoglobulin classes. IgG constitutes the major class in man and the eutherian mammals studied thus far. The preceding data clearly establish that vertebrates ranging from Dipnoi to birds possess low molecular weight immunoglobulins that contain heavy chains distinct from $\mu$ chains. The above evidence does not however, provide support for direct homology between any of the IgG-like molecules and the IgG class of man. Lower mammals such as monotremes and marsupials are located at a pivotal juncture in the evolution of $\gamma$ chains.

Diener and his colleagues (Diener, Ealey, and Legge 1967; Diener, Wistar, and Ealey 1967) carried out a series of investigations designed to determine the characteristics of the lymphatic system of the echidna and the capacity of this monotreme to form antibodies. In this series of experiments the animals responded well to *S. adelaide* flagellar antigens, producing high titer antibodies characterized by high (19S) and low (7S) molecular weights (Diener, Ealey, and Legge 1967). Moreover, a typical secondary response was induced by repeated injections of this antigen. Atwell, Marchalonis, and Ealey (1973) have recently isolated immunoglobulins from echidnas and shown that the two major classes of immunoglobulin closely resemble the IgM and IgG classes of man in size and polypeptide chain structure. Figure 42 presents a comparison by polyacrylamide gel electrophoresis in acid urea of reduced alkylated immunoglobulins of man and echidna. In contrast to the patterns obtained with low molecular weight antibodies of species more primitive than mammals, the heavy chain of the echidna antibody (M. W. 150,000) corresponds exactly in mobility to that of human $\gamma$ chain. The pattern regarding light chains and $\mu$ chains reflects that observed in Chapter 6. The electrophoretic similarities described here were corroborated by the additional finding that the molecules possess identical molecular weights and carbohydrate compositions (Atwell and Marchalonis 1975). Moreover, preliminary evidence was obtained which suggests an immunological cross reaction between echidna IgG and human IgG. One rabbit antiserum to echidna IgG showed a weak cross reaction with human IgG but not IgM when assayed using a sensitive test that entailed use of unlabelled human proteins to inhibit the binding of $^{125}$I-labelled echidna IgG to the antiglobulin (Marchalonis and Atwell 1972). Another finding which supports the identification of the IgG-like immunoglobulin of the echidna as a homologue of IgG of higher mammals results from work of Kronvall et al. (1970). These workers found that the A protein of *Staphylococcus*, which binds specifically to the Fc portion of $\gamma$ chains of eutherian mammals, reacted with immunoglobulin of the echidna. This reagent did not react with immunoglobulins of chickens, reptiles, or amphibians.

Approximately 15 mg per ml of echidna serum consists of immunoglobulin, and IgG molecules comprise 85-90% of this mass. From this standpoint, the echidna resembles the rabbit, but further work is required to determine if classes distinct from IgG are present in the non-IgM fraction. Furthermore, the existence of IgG subclasses remains an open question.

Since a prototherian species, the echidna, possesses immunoglobulins comparable to those of eutherian mammals, it is reasonable to propose that metatherians should present a similar situation. At this stage,

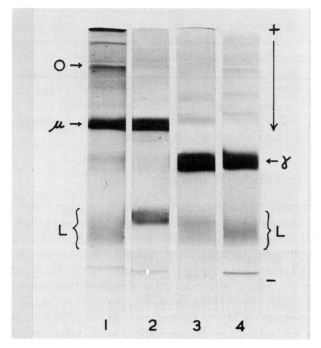

Fig. 42. Comparison by polyacrylamide gel electrophoresis in acid urea of polypeptide chains of immunoglobulins of man and echidna. Cell samples were reduced and alkylated in the presence of 9M urea. Origins: (1) echidna IgM, (2) human IgM myeloma protein, (3) echidna IgG immunoglobulin, (4) human IgG immunoglobulin. Notice the exact correspondence between the gamma chains of man and echidna. (From Atwell, Marchalonis, and Ealey 1973).

unfortunately, no detailed studies of the subunit structure of isolated marsupial immunoglobulins have been carried out, so the truth of this prediction cannot be tested. Rowlands and Dudley (1968) have shown that the American opossum can produce 19S and 7S antibodies to bacteriophage. In addition, Thomas et al. (1972) described the presence of antibody activity in the quokka, a small kangaroo, in fractions obtained by gel filtration on Sephadex G-200. These fractions had molecular weights of about 900,000 (void volume), 160,000 and 80,000. The former two molecular species may be comparable to IgM and IgG immunoglobulins of other species. The low molecular weight component of the quokka is too small to represent an intact immunoglobulin comprised of two pairs of light and heavy chains. This protein might be a "half-molecule" containing one heavy and one light chain. Marchalonis and Atwell (1972) have carried out preliminary studies of immunoglobulins of the brush-tailed opossum (*Trichosurus vulpecula*) and found that this marsupial possesses a major immunoglobulin containing heavy chains very similar to the $\gamma$ chains of man. In any case, marsupials possess immunoglobulins resembling the IgM and IgG molecules of other mammalian species. In addition, it was recently shown that the quokka possesses homocytotropic antibodies similar to IgE (Lynch and Turner 1974) and immunoglobulins comparable to IgA (Bell, Stephens, and Turner 1974) in its milk. Detailed chemical characterizations of these molecules have not yet been performed.

## Functions of non-IgM immunoglobulins

The emergence of immunoglobulin classes distinct from IgM-like molecules necessitated the generation of genes encoding heavy chains which differed from the $\mu$ chain. If this event occurred as part of a standard evolutionary process, some selective pressures must operate to maintain the existence of a separate heavy-chain gene. The question then arises as to the nature of the selective pressures that engendered the origin of distinct heavy chains and, therefore, new immunoglobulin classes. Some clues to this puzzle might be obtained from studies of the functions of the different immunoglobulin classes. A few generalizations can be made concerning functional differences among the major classes of vertebrate immunoglobulins, but a functional distinction among IgG subclasses, for example, is difficult to detect (Natvig and Kunkel 1973). IgM immunoglobulin is extremely efficient in the fixation of complement (Metzger 1970) and is very effective in agglutination reactions (Dreesman et al. 1965). In both of these properties, it surpasses IgG. IgG antibodies can carry out functions which cannot be fulfilled by IgM. For example, IgG molecules in man can pass through the placenta from mother to fetus, whereas IgM molecules are restricted to the maternal circulation (Metzger 1970). Another property in which IgG antibodies differ from IgM molecules is that the affinities of IgG for haptens increase with time following immunization, while those of IgM antibodies remain relatively low through the course of immunization. Immunoglobulins of other classes can differ from both IgM and IgG in distribution within the body and in binding properties. IgA, for example, tends to be concentrated in external secretions rather than in serum (Cohen and Porter 1964). IgE adheres to mast cells via its Fc piece, and combination of antigen (allergen) with cell-bound IgE initiates an anaphylactic reaction (Bennich and Johansson 1971).

Another selective influence that might have played an important role in the development of non-IgM antibodies could be related to recent observations that the production of IgM antibodies does not necessarily require cooperation between T and B cells (Mitchell 1974). In contrast,

IgG, IgA, and IgE are strongly dependent upon this cooperative interaction and might be called thymus-dependent antibodies. The primordial IgM immunoglobulins may have arisen in species where the fine control imparted by collaboration between T and B cells was not yet optimal. As the capacity for this collaboration intensified, B cells might have differentiated to the stage at which they could form IgM antibodies and, in addition, produced other classes, possibly of higher binding affinity, as a result of certain interactions with T cells.

Bearing these functional distinctions in mind, we can inquire whether the non-IgM antibodies of lower vertebrates differ from the IgM molecules in function and distribution. IgN of the turtle fixes complement, as does IgM, but it is laid down in the egg whereas IgM is not (Chartrand et al. 1971). These properties are shared by the IgN molecule of the duck (Grey 1969) and the IgG (Y) of the chicken (Orlans 1967). In the case of the chicken, another immunoglobulin class which resembles IgA occurs in the bile duct (Lebacq-Verheyden, Vaerman, and Heremans 1972). I would mention here that in species such as the plaice, a teleost that lacks non-IgM classes, IgM constitutes the immunoglobulin class present in the gut mucus (Fletcher and White 1973). The presence of IgE molecules in non-mammalian species can be inferred only from functional tests, such as induction of anaphylaxis, and no conclusive studies designed to establish its existence have been reported. Immediate hypersensitivity to fungal extracts has been observed in teleosts such as plaice and flounder. This phenomenon in mammals generally indicates the presence of IgE, but such antibodies were not shown in these teleosts (Fletcher and Baldo 1974). In any case, certain functional differences between IgM and non-IgM molecules of lower species have been observed, but a great deal of further study is required to provide complete information on distinctive functions of the latter molecules.

## Sequence comparison of mammalian mu and gamma chains

Comparisons among amino acid sequences obtained for the carboxyl-terminal residues of human $\mu$ chain, human $\gamma1$ chain and rabbit $\gamma$ chain are made in Table 16. Although the number of residues considered here represents only about 20% of the constant region sequence, the pattern that emerges reflects that shown by the entire constant portion. Human $\gamma1$ and rabbit $\gamma$ are identical in approximately 71% of the residues considered, whereas both $\gamma$ chains differ from the $\mu$ chain by 60-65%. This observation is consistent with an evolutionary scheme in which the $\mu$ chain and the common ancestor of the rabbit $\gamma$ and human

TABLE 16: Carboxyl terminal sequences of human $\mu$ chain, $\gamma_1$ chain, and rabbit $\gamma$ chain.

| Chain | | | | | | | | Residue |
|---|---|---|---|---|---|---|---|---|
| | 383 | 384 | 385 | 386 | 387 | 388 | | |
| $\gamma_1$ | Ser | Asn | Asp | Gly[b] | Glu | Pro | - | - |
| R $\gamma$ | Lys | Asp | Gly | Lys | Ala | Glu | - | - |
| $\mu$ | Met | Gln | Arg | Gly | Glu | Pro | Leu | Ser |
| | 396 | 397 | 398 | 399 | 400 | 401 | 402 | 403 |
| $\gamma_1$ | Pro | Val | Leu | Asp | Ser | Asp | Gly | Ser |
| R $\gamma$ | Ala | Val | Leu | Asp | Ser | Asp | Gly | Ser |
| $\mu$ | Pro | Glu | Pro | Gln | Ala | Pro | Gly | Arg |
| | 414 | 415 | 416 | 417 | 418 | 419 | 420 | 421 |
| $\gamma_1$ | Lys | Ser | Arg | Trp | Gln | Glu | Gly | Asn |
| R $\gamma$ | Thr | Ser | Glu | Trp | Gln | Arg | Gly | Asp |
| $\mu$ | Glu | Glu | Glu | Trp | Asn | Thr | Gly | Glu |
| | 432 | 433 | 434 | 435 | 436 | 437 | 438 | 439 |
| $\gamma_1$ | Leu | His | Asn | His | Tyr | Thr | Gln | Lys |
| R $\gamma$ | Leu | His | Asn | His | Tyr | Thr | Gln | Lys |
| $\mu$ | Leu | Pro | Asn | Arg | Val | Thr | Glu | Arg |
| | 450 | 451 | 452 | 453 | 454 | 455 | 456 | 457 |
| $\mu$ | Leu | Tyr | Asx | Val | Ser | Leu | Val | Met |

$\gamma$ chain diverged early in vertebrate evolution. Moreover, Putnam, Shimizu, and Shinoda (1972) have emphasized that the human $\mu$ chain is essentially as similar to the human $\lambda$ chain as it is to the $\gamma$ chain in constant sequence. This suggests that the genes encoding the constant regions of $\mu$ and $\gamma$ chains must have diverged early in vertebrate evolution soon after the separation of primitive light- and heavy-chain genes. This event clearly occurred before the divergence of primates from other eutherian mammals. It is difficult to make a precise estimate of the time of the divergence of $\mu$ chain and $\gamma$ chain from their common ancestor, because, as Smith (1973) suggests, the marked differences between human C$\mu$ and the C$\gamma$ regions of man and other mammals might reflect an extraordinarily large accumulation of mutation in the C$\mu$ of man, rather than a distant evolutionary relationship. Hill and his coworkers (1966) found that human and rabbit $\gamma$ chains are 76% identical in the Fc piece. Since human and rabbit hemoglobins are more than 80% identical, the possibility has been raised that immunoglobulin evolution may proceed at a faster rate than hemoglobin evolution (Grey 1969), at least during mammalian evolution. Such conclusions are prenumber [a]

| | 389 | 390 | 391 | 392 | 393 | 394 | | 395 | |
|---|---|---|---|---|---|---|---|---|---|
| - | Glu | Asn | Tyr | Lys[c] | Thr | Thr | - | Pro | - |
| - | Asp | Asp | Tyr | Lys | Thr | Thr | - | Pro | - |
| Pro | Gln | Lys | Tyr | Val | Thr | Ser | Ala | Pro | - |

| 404 | 405 | 406 | 407 | 408 | 409 | 410 | 411 | 412 | 413 |
|---|---|---|---|---|---|---|---|---|---|
| Phe | Phe | Leu | Tyr | Ser | Lys | Leu | Thr | Val | Asp |
| Trp | Phe | Leu | Tyr | Ser | Lys | Leu | Ser | Val | Pro |
| Tyr | Phe | Ala | His | Ser | Thr | Leu | Thr | Val | Ser |

| 422 | 423 | 424 | 425 | 426 | 427 | 428 | 429 | 430 | 431 |
|---|---|---|---|---|---|---|---|---|---|
| Val | Phe | Ser | Cys | Ser | Val | Met | His | Glu | Ala |
| Val | Phe | Thr | Cys | Ser | Val | Met | His | Glu | Ala |
| Thr | Tyr | Thr | Cys | Val | Val | Ala | His | Glu | Ala |

| 440 | 441 | 442 | 443 | 444 | 445 | 446 | 447 | 448 | 449 |
|---|---|---|---|---|---|---|---|---|---|
| Ser | Leu | Ser | Leu | Ser | Pro | Gly | - (COOH) | | |
| Ser | Ile | Ser | Arg | Ser | Pro | Gly | - (COOH) | | |
| Thr | Val | Asp | Lys | Ser | Thr | Gly | Lys | Pro | Thr |

| 458 | 459 | 460 | 461 | 462 | 463 | 464 | 465 | |
|---|---|---|---|---|---|---|---|---|
| Ser | Asp | Thr | Ala | Gly | Thr | Cys | Tyr | - (COOH) |

*Source:* Data compiled by Gally and Edelman 1972, $\gamma_1$ chain of myeloma protein Eu; Putnam et al. 1973, $\mu$ chain of myeloma protein Ou; Hill et al. 1966, $\gamma$ chain of normal rabbit IgG immunoglobulin.

a. Numbering system is based upon the human $\gamma_1$ chain. The $\mu$ chain has 576 residues, but its carboxyl terminal segment is aligned with corresponding stretches of $\gamma_1$ chain sequence.

b. Line above indicates identical residues in human $\mu$ and $\gamma_1$ chains.

c. Line below indicates identical residues in human $\gamma_1$ chain and rabbit $\gamma$ chain.

mature because the C-region immunoglobulin pattern may be obscured to some degree by the presence of subclasses and allotypes. More extensive sequence data for immunoglobulins of nonmammalian vertebrates are required in order to resolve this issue. Recent sequence comparisons among immunoglobulins of mammalian species suggest an observed rate of evolution of 20-25 residues per 100 residues per $100 \times 10^6$ years (Chuang, Capra, and Kehoe 1973). The present data are sufficient to establish the homology among immunoglobulin light chains and heavy chains, thereby indicating that such molecules are derived from the same ancestral gene.

## Classification of non-IgM immunoglobulins of lower species

In this chapter the evidence has been reviewed establishing that the tetrapod vertebrates and their close relatives, the Dipnoi, possess low molecular weight immunoglobulins distinct from the IgM class, which is ubiquitous in distribution among vertebrate species. Although the final assessment of evolutionary homology among these immunoglobulins and those of higher mammals must entail comparisons of amino acid sequences of heavy chains from the diverse species, sufficient data presently exist to allow some general conclusions to be drawn regarding the non-$\mu$ heavy chains. Figure 43 depicts a selection of reduced low molecular weight immunoglobulins analyzed by polyacrylamide gel electrophoresis in buffer containing SDS. Marked differences in heavy chain mobilities can be observed, although light chains of all species possess similar mobilities. The heavy chains of the 7S immunoglobulins of the toad and chicken penetrated the gel to an extent intermediate between those mobilities characteristic of the $\mu$ chain marker (M. W. 70,000) and the $\gamma$ chain marker (M. W. 50,000). The heavy chain of the 5.9S immunoglobulin of the lungfish possesses a faster mobility than that of the human $\gamma$ chain. This result is consistent with its molecular weight of 38,000 (Marchalonis 1969) which is substantially lower than the mass of 50,000 characteristic of $\gamma$ chains. Heavy chains identical in mobility to the human $\gamma$ chain marker were present in 7S immunoglobulins of both lower mammalian species the echidna (a monotreme) and the brush-tailed opossum (a marsupial). On the basis of this type of analysis, three categories of heavy chains can be discerned among lower vertebrate immunoglobulins.

The first category is comprised of heavy chains characterized by a mobility identical to that of human $\gamma$ chain. The second consists of molecules substantially lower in molecular weight than the $\gamma$ chain. These molecules, therefore, penetrated substantially more deeply into the gels in Figure 43 than did those of the $\gamma$ chain. The third category consists of heavy chains which are retarded in mobility relative to the $\gamma$ chain and possess higher molecular weights than does the $\gamma$ chain. More detailed analyses of the non-IgM immunoglobulins of lower species justify this classification. It is interesting that, although immunoglobulins comparable in size to the human IgG molecule are found in certain anuran amphibians, their heavy chains possess mobilities distinct from that of the human $\gamma$ chain where analyzed by polyacrylamide gel electrophoresis in acid urea (Marchalonis and Edelman 1966b) or in buffers containing SDS (Hadji-Azimi 1971; Geczy, Green, and Steiner 1973). The latter results probably reflect differences in carbohydrate content of the heavy chains, a condition which is particularly critical

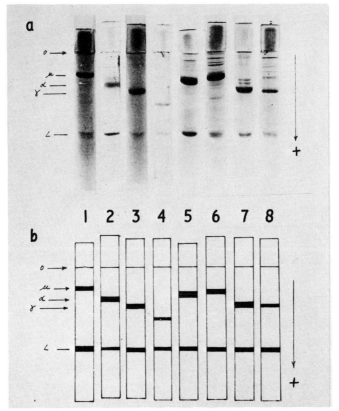

Fig. 43. a. Polyacrylamide gel electrophoresis in discontinuous sodium dodecylsulfate-containing buffers of reduced low molecular weight immunoglobulins from diverse vertebrate species. From left to right: (1) Human IgM, (2) Mouse IgA, (3) Human IgG, (4) Lungfish IgN, (5) Toad IgG-like, (6) Chicken IgG-like, (7) Echidna IgG, (8) Possum IgG. O indicates origin of running gel and ⟶ + shows the direction of migration of the protein bands. b. Tracings of gels shown.

in the determination of molecular weights by gel electrophoresis using buffer containing SDS (Glosmann and Neville 1971; Segrest et al. 1971).

Using terminology presented earlier in this chapter, the three categories of non-IgM immunoglobulins can be summarized as follows. (1) Immunoglobulin molecules that are similar to IgG of man in all general properties studied can be considered homologues of IgG. These occur only in mammalian species. (2) Immunoglobulins characterized by a mass 120,000 daltons and containing heavy chains of mass 38,000 have been termed IgN immunoglobulins (Marchalonis 1970). This group was the first distinct class to diverge from the originally existing IgM group. IgN-like molecules occur in Dipnoi, reptiles, birds, and possibly mammals. (3) IgG-like molecules were found in anuran amphibians, reptiles and birds. Although certain of these molecules exhibit definite resemblances to IgG immunoglobulin of eutherian mammals, none of them show identity in all properties tested. For example, heavy chains of all of the proteins in this category are distinct from true γ chain in mobility when analyzed by polyacrylamide gel electrophoresis in SDS-containing buffers (Figure 43). The IgG-like proteins of the chicken, moreover, differ substantially in molecular weight from IgG (Leslie and Clem 1969) and attempts by Sanders, Travis, and Wiley (1973) to find amino acid sequence homology between the chicken heavy chain and human γ chain did not provide evidence indicating marked evolutionary relatedness of the two proteins. Leslie and Clem (1969) termed the

TABLE 17: Properties of non-IgM immunoglobulins of lower vertebrates.

| Species | Class | Intact | Light |
|---------|-------|--------|-------|
| Lungfish[1] | Dipnoi | 120,000 | 22,000 |
| Bullfrog[2] | Amphibian | (150,000) | 22,000 |
| Marine[3] toad | Amphibian | 160,000 | 22,500 |
| *Xenopus*[4, 5] | Amphibian | N.D.[b] | 22,000 |
| | | | 26,700 |
| Turtle[6] | Reptile | 180,000 | 22,500 |
| | | 120,000 | 22,500 |
| Sleepy lizard[7] | Reptile | 151,000 | 22,400 |
| Duck[8] | Avian | 178,000 | 22,400 |
| | | 118,000 | 23,000 |
| Chicken[9] | Avian | 174,000 | 22,500 |
| | | N.D. | N.D. |
| Echidna[10] | Monotreme mammal | 150,000 | 22,500 |

chicken 7.1S molecule IgY. This notation has been incorporated into Table 17, where IgG-like immunoglobulins are designated either IgG(Y) for reptiles and birds and IgG(?) in the case of amphibians and the sleepy lizard. This terminology takes into account the possibility that the IgG(Y) of reptiles and birds and the IgG(?) of amphibians might not be directly homologous molecules, but that the genes encoding the various heavy chains could have arisen independently during amphibian and reptilian evolution.

The data considered here will be given a molecular and genetic interpretation in Chapter 17, where the general evolution of immunoglobulin polypeptide chains will be discussed.

| Heavy | CHO (%)[a] | Designation | Formula |
|---|---|---|---|
| 38,000 | N.D. | IgN | $L_2 \nu_2$ |
| 53,000 | 2.0 | IgG(?) | $L_2 \gamma(?)_2$ |
| 53,000 | 4.2 | IgG(?) | $L_2 \gamma(?)_2$ |
| 53,000[c] | N.D. | IgG(?) | $L_2 \gamma(?)_2$ |
| 64,500[d] | N.D. | IgG(?) | $L_2 \gamma(?)_2$ |
| 67,500 | | IgG(Y) | $L_2 \gamma(Y)_2$ |
| 38,000 | 0.9 | IgN | $L_2 \nu_2$ |
| 51,000 | N.D. | IgG(?) | $L_2 \gamma(?)_2$ |
| 68,000 | 5.0 | IgG(Y) | $L_2 \gamma(Y)_2$ |
| 35,000 | 0.6 | IgN | $L_2 \nu_2$ |
| 67,500 | 2.2 (HEX) | IgG(Y) | $L_2 \gamma(Y)_2$ |
| N.D. | N.D. | IgA | $(L_2 a_2)$ |
| 49,000 | 2.0 (HEX) | IgG | $L_2 \gamma_2$ |

*Sources:* (1) Marchalonis 1969; Litman et al. 1971; (2) Marchalonis and Edelman 1966b; (3) Acton et al. 1972a; (4) Marchalonis, Allen, and Sarni 1970; (5) Haji-Azimi 1971; (6) Leslie and Clem 1972; (7) Wetherall 1969; (8) Zimmerman, Shalatin, and Grey 1971; (9) Leslie and Clem 1969; (10) Atwell, Marchalonis, and Ealey 1973.

a. CHO, carbohydrate content of intact molecule as a percentage of the total mass.

b. N.D. = no data.

c. Molecular weights obtained by polyacrylamide gel electrophoresis under varying concentrations.

d. Molecular weights obtained by polyacrylamide gel electrophoresis in buffer containing SDS.

# 9 Emergence of Variable Regions

## *Variable region structure and function*

Information in Chapter 7 establishes that the variable regions of light chains of immunoglobulins of primitive vertebrate species are directly homologous to the light chains of mammalian antibodies. The available data, however, were insufficient to determine whether such molecules resembled the kappa and lambda chains of mammals, which are encoded by distinct gene clusters and possess markedly divergent primary sequences. Heavy chains of mammalian species are the products of closely linked genes and share a common pool of variable region sequences (Gally and Edelman 1972). Therefore, variable region sequences of heavy chains are not useful in making a distinction among immunoglobulin classes (e.g., IgM versus IgG or IgA) but provide evidence on phylogenetic divergence or homology of variable regions. Since the combining sites for antigen are formed by interaction of the variable regions of light and heavy chains, another means of assessing homology of variable regions of diverse vertebrate species is to characterize quantitatively their binding properties for antigens. The most suitable antigens for such quantitative studies are haptens such as dinitrophenol, because a series of precise thermodynamic measurements can be readily carried out (see Chapter 2). I will now discuss studies of lower vertebrates which bear upon the problem of V-region evolution within heavy chains and upon comparative studies of the properties of the combining site for antigen.

TABLE 18a: Amino terminal amino acid sequences of heavy chains of immunoglobulins of diverse species.

| Species | Vertebrate class | Immunoglobulin class or V-designation | Residue numbers | | | | | |
|---|---|---|---|---|---|---|---|---|
| | | | 1 | 2 | 3 | 4 | 5 | 6 |
| Man[1,2,3] | Mammal (Primate) | $V_H$ I | PCA | Val | Gln | Leu | Val | Gln |
| | | $V_H$ II | PCA | Val | His | Leu | Lys | Gln |
| | | | - | - | Thr | - | Thr | - |
| | | | - | - | - | - | Arg | - |
| | | $V_H$ III | Glu | Val | Gln | Leu | Leu | Glu |
| | | | - | Ile | - | - | Val | - |
| Cat[4] | Mammal (Carnivore) | $V_H$ III | Asp | Val | Gln | Leu | Val | Glu |
| Opossum[5] | Mammal (Marsupial) | $V_H$ III | Glu | Ile | Gln | Leu | Val | Glu |
| Echidna[6] | Mammal (Prototherian) | $V_H$ III gamma | Glu | Val | Glx | Leu | Val | - |
| | | | - | Ile | - | - | Met | - |
| | | | - | - | - | - | Leu | - |
| Chicken[7] | Avian | Anti-DNP | Ala | Val | Thr | Leu | Asp | Glu |
| Gar[8] | Holostean | mu | Asp | Ala | Val | - | - | - |
| | | | - | Val | Ile | Val | - | - |
| | | | - | Leu | Leu | - | - | - |
| Paddlefish[9] | Holostean | mu | Asp | Ile | Val | Ile | Thr | - |
| Leopard shark[2] | Elasmobranch | HS1 | PCA | Val | Pro | Gly | - | Gln |
| | | HS2 | PCA | Asp | Leu | Pro | Thr | Pro |
| | | $V_H$(ub) | Glu | Ile | Val | Leu | Thr | Gln |
| | | | - | Val | - | - | - | - |
| Lamprey[10] | Cyclostome | mu | Asp | - | - | - | - | - |

*Sources:* (1) Hood and Talmage 1970; (2) Wang et al. 1970; (3) Kubo, Rosenblum, and Benedict 1970; (4) Kehoe, Hurwitz, and Capra 1972; (5) Wasserman, Kehoe, and Capra 1974; (6) Atwell and Marchalonis (unpublished observation); (7) Kubo, Rosenblum, and Benedict 1971; (8) Acton et al. 1971a; (9) Pollara et al. 1968; (10) Marchalonis and Edelman 1968b.

## Variable region sequences of heavy chains

The data listed in Table 18a show that obvious homologies exist among the $V_H$ regions of diverse vertebrate species. Human $V_H$ regions can be classified into three subgroups on the basis of amino acid sequence. The $V_H$ III subgroup does not have a blocked N-terminal residue and, consequently, has been relatively easy to investigate using pooled immunoglobulins of a number of species (Wasserman, Kehoe, and Capra

1974). The data of Table 18a indicate that cats (eutherians), opossums (metatherians) and echidnas (prototherians) all possess heavy chains with V-region sequences corresponding to the $V_H III$ subgroup of man. Wasserman, Kehoe, and Capra (1974) have found that this subgroup is the predominant one in opossums, dogs, minks, cats, sea lions, and seals. In rodents (guinea pig, mouse, and rat) and primates (green monkey, rhesus monkey, man) the $V_H III$ subgroup comprises 20-30% of total $V_H$ sequence as judged by quantitative application of the automated amino acid sequencer. The $V_H III$ subgroup was either absent or present in very small quantities among heavy chains of perissodactyls (horses), artiodactyls (pig, cow, sheep, goat, mouse), lagomorphs (rabbit), and cetaceans (white whale and fin whale). The presently available data indicate that the establishment of heavy chain V-region structure, notably that of the $V_H III$ subgroup, preceded the divergence of mammals into prototherians, metatherians and eutherians. The ancient origins of the $V_H III$ subgroup were established by recent studies showing the presence of extremely similar sequences in avian species (Wasserman, Kehoe, and Capra 1974) and sharks (Sledge, Clem, and Hood 1974). The sequences of the first twenty-four residues of avian and mammalian $V_H III$ prototypes and the heavy chain of shark antipolysaccharide antibody are compared in Table 18b. Close homology (identity of 17 of 24 residues) exists between avian and mammalian $V_H III$ sequences. The similarities between the shark sequence and those of mammals (11 of 24) and birds (8 of 24) were less striking.

The sequence data for lower species are too fragmentary to allow unequivocal identification of the subgroups present in all vertebrate classes. Three important conclusions merit comment, however. (1) Partial N-terminal sequences of heavy chains of all species exhibit clear-cut homology. (2) Unblocked amino-terminal sequences occur in a variety of lower species, although exceptions to this rule have been reported. Acton et al. (1970) found that the heavy chains of immunoglobulins of the marine toad (*B. marinus*) had blocked N-terminals and were not amenable to direct sequence analysis. (3) V-region subclasses are present in the leopard shark. This important observation, make by Klaus, Nitecki, and Goodman (1971) suggests that the genetic events underlying the formation of subclasses, probably duplications of germ-line genes, must have occurred in vertebrate evolution. If, in fact, there exist multiple copies of germ-line genes coding for V-region subclasses, phylogenetic divergence may have included duplication or deletion of different primordial germ-line genes (Klaus, Nitecki, and Goodman 1971). Thus, different germ-line genes than those that led to the human $V_H$ sequences may have duplicated during the intraspecies evolution of other species such as the shark. This particular hypothesis would suggest

TABLE 18b: Comparison of V$_H$ III-like sequences of mammalian, avian, and elasmobranch species.

| | Residue number | | | | | | | | | | | |
|---|---|---|---|---|---|---|---|---|---|---|---|---|
| | 1 | 2 | 3 | 4 | 5 | 6 | 7 | 8 | 9 | 10 | 11 | 12 |
| V$_H$ III Prototype (Mammal) | Glu[a] | VAL[b] | Gln[c] | LEU | Val | GLU | Ser | Gly | Gly | Gly | Leu | Val |
| V$_H$ III Prototype (Avian) | Ala | VAL | Gln | LEU | Asp | GLU | Ser | Gly | Gly | Gly | Leu | Val |
| Nurse shark | Glu | VAL | Thr | LEU | Thr | GLU | Pro | Glx | Ala | Glx | Asp | Ser |

| | Residue number | | | | | | | | | | | |
|---|---|---|---|---|---|---|---|---|---|---|---|---|
| | 13 | 14 | 15 | 16 | 17 | 18 | 19 | 20 | 21 | 22 | 23 | 24 |
| V$_H$ III Prototype (Mammal) | Gln | PRO | GLY | Gly | Ser | LEU | Arg | LEU | Ser | CYS | Ala | Ala |
| V$_H$ III Prototype (Avian) | Gly | PRO | GLY | Val | Gln | LEU | Arg | LEU | Val | CYS | Gly | Ala |
| Nurse shark | Glx | PRO | GLY | Gly | Ala | LEU | Thr | LEU | Thr | CYS | Glx | Val |

*Source:* Wasserman, Kehoe, and Capra 1974.
a. Line above indicates identity between mammalian and shark sequence.
b. The residues common to all three groups are in capitals.
c. Line below the amino acid indicates identity between avian and mammalian sequence.

that $V_H$ regions of shark and man, although arising from a common set of germ-line genes, might bear little resemblance to each other at this time. Unfortunately sufficient data do not exist to enable the formulation of a detailed picture of V-region evolution at this time.

## *Binding affinities of antibodies of lower vertebrates*

The above partial sequence information, in conjunction with that cited for light chains (Table 10), indicates that the N-terminals of the V regions of antibodies of all species share considerable evolutionary homology. Unfortunately, the studies are not yet of sufficient extent to probe the combining-site regions that are composed of residues associated with the so-called hypervariable regions of light and heavy chains. Indirect evidence on the phylogeny of the combining site can be obtained by studying the binding of anti-hapten antibodies of lower species to the homologous hapten. Clem and Small (1970) studied the thermodynamic properties of specifically purified antibodies of the grouper (a teleost fish) to the DNP-hapten. Using the technique of equilibrium dialysis and the mathematical analysis outlined in Chapter 2, they measured binding affinities of the 16S and 6.2S immunoglobulins for this hapten. The 16S antibodies possessed two populations of binding sites; four strong sites characterized by an intrinsic binding constant of $2.2 \times 10^6 M^{-1}$ and four sites exhibiting weaker affinities of $8 \times 10^4 M^{-1}$ (see Chapter 2 for a discussion of the measurement of binding affinity). This situation parallels that of rabbit IgM antibodies where 5 sites show high affinity combination with antigen and 5 sites possess lower binding constants (Onoue et al. 1968; Merler, Karlin, and Matsumoto 1968). The 6.3S antibody showed an intrinsic affinity of $3 \times 10^4 M^{-1}$. The molecule seemed to possess only one combining site for antigen instead of the expected two. This discrepancy may result from a variety of causes including possible lability of the molecule during fractionation (Underdown, Simms, and Eisen 1971). In addition to determining association constants, Clem and Small (1970) calculated the thermodynamic parameters free energy ($\Delta F$), enthalpy ($\Delta H$) and entropy ($\Delta S$). Their values for grouper antibodies corresponded closely to those observed for rabbit antibodies (Table 19), prompting them to conclude that the antibody combining site may have remained relatively unchanged during the time required for the emergence of teleosts and mammals (Clem and Small 1970). I would comment that even the larval lamprey possesses cells bearing receptors which specifically recognize haptens. If the binding studies could be performed, it is likely that such molecules would show affinities comparable to those observed for grouper and rabbit antibodies. This conservatism within the variable

TABLE 19: Thermodynamic properties[a] of rabbit and grouper antibodies to the dinitro-phenyl hapten.

| Species | Hapten | Temperature °C | $\Delta F$ (K cal/mole) | $\Delta H$ (K cal/mole) | $\Delta S$ (eu/mole) |
|---------|--------|----------------|-------------------------|-------------------------|----------------------|
| Rabbit (IgG) | -DNP-1 lysine | 12.8 | -10.9 | -19.6 | -30.4 |
| | | 22.2 | -10.6 | | |
| | | 39.5 | -10.1 | | |
| Grouper (6.4S) | -DNP-amino caproic acid | 2.5 | -10.0 | -15.6 | -22.8 |
| | | 36.5 | - 9.3 | | |
| Grouper (16S) | -DNP-amino caproic acid | 2.5 | - 9.5 | -15.7 | -22.8 |
| | | 18.5 | - 9.4 | | |
| | | 36.5 | - 9.2 | | |

a. The thermodynamic parameters for the binding of antibody and hapten given here are calculated as follows:

The Gibbs free energy is computed from the association constant according to the equation $\Delta F° = - RT \ln K$, when R is the gas constant and T is the absolute temperature at which the binding is carried out.

The enthalpy $\Delta H°$ is computed by measuring the affinity constant at two different temperatures and solving the equation

$$\frac{\ln K_2}{K_1} = \frac{\Delta H°}{R} \left(\frac{1}{T_2} - \frac{1}{T_1}\right)$$

where $T_1$ and $T_2$ are the different temperatures and $K_1$ and $K_2$ are the corresponding association constants.

The entropy $\Delta S$ is then computed from the equation $\Delta F = \Delta H - T\Delta S$.

region has also been substantiated by amino acid sequence data for the amino-terminal portions of polypeptide chains of Chondrichthyes and Osteichthyes.

## *Affinity variation of antibodies of vertebrates*

Critical measurements by Eisen and Siskind (1964) established that the binding affinities of IgG antibodies increase markedly with time after immunization and as a consequence of secondary or booster injections of the same hapten. It has been difficult to perform comparable detailed studies of IgM antibodies in mammals because these constitute a small fraction of serum immunoglobulin, and purified antibodies must be used in a precise assay such as equilibrium dialysis. Sarvas and Mäkelä (1970) have obtained evidence showing that serum IgM antibodies of mammals do not exhibit increasing affinity with time following immunization.

Analyses of antibodies to DNP synthesized by lower species provide a useful approach to examining variation of IgM binding affinities for hapten in primary and secondary responses. Since species more primitive than Dipnoi possess only IgM antibodies, a clear-cut experimental system for study of the affinity properties of an IgM response is available. Furthermore, the existence in reptilian species of low molecular weight immunoglobulins distinct from IgG antibodies of mammals provides a further opportunity to determine if increasing affinity is a general property of all antibody classes.

The profound increase in affinity of rabbit IgG antibodies to DNP is illustrated in Table 20, which also presents comparable data for the grouper, carp, and tortoise. The IgM antibodies of both teleost fish failed to increase significantly either with increasing time or after secondary injections. Even following multiple booster injections, the affinities of grouper antibodies actually decreased (Clem and Small 1970). The binding affinities of carp and turtle IgM antibodies remained roughly constant over a period of seven months. In contrast, IgN antibodies of the turtle (Ambrosius and Fiebig 1972) and IgG antibodies of the rabbit showed greater than 100-fold increases in affinity during the immunization courses. These results indicate that the primitive IgM antibody forming system does not usually improve with experience. The quantitative increase in titers of specific IgM antibody following immunization most probably results only from an increase in the number of cells capable of secreting these molecules. The increased affinities exhibited by the IgG and IgN antibody populations suggest that selection for better binding continued as the specific antibody-forming cells divided after stimulation by antigen. This observation may account in part for the evolutionary impetus that prompted the emergence of immunoglobulin classes distinct from the ancient IgM molecules. Similar studies are required to determine if the IgG-like antibodies of amphibians also are the product of an immunological learning system.

## Dimensions of the combining site

Another parameter indicating similarities in the combining sites of anti-DNP antibodies of trout and rabbits was obtained by Roubal, Etlinger, and Hodgins (1974) who used spin-labelled haptens and paramagnetic resonance spectroscopy to determine the dimensions of the combining sites. These workers found that the average depth of the antigen combining sites in both trout and rabbit was 11.5Å. This result agrees quite closely with that of Padlan et al. (1974) who found, by crystallographic analysis, that the depth of the phosphorylcholine binding site of the mouse IgA myeloma McPc 603 was 12 Å (see Chapter 2).

TABLE 20: Variation of binding affinities of antibodies to DNP with time after immunization.

| Species | Immunoglobulin | Average intrinsic association constant (liter/mole x $10^{-6}$) | |
|---|---|---|---|
| | | Early antibodies | Late antibodies |
| Grouper[1] | 16S IgM | 2.1 (30 days) | 0.06 (365 days) |
| | 6.4S IgM$_F$ | 3.0 (30 days) | 0.40 (365 days) |
| Carp[2] | 16S IgM | 0.08 (50 days) | 0.15 (125 days)[a] |
| Tortoise[3] | 19S IgM | 0.06 (45 days) | 0.05 (162 days)[a] |
| | | | 0.08 (203 days)[a] |
| | 5.7S IgN | 0.20 (45 days) | 1.00 (162 days)[a] |
| | | | 60.0 (203 days)[a] |
| Rabbit[4] | IgG | 0.14, 0.3[b] | 34.0, 84.0 (8 days |
| | | 0.20, 0.15[b] | 500.0 after secondary injection) |

*Sources:* (1) Data of Clem and Small 1970; (2) Data of Ambrosius and Fiebig 1972; (3) Data of Eisen, Little, Steiner, and Simms (quoted by Eisen in Davis et al. 1969).

a. Values cited were obtained for animals which had received secondary or booster injections.

b. Data for 4 individual animals at day 14.

## *Synopsis*

Comparative amino-terminal sequence data establish that considerable homology exists among immunoglobulin chains of diverse vertebrate species. Moreover, studies of the thermodynamic properties of the combination of the dinitrophenyl hapten with antibodies of fish, reptiles, and mammals establish that the energetic parameters of this binding are similar for all vertebrate antibodies. This finding provides additional support for the conclusion that substantial conservatism has characterized the evolution of the variable regions of immunoglobulin light and heavy chains. Furthermore, measurements of binding affinities as a function of time after immunization indicate that a clear-cut rise in affinities does not occur in fish that possess only IgM antibodies. This increase in affinity was definitely observed in the tortoise and the rabbit, which exhibit immunological memory and express classes of immunoglobulin distinct from IgM antibodies.

# 10 Emergence of Lymphoid Cells and Organs

". . . the inference is drawn, that in later life the [thymus] ceases to exist. It no more ceases to exist than would the Anglo-Saxon race disappear, were the British Isles to sink beneath the waves.

. . . for just as the Anglo-Saxon stock has made its way from its original home into all parts of the world, and has there set up colonies for itself and for its increase, so the original leucocytes, starting from their birthplace and home in the thymus, have penetrated into almost every part of the body, and have there created new centres for growth, for increase, and for useful work for themselves and for the body."
[J. Beard, *The Source of Leucocytes and the True Function of the Thymus*, 1900.]

This chapter deals with the phylogenetic distribution of lymphoid cells and organs. The lymphoid cells considered here are those which have been shown to bind antigen and to synthesize antibodies. I will place major emphasis here on primary lymphoid organs, in particular the thymus, which arise early both in phylogeny and in ontogeny. In addition, I will use the embryonic chicken as a model system to document the development of thymus and bursa, the larval frog to illustrate the role of the thymus in the ontogeny of amphibian immunity, and the fetal or pouch-young opossum as a model for the development of mammalian primary and secondary lymphoid organs. All three lower vertebrates possess special advantages which facilitate detailed study of the ontogeny of the cells and organs of the immune system.

## Phylogenetic origins of antibody-forming cells

Plasma cells, which are the major secretors of antibody in higher vertebrates, are the end stage in the antigen-directed maturation of B lymphocytes. Although cyclostomes and primitive elasmobranchs can synthesize antibodies that resemble the immune macroglobulins (IgM) of man, cells possessing the morphology of plasma cells have not been found in these species (Good et al. 1966). The cells which produce antibodies in these species resemble small to large lymphocytes (Hildemann 1972; Good et al. 1966). Plasma cells are present in advanced sharks such as the smooth dogfish, *Mustelus canis* (Engle et al. 1958), and in all higher forms investigated. Studies have been performed at the level of electron microscopy which clearly establish the similarities of plasma cells of the paddlefish, a chondrostean (Clawson, Finstad, and Good 1966), and various amphibians (Cowden and Dyer 1971) to those of mammals. In addition, all species of vertebrates possess circulating cells which have been described as lymphocytes (Hildemann 1972).

Investigation of the morphological differentiation of antibody-forming cells of the marine toad (*B. marinus*), an anuran amphibian, serves to illustrate the cellular transitions involved in the immunological progression of antigen recognition to immunoglobulin synthesis.

The immunocytoadherence technique used to enumerate antibody-forming cells (AFC) lends itself readily to the identification of the cells involved. Figure 44 presents a selection of AFC as they appear in histological smears. The AFC are distinguished by their coronas of adherent bacteria. The variety of cell types resembles that observed in mice. The AFC population in toads consists of lymphocytes, immunoblasts, and plasma cells. The lymphocytes constitute a large fraction of the AFC population at early stages of the response (less than one week) and the larger cells predominated at later times (10-28 days). Studies involving uptake of tritiated thymidine into AFC indicate that the exponential rise in numbers of these cells is due chiefly to cell division. A dividing immunoblast which has incorporated ³H-thymidine is shown in Figure 44c. Cowden and Dyer (1971) have confirmed the identification of the variety of antibody-forming cells of amphibians by means of electron microscopy.

The differentiation of antibody-forming cells has also been followed by separation on the basis of density. Shortman (1972) has devised a procedure for separating lymphocytes by centrifugation through density gradients formed from bovine serum albumin. Kraft, Shortman, and Marchalonis (1971) used this procedure to assess the density heterogeneity of lymphocytes from the spleen of rats which formed antibodies to *S. adelaide* in the primary response to this antigen. The pattern that

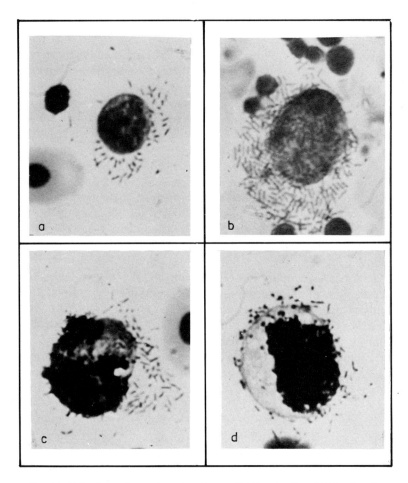

Fig. 44. Cells from *B. marinus* forming antibodies to *S. adelaide* flagella. Antibody secreted to the surface of the cells causes adherence of *Salmonella* bacteria. (a) Lymphocyte; (b) immunoblasts; (c) dividing immunoblast labeled with tritiated thymidine (autoradiography); (d) immature plasma cell with nucleus labeled with tritiated thymidine (autoradiography) (x 1,500). (From Marchalonis 1971b).

was obtained was complicated, even extremely early in the response. It was initially hoped that cells responding early would represent a single component on the density gradient and that heterogeneity would emerge later as the result of differentiation and division. Studies were performed to compare the AFC in the primary response of the toad to those of the rat (Kraft, Shortman, and Marchalonis 1971). The situation with immunocytoadherent lymphocytes from toad spleen was much less complex than that of the rat (Figure 45). Initially, the toad AFC were restricted primarily to one peak of relatively low density, and the pattern became complicated as time after immunization increased. Unfortunately, the density of cells within the density gradient cannot be correlated in a simple fashion with cell morphology, because the physiological state of the cell also affects the density (Shortman 1972). The relatively simple pattern obtained for the toad primary response relative to that of the rat is due in a large part to the fact that the toad produced only IgM antibodies to flagella (see Chapter 12), whereas the rat synthesizes IgM and IgG antibodies.

## Primary and secondary lymphoid organs

Since antibody production is a property of all vertebrate classes, the cellular machinery required to recognize foreign antigens and synthesize immunoglobulins must be ubiquitous among vertebrates. Critical investigation of the patterns of lymphoid structure present in diverse vertebrate species can provide information about the minimal structural requirements for humoral and cellular immunity. These studies should illustrate evolutionary analogies in which disparate groups of vertebrates have evolved different lymphoid structures that fulfill common roles. Aggregates of lymphocytes and phagocytic cells tend to be associated with sites where potentially pathogenic organisms might be trapped. Larval lampreys, for example, may represent the primitive condition in which clusters of lymphocytes are found within the gills and along the gut (Good et al. 1966; Rowlands 1969). All placoderm-derived vertebrates possess certain discrete lymphoid organs, such as the thymus and spleen, which probably developed over evolutionary time from such primitive collections of lymphoid cells.

The lymphoid architecture of mammals is quite complex. The roles of various discrete organs (including the thymus, lymph nodes, bone marrow, and appendix) in the trapping of antigen and the formation of antibodies have been investigated in great detail (Nossal and Ada 1971). Recent studies have established that lymphoid organs of mammals and birds can be separated into two categories, namely, primary and secondary lymphoid organs. The distinguishing features of these two classes of lymphoid structures are outlined in Table 21. In essence, primary lymphoid organs appear early in ontogenetic development and have a high turnover of lymphocytes. This proliferation of lymphocytes is independent of exogenous antigen, and the primary organs do not under normal circumstances contain antibody-forming cells. Moreover, the primary organs tend to involute as the animal reaches maturity. Secondary lymphoid organs, in contrast, develop later in ontogeny, remain throughout the life of the animal, require antigen to maintain lymphopoietic activity, and contain antibody-forming cells or other end cells of immunocyte differentiation. In the consideration of the lymphoid organs of lower vertebrates, I shall not dwell on attempts to describe the exact homology of structures such as the procoracoid bodies of anurans (Baculi, Cooper, and Brown 1970; Cooper, Brown, and Baculi 1971) to definite lymphoid structures of mammals, but will ask whether they are primary or secondary organs. Another major issue to be discussed here is the existence of T cells and B cells in lower species. The derivation of these terms reflects the names of the primary lymphoid organs, the thymus, which is the source of cells mediating cellular immunity, and the bone marrow or bursa of Fabricius, which provide the cells which actually synthesize antibodies. The issue of the phylogenetic distribution of T cells and B cells is a crucial one to present day cellular immunology, which is deeply concerned with the mechanisms by which such

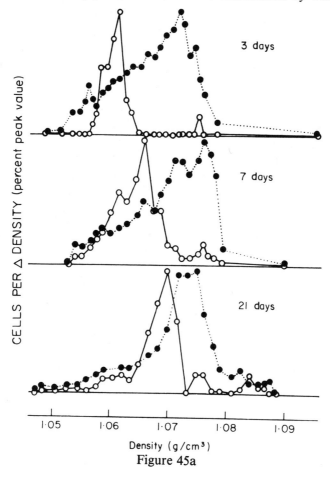

Figure 45a

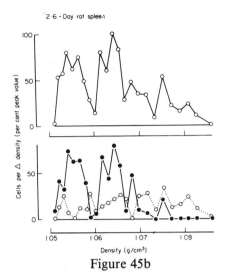

Figure 45b

Fig. 45. Studies by density gradient resolution of antibody-forming cells (immunocytoadherent) to *S. adelaide* flagella in the toad and the rat. (A) Density distribution of toad AFC (o——o) and total spleen cells (•---•) as a function of time after immunization. AFC initially appear in a fraction of light density. The pattern becomes more complex with time after immunization. (B) Density distribution of rat spleen lymphocytes forming different classes of antibody. The upper curve gives the density distribution of all AFC. The lower curve resolves this curve into IgM producers (•——•) and non IgM producers (o--o). Notice the increased complexity of these rat patterns relative to the distribution of AFC in toad spleen at days 3 and 7. (From Kraft, Shortman, and Marchalonis 1971).

cells recognize antigen and interact. Because a clear-cut distinction between these cells can be made only in a few mammalian species such as the mouse, questions of the existence of T and B lymphocytes in lower species must be based upon functional properties. Therefore, answers to questions involving the homology among putative T or B cells of lower vertebrates and man cannot be taken with the same assurance as can the results of comparative studies of antibody structure. Moreover, one must be careful to avoid forcing immunological phenomena occurring in various species into the theoretical mold devised for mice, which might not be a universal one.

## Phylogenetic emergence of lymphoid tissues

All true vertebrates possess the capacity to reject allografts (Hildemann 1972) and to form circulating antibodies to a variety of antigens

TABLE 21: Properties of primary and secondary lymphoid organs.

| Property | Primary lymphoid organ[a] | Secondary lymphoid organ[b] |
|---|---|---|
| Embryonic origin of stromal cells | Epithelial (Ectoendodermal junctions) | Mesenchymal |
| Appearance of lymphopoiesis | Early | Late |
| Lymphopoietic activity | High, about 1% | Low, approx. 0.1% |
| Antigen-dependence of lymphopoiesis | No | Yes |
| Lymphopoiesis in germ-free animals | Normal | Reduced |
| Effects of embryonic or neonatal removal | Profound immunological deficiency | Slight local effects |
| Persistence in life | Involution at maturity | Retained throughout life |
| Contain antibody-secreting cells[c] | No | Yes |

*Source:* Adapted from Warner 1972.
a. Examples, thymus and bursa of Fabricius.
b. Examples, spleen and lymph nodes.
c. Following normal immunization procedures.

(Grey 1969; Clem and Leslie 1969; Marchalonis and Cone 1973b). Although all species studied contain circulating lymphocytes, a progressive development of lymphoid organs can be discerned. The minimal requirements for the elaboration of an immune response appear to be the presence in the blood of lymphocytes and diffuse clusters of lymphatic tissue in the gill regions. Larval lampreys possess such collections of lymphocytes in the gills, which may constitute a primitive homologue of the thymus (Good et al. 1966), the organ that plays the major role in the development of lymphocytes mediating cellular immunity in mammals and birds. The foci of lymphoid cells are found beneath the epithelium of the pharyngeal gutter between the second to fifth pharyngeal pouches (Good et al. 1966; Rowlands 1968). The identification of such collections of lymphocytes as homologues to the thymus is reasonable, because the thymus in ontogeny develops from epithelial and mesenchymal rudiments in the region of these pharyngeal pouches. The lamprey also possesses a collection of lymphoid cells and hematopoietic cells in an invagination of the anterior gut which is probably homologous to the spleen of higher vertebrates (Good et al. 1966; Rowlands 1968). Although adult hagfish form antibodies and reject allografts (Theones and Hildemann 1970), they do not possess any definitive aggregation of lymphoid tissue. Studies of larval hagfish, if it is possible to obtain them, should provide evidence regarding lymphoid tissue in these extremely primitive vertebrates.

Cartilaginous fishes possess thymuses, like those of mammals, which are organized histologically into discrete cortex and medulla. Although the role of the thymus in mammals was not conclusively established until the 1960's, the Scottish anatomist Beard (1900) earlier reasoned shrewdly from observations of the thymus in embryonic rays that the probable function of this organ was to produce and export lymphocytes. All vertebrates higher than cyclostomes possess a discrete thymus at early stages of their life history, although this key organ may be lacking in adult forms. McKinney and Sigel (personal communication), for example, did not observe a thymus in the gar (*Lepisosteus platyrhincus*, a holostean), while Marchalonis, Ealey, and Diener (1969) found no evidence of a thymus in the adult *Sphenodon*. In addition to thymuses, all placoderm-derived vertebrates contain discrete spleens, which are organized into red pulp and white pulp. The red pulp is rich in blood and in red cells undergoing destruction, and the white pulp consists primarily of packed lymphocytes.

## Lymph nodes and lymph-node-like structures

True lymph nodes, containing a lymphoid cortex and a medulla rich in macrophages, do not occur below the phylogenetic level of marsupial mammals (Good et al. 1966). Amphibians, particularly anurans such as the marine toad, *Bufo marinus* (Kent, Evans, and Attleberger 1964; Diener and Nossal 1966), birds (Jordan 1938), and even primitive monotremes, such as the echidna (Diener, Ealey, and Legge 1967), possess tightly packed small clusters of lymphocytes occurring within lymphatic vessels. These structures have been termed jugular bodies in toads (Kent, Evans, and Attleberger 1964; Diener and Nossal 1966) and are more similar to naked lymphoid follicles than to intact lymph nodes of mice or man (Diener and Nossal 1966). The kidneys of fish (Chiller et al. 1969) and amphibians (Diener and Nossal 1966) and the urinary bladders of turtles (Lefevre et al. 1973) also contain clusters of lymphatic tissue. Anuran amphibians possess a variety of additional lymphoid structures which have been described in detail for frogs by Cooper and his colleagues (Baculi, Cooper, and Brown 1970). Adult bullfrogs, for example, possess a variety of lymphoid organs in addition to the thymus. These include a spleen, jugular bodies, bone marrow and a variable number of paired lymph-node-like organs in the ventral region of the neck. These lymph-node-like structures are not part of the lymphatic system, but filter blood. They consist of encapsulated aggregates of lymphoid cells, which are separated into lobules by sinusoids lined with reticuloendothelial cell elements. Not all anuran amphibians possess such structures; the toad, *Bufo boreas*, for example, lacks all of these additional lymphoid organs, including jugular bodies. Moreover, the primitive Pipidae, represented by the clawed toad *X. laevis*, lack the tentative lymph nodes described for *Rana* and *B. marinus*. Their major lymphoid organs are the thymus, spleen, bone marrow and kidney (Manning and Horton 1969). Unlike the cyclostomes, elasmobranchs, and teleosts, adult anuran amphibians contain a bone marrow which is lymphopoietic as well as myelopoietic and erythropoietic (Jordan 1938; Baculi, Cooper, and Brown 1970). Larval anurans lack this bone marrow.

The fact that anuran amphibians possess primitive lymphoid nodules whereas urodeles do not (Good et al. 1966) suggests that the emergence of these structures in anurans may represent an evolutionary event unrelated to their appearance in birds and in lower mammals. This line of argument gains support from the observation that alligators (Good et al 1966) and *Sphenodon* (Marchalonis, Ealey, and Diener 1969) do not possess lymphoid nodules. Both species are related to the archosaurian stream of reptilian evolution, which was ancestral to birds. Figure 46 presents a schematic comparison of lymphoid nodules of a prototherian mammal (the spiny anteater, or echidna) and the true lymph node of a eutherian mammal, the rat. Diener and his colleagues (Diener, Ealey, and Legge 1967; Diener, Wistar, and Ealey 1967) carried out a series of studies designed to ascertain the properties of the lymphatic system of this key species. These workers found that adult echidnas possessed characteristic thymuses, spleens, and gut-associated lymphoid tissues. The lymph nodes were surprisingly primitive, however, consisting of small nodules of lymphocytes occurring within the main lymphatic vessels. The situation resembled that which occurs in amphibians and fowls rather than the multifollicular lymph nodes characteristic of eutherian mammals. The echidna lymph nodule shows some sophistication with respect to the jugular bodies of the toad, because the former structure shows differentiation into a peripheral area of small lymphocytes surrounding a central area consisting mainly of pyroninophilic cells. Diener and his colleagues, furthermore, showed that [125]I-labelled *S. adelaide* flagellar antigens were trapped in the lymphoid nodules, spleen, appendix, Peyer's patches, and even within the Hassall's corpuscles of the thymus.

Studies of the localization of antigens labelled with radioactive iodine have been performed on the lymphomyeloid organs of anuran am-

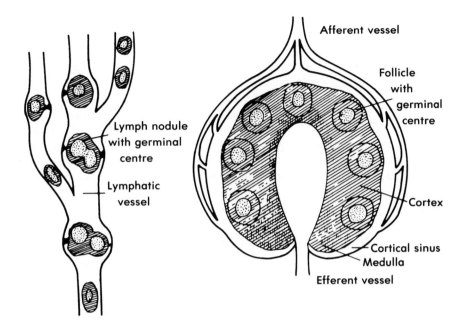

Fig. 46. Schematic comparison between the lymphoid nodules of the echidna
(left) and the true lymph node (right) of a eutherian mammal. (From
Diener and Nossal 1966).

phibians, which bear certain similarities to the lymph nodes of mammals
(Evans et al. 1966; Diener and Nossal 1966; Baculi, Cooper, and Brown
1970). *Bufo marinus*, in particular, possesses lymphoid organs termed
jugular bodies (Evans et al. 1966; Diener and Nossal 1966), which have
been studied in some detail. These small lymphoid structures, which are
attached to the main blood vessels close to the pericardium, appear
histologically to resemble lymphoid follicles (Diener and Nossal 1966).
The structures are small and contain no evidence of medullary organization characteristic of true lymph nodes. Similar lymphoid nodules
occur in the mammalian monotreme, the echidna, which likewise lacks
classical lymph nodes. The presence of small lymphoid nodules within
vessels obtains also for anuran amphibians and for birds. Diener and
Marchalonis (1970) injected *B. marinus* with [125]I-labelled flagella in
order to determine the mode of retention of antigen in the jugular nodules. In higher mammals two types of antigen localization occur in
lymph nodes (Nossal 1967). One type of antigen localization, which predominates in the medulla, is characterized by ingestion of antigen by

macrophages. The second type, termed follicular localization, involves attachment of antigen to the surfaces of reticular cells in the lymphoid follicles. Electron microscopic autoradiography of labelled toad jugular bodies disclosed only the latter process to be involved in retention of flagellar antigen (Diener and Marchalonis 1970). In this study, auto-radiography showed that [125]I-labelled flagella were distributed along the surfaces of jugular body macrophages taken one day after injection of antigen. The localization of antigen on the surfaces of dendritic macrophages requires the presence of antibodies which adhere to the cell surface via their Fc pieces, thereby leaving the binding sites free for combination with antigen (Herd and Ada 1969). This mechanism of anti-gen retention is responsible for the intense uptake of antigen by ger-minal centers of mammals (Nossal and Ada 1971). Pollara et al. (1969) have obtained histological evidence for the existence of lymphoid col-lections resembling germinal centers in the spleen of the bullfrog, *Rana catesbeiana.*

Since the lymphoid system of the chicken presents an extremely clear-cut example of a demarcation between organs generating thymus-derived and bone-marrow-derived lymphocytes, I shall consider the bursa in more detail below. Alligators (Good et al. 1966) and the adult *Sphenodon punctatum* (Marchalonis, Ealey, and Diener 1969) do not possess a recognizable bursa. Because the animals studied were adults, the possibility cannot be discounted that such organs were present in early development but involuted with aging. Alligators do, however, possess lymphoid tissue in the pharyngeal region, which has been com-pared to the tonsils of mammals (Good et al. 1966). Moreover, although the snapping turtle has discrete lymphoid tissue associated with its cloaca, this does not function as a primary lymphoid organ (Sidky and Auerbach 1968) in which division of lymphocytes occurs in the absence of antigenic stimulation. It has been argued that since mammals possess B lymphocytes they must contain a lymphoid organ equivalent to the bursa (Cooper, Gabrielson, and Good 1967). Chief candidates for this equivalent in the rabbit have been gut-associated lymphoid structures such as the appendix (Archer, Sutherland, and Good 1964) and Peyer's patches (Cooper et al. 1966). Since bone marrow cells act as a source of B lymphocytes (Armstrong, Diener, and Shellam 1969) in experimental reconstruction of the immune response to sheep red blood cells, using thymus cells or thoracic duct lymphocytes to function as a T-cell source, it can be considered that the functional equivalent of the bursa is dif-fusely spread throughout the bone marrow of mammals (Abdou and Abdou 1972). The lymphoid tissues of representative lower vertebrates are tabulated in Table 22.

The bursa of Fabricius, which is clearly the source of antibody-forming B cells in birds (Szenberg and Warner 1962; Warner 1972; Kincade and Cooper 1971), may constitute an adaptation of the lymphoid system which arose in the evolution of this class of vertebrates subsequent to their divergence from reptiles.

## *The bursa of Fabricius*

The chicken possesses a second major lymphoid organ which has been classed as a primary structure. This is the bursa of Fabricius, which appears to be unique to birds. It is located on the distal side of the cloaca and is connected to the cloaca by a short bursal duct. The location of this organ and the thymus of a young chick are illustrated in Figure 47. The thymus consists of a number of lobes arranged linearly along the neck. Definite epithelial rudiments of the thymus and bursa are present in a six-day-old chick embryo. The spleen rudiment does not appear until day 8, which is consistent with its function as a secondary lymphoid organ. Large dividing lymphocytes occur in the thymus at day 9. The bursa becomes completely lymphoid in a period of about two days, which occurs between days 12 and 14 of embryonic life. The spleen does not become lymphoid until day 19 (Metcalf and Moore 1971). Division of lymphocytes in the thymus and bursa is independent of foreign antigen, and these organs involute after the chicken hatches and becomes mature. It has been shown (Warner, Szenberg, and Burnet 1962; Szenberg and Warner 1962) that extirpation of the thymus suppresses the capacity of chickens to give cell-mediated immune responses. Immunoglobulin synthesis in thymectomized chickens is largely unimpaired, however (Warner 1972). In contrast, removal of the bursa did not affect cell-mediated immunity but eliminated the capacity of such chickens to synthesize immunoglobulins (Warner et al. 1969). In the original experiments of Glick, Chang, and Jaap (1956), the bursa was removed surgically and variable results were obtained. Complete suppression of bursal development can be achieved by injection of 19-nortestosterone into five-day chick embryos (Warner et al. 1969) or by surgical removal followed by x-irradiation (Cooper et al. 1966). These rigorous treatments eliminated the antibody-forming capacity of the chickens. This situation represents the converse of thymectomy; the bursectomized animals cannot produce antibodies but their cellular immune system is normal.

The concept of a dichotomy of immune responses has been confirmed in mammalian systems (Cooper, Gabrielsen, and Good 1967) and is also evident in many clinical syndromes involving depression of various aspects of the immune system. Classical sex-linked agammaglobulinemia is a condition in which male children are susceptible to bacterial

TABLE 22: Phylogenetic distribution of lymphoid organs.

| Species | Primary | | |
|---|---|---|---|
| | Thymus | Bursa | Spleen |
| Lamprey | Diffuse | - | Diffuse |
| Hagfish (adult) | ? | - | Hematopoietic foci in lamina propria of gut wall (splenic precursor?) |
| Sharks (many species) | Discrete (cortex, medulla, Hassall's corpuscles) | - | Discrete (red pulp; white pulp) |
| Chondrostei (paddlefish) | Discrete | - | Discrete |
| Holostei (bowfin; gar) | Discrete | - | Discrete |
| Teleosteii (many species) | Discrete | - | Discrete |
| Urodelea (salamanders) | Discrete (cortex medulla, not clearly demar-arated) | - | Discrete |
| Rana pipiens | Discrete | - | Discrete |
| X. laevis | Discrete | - | Discrete |
| Snapping turtle | Discrete | - | Discrete |
| Alligator | Discrete | - | Discrete |
| Chicken | Discrete | Discrete | Discrete |
| Echidna | Discrete | Equivalent(?) | Discrete |
| Mouse | Discrete | Equivalent(?) | Discrete |
| Rabbit | Discrete | Equivalent (Appendix) ? | Discrete |

| Secondary | | Does thymus involute? |
|---|---|---|
| Gut-associated | Other | |
| ? | - | ? |
| - | - | ? |
| Lymphoid accumulations | Parenchyma of gonads; kidneys (some elasmobranchs) | Advanced species, yes; primitive species, no |
| Lymphoid nodules in region of ileocecal valve and in the gut | Hematopoietic tissue overlaying the base of the heart (primary?) | ? |
| Not done | Not done | ? |
| Not done | Pronephros | Yes |
| Lamina propria contains lymphoid accumulations and plasma cells | Buccal region; Perihepatic area | Yes |
| Yes | Bone marrow; jugular bodies; propericardial bodies; epithelial body (near parathyroids) | Yes |
| Yes | Bone marrow; kidney | Yes |
| Pharynx; intestine | Cloaca; bladder; kidney; lung | Yes |
| Pharynx; gut | Lymphatic plexus; cloaca | ? |
| Yes | Lymphoid nodules | Yes |
| Yes | Lymphoid nodules | Yes |
| Yes | Lymph nodes | Yes |
| Yes | Lymph nodes | Yes |

Fig. 47. Sketch of newly-hatched chicken showing location of multi-lobed thymus (T) and bursa (B).

infection but generally respond well to viral infections and can exhibit cellular immune reactions (Cooper, Gabrielsen, and Good 1967). Children having this disease were found by Bruton (1952) to lack serum immunoglobulins. In a general sense, bursectomized chickens can be considered analogous to this disease situation. The converse phenomenon, the lack of a thymus, also presents an immunodeficiency syndrome in man (Cooper, Gabrielsen, and Good 1967). Naturally, the clinical manifestations are not as readily interpretable as are the experimentally-induced dysfunctions. However, the validity of the experimental comparison of thymus-dependent and bursa-dependent immunity as a model for the understanding of immune deficiency diseases is well established.

Since the bursa is required for antibody production, lymphocytes of this organ should possess the capacity to synthesize immunoglobulins. Thorbecke et al. (1968) cultured fragments of embryonic thymus, bursa, or spleen *in vitro* in the presence of $^{14}$C-labelled amino acids. When the organ fragments were obtained from 18-day embryos, only the bursa fragments synthesized immunoglobulins. The immunoglobulin synthesized was exclusively of the IgM class. Bursal fragments taken from 25-day embryos synthesized IgG(Y) as well as IgM. This sequence of IgM to IgG(Y) production was found in germ-free chick embryos, thereby suggesting that immunoglobulin synthesis in the bursa is independent of stimulation by exogenous antigens. These results were confirmed and extended using an even more sensitive system consisting of the use of fluorochrome-labelled goat antibodies to chicken $\mu$, $\gamma(Y)$

and light chains as a means of detecting immunoglobulin-containing cells. IgM-containing cells occur in the bursa of 14-day embryos (Kincade and Cooper 1971). Cells containing IgG(Y) appeared in the bursa about six days later. Very few immunoglobulin-containing cells could be detected in other parts of the chick until after hatching.

Szenberg (personál communication) has recently completed a detailed analysis of the emergence in the bursa of lymphocytes possessing immunoglobulin on their surfaces. He employed an extremely sensitive technique involving $^{125}$I-labelled antibodies to chicken light chains, $\mu$ chains, and $\gamma$(Y) chains and detected cells displaying surface immunoglobulins by autoradiography. The number of such cells was at background level at day 12, increased slightly by day 13 and accounted for 50% of the bursal lymphocytes at day 15. By the eighteenth day of embryonic development, over 80% of the bursal lymphocytes contained surface immunoglobulin. IgM immunoglobulin alone was detectable at day 14 and IgG(Y) was observed by day 15. An unusual finding was disclosed. The sum of the percentage of cells displaying $\mu$ chains plus the percentage of cells displaying $\gamma$(Y) chains was substantially greater than the value obtained for cells expressing surface light chains. The latter value is considered to be the 100% level because both IgM and IgG(Y) molecules contain the same light chains. This superaddition phenomenon was interpreted as suggesting that some cells contained both IgM and IgG(Y) simultaneously upon their surfaces.

The events underlying the development of the bursa-dependent immune system of chickens can be outlined as follows: (a) The epithelial anlage of the bursa appears at about the 5th day of embryonic development. (b) Lymphopoietic stem cells originating in the yolk sac begin to populate this rudiment on the eleventh day. This influx of stem cells reaches its peak on days 12 and 13. (c) Cells possessing surface immunoglobulin are definitely identifiable at the fourteenth day, and over 80% of the lymphocytes possess immunoglobulin by the eighteenth day of embryonic life. (d) Individual cells undergo a switch from IgM synthesis to IgG(Y) synthesis within the bursa (Kincade et al. 1970) during the period of rapid bursal expansion and differentiation. The appearance of immunoglobulin and the IgM to IgG(Y) switch occur independently of the presence of exogenous antigen.

This period of antigen-independent proliferation and differentiation of immunologically competent lymphocytes may prove to be a crucial phase in the elaboration of an individual's set of antibody-forming cells. Furthermore, careful analyses of immunoglobulins present during this critical four days of bursal development may provide a critical test of the present hypotheses for the generation of V-region diversity (see Chapter 17). If a multiple gene germ-line theory is the correct one, the first immunoglobulin to appear should be as heterogeneous in amino acid sequence as that produced later when the immune system has matured. If, however, a somatic theory for the generation of diversity is correct, the early immunoglobulins would reflect only the small subset of V-region genes which are carried in the germ line. Subsequent division of lymphocytes incorporating somatic mutation and/or recombination of V-region genes would create an increasingly varied pool of these genes. In this case, we would expect the early bursal immunoglobulin chains to show restricted heterogeneity in sequence and in properties such as electric charge, which reflect the primary sequence patterns. As in all gedanken-experiments, confrontation with the real situation may lead to serious complications. The simplisitic alternatives outlined here are subject to the operation of differential gene activation; e.g., sequential activation of individual genes from a large pool of germ-line genes would give results equivalent to those predicted for somatic diversification. Only one possible result of such an experiment can be interpreted unambiguously and that is the situation in which all immunoglobulin variants are expressed from the first appearance of immunoglobulin.

The chicken bursal system has provided an extremely valuable model for detailed investigations of the origins and functions of lymphocytes responsible for antibody synthesis and secretion (B cells). It is very tempting to extrapolate this finding to man, but mammals do not possess a discrete lymphoid structure which can be considered formally equivalent to the bursa. This organ would be a primary lymphoid organ which would generate lymphocytes destined to produce antibodies. Clearly, thymus-dependent cellular immune systems and thymus-independent antibody-forming systems exist in mammals. The mouse has been intensively studied from this standpoint. This common rodent possesses T lymphocytes analogous to those of the chicken and B lymphocytes, arising in the bone marrow, which function similarly to the bursa-derived lymphocytes of chickens. Roitt et al. (1969) have formalized the analogy between the corresponding lymphocyte classes of birds and mice, coining the T and B nomenclature.

I use the term analogy rather than homology because phylogenetic relationships between birds and mammals are not direct. As shown in Figure 26, birds were a relatively late offshoot of the ruling reptile branch, whereas mammals diverged early from stem reptiles. Moreover, lymphoid structures similar to the bursa have not been found in the few reptilian species which have been examined in detail (Good et al. 1966; Marchalonis, Ealey, and Diener 1969; Sidky and Auerbach 1968). The lack of bursa-like structures in adult animals such as *Sphenodon* (Marchalonis, Ealey, and Diener 1969) may be explained by the general observation that primary lymphoid organs tend to involute as the animal matures. Sidky and Auerbach (1968) have reported that newly-hatched snapping turtles possess a lymphoid organ located in close association to the cloaca. However this organ has the properties of a secondary lymphoid organ, so its functional relationship to the avian bursa is uncertain. The dichotomy between cellular (T) and humoral (B) immunity in the chicken has provided a model system in experimental immunology, but it must be remembered that observations made in one species cannot be directly extrapolated to other species representing divergent vertebrate classes.

## Development of immunity in larval amphibians

In the preceding sections, I described the ontogenetic emergence in the chicken of the bursa, a primary lymphoid organ producing B lymphocytes. I will now consider the development of immunity in larval amphibians as a model for the ontogeny of the thymus. The thymus of larval amphibians provides an extremely interesting system for the study of immunoglobulins associated with thymus lymphocytes. The free-living larval stages (tadpoles) of anurans allow the opportunity to perform detailed investigations on animals in which the lymphoid system is extremely primitive (Manning and Horton 1969). Larval frogs are capable of rejecting homografts and producing circulating antibodies (Hildemann and Haas 1959; Cooper and Hildemann 1965). Furthermore Du Pasquier (1970) has recently shown that *Alytes obstetricans* tadpoles are immunologically competent at a stage in which they possess less than one million lymphocytes. This system contrasts markedly with a usual immunological model such as the adult mouse, which contains from $10^8$ to $10^9$ lymphocytes.

The determination of the types of immunoglobulin present in larval anurans presented an experimental challenge because of the small size of early tadpoles and the minute quantities of sera available. This problem was approached by labelling tadpole serum proteins with $^{125}$I, and adding the labelled protein to sufficient quantities of frog serum to enable standard fractionations of immunoglobulins. The application of techniques involving electrophoresis and gel filtration showed that *R. pipiens* tadpoles possessed molecules resembling the IgM immunoglobulins of the adult frog in mobility and size (Marchalonis 1971a, 1971b). Furthermore, rabbit antisera to frog IgM made possible the positive identification of IgM immunoglobulin in tadpole serum (Figure 48). Under conditions which were capable of resolving 1 to 10 nanograms of protein, IgM molecules were first detected in tadpoles at developmental stage 25 (Witschi 1957). At this stage of development, the only

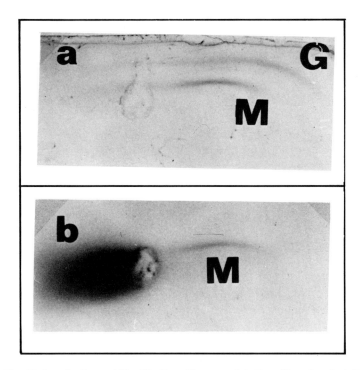

Fig. 48. Localization and identification of immunoglobulins of larval and adult
*R. pipiens* by immunoelectrophoresis against rabbit antiserum to adult
IgM immunoglobulin. A trace amount of [125]I-labeled tadpole serum
(Stage 27 tadpoles) was added to adult serum prior to electrophoresis.
(a) The precipitin arcs formed by adult immunoglobulin detected by
staining. (b) The precipitin arc of tadpole immunoglobulin developed
by autoradiography. M refers to the position of frog IgM immunoglobulin; G indicates the position of frog IgG-like immunoglobulin.
The anode is at the left. The correspondence of the stained frog IgM
line with the autoradiographic localization of tadpole immunoglobulin shows that the tadpole contains IgM immunoglobulin.

lymphoid structure in the tadpole was the thymus, which possessed a
characteristic histological structure (Figure 49). The spleen becomes
lymphoid approximately 4 days later in this species (Horton 1971).
Circulating lymphocytes were present in the blood stream at the time
immunoglobulin first appeared. Although low molecular weight immunoglobulins were sometimes observed in this study, they were the
exception, rather than the rule, and were considered artifacts of the
iodination procedure used (Marchalonis 1971b). More recently, Geczy,
Green, and Steiner (1973) reported that IgG-like immunoglobulins, as
well as the IgM molecules, are found in *Rana catesbeiana* tadpoles.
Du Pasquier (personal communication) also thinks that low molecular

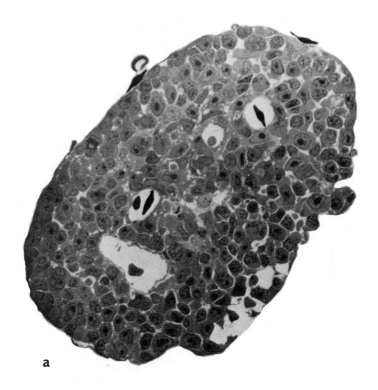

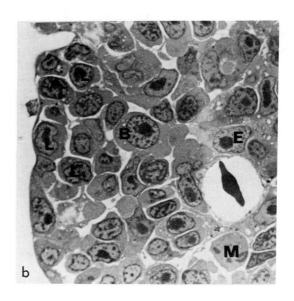

Figure 49a and 49b, see caption p. 163.

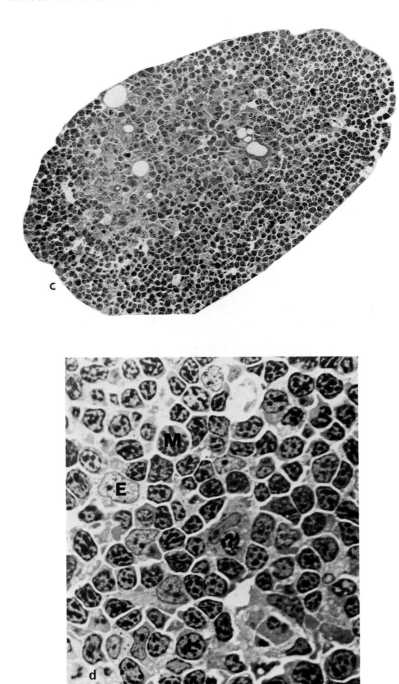

Figure 49c and 49d.

weight immunoglobulins, possibly 7S IgM proteins, are detectable in *Xenopus* larvae. In any case, it is clear that tadpoles possess immunoglobulin which is identical to a major antibody class (or classes) of the adult. This observation is in contrast to results of recent studies of other biochemical systems in *R. catesbeiana* larvae. Glutamate dehydrogenase (Cohen 1970) and hemoglobin (Maniatis, Steiner, and Ingram 1969; Maniatis and Ingram 1971) of larval bullfrogs differ markedly from those of the adult frog. The discontinuity in these systems probably results from the fact that anuran metamorphosis effects a transition from an aquatic mode of life to a terrestrial one, in which problems of respiration and nitrogen excretion are markedly divergent. The immune system, however, emerged early in ontogeny to protect the free-living larva and apparently serves the amphibian equally well in his later mode of existence.

Larval amphibians may provide an important model for the examination of basic aspects of the development of immunity and tolerance. One question to be considered is the chemical nature of the T-lymphocyte receptor for antigen. Although the obvious candidate for this receptor is immunoglobulin (Mitchison 1971b; Greaves and Hogg 1971), certain negative results in attempts to block cellular immune phenomena by antisera directed against immunoglobulins have brought this hypothesis into question (Crone, Koch, and Simonsen 1972). Clearly, the mere presence of immunoglobulin on the surface of a cell does not guarantee that this molecule functions as the receptor for antigen, but a definite absence establishes that it cannot fulfill this role. The study of *R. pipiens* tadpoles (Marchalonis 1971b) provided evidence that IgM immunoglobulin can be detected at stages when the thymus was the only discrete lymphoid organ present. Since the experiments were based upon analysis of serum immunoglobulin, it was not possible to establish that this organ was, in fact, the source of the immunoglobulin. However, it was known that the immunoglobulin was not of

---

Fig. 49. Two stages of the development of the thymus of tadpoles of the Australian tree frog *Hyla ewingi* (development is comparable to that of *Rana pipiens*). Figures a and b were taken from animals sacrificed at five days posthatching (20°C). (a) Cross section of whole thymus (x 560). The thymus is vascularized, undifferentiated and contains blast-like lymphoid cells. (b) Detail of (a) (x 1,190). Blast cell (B), epithelial cell (E), mitotic figure (M), lymphocyte (L). Figures c and d were obtained from individuals, taken from the same group of animals, that were sacrificed at 13 days posthatching. (c) Cross section of whole thymus (x 280) showing the presence of a cortex rich in small lymphocytes and a medulla containing a few lymphocytes. (d) Detail of (c) (x 1,360) showing cortical small lymphocytes (L), epithelial cells (E), and mitotic figures (M). (Photographs courtesy of Dr. T. Mandel).

maternal origin. Du Pasquier, Weiss, and Loor (1972) used the technique of immunofluorescence to establish that thymus lymphocytes of *Xenopus* larvae possess surface immunoglobulin. These workers prepared rabbit antiserum to the IgM immunoglobulin of the adult (the antiserum reacted with light chains and $\mu$ chains) and isolated IgG immunoglobulin from the rabbit serum. They coupled the fluorescent dye fluorescein to this rabbit globulin, which then served as their reagent to ascertain the presence of *Xenopus* immunoglobulin on lymphocytes. The results of this study are plotted in Figure 50. The percentage of immunoglobulin-containing lymphocytes in the thymus rises from low levels to over 80% between days 12 and 15. This high level remains until after metamorphosis. The spleen becomes lymphoid 3 days later than the thymus and possesses its full complement of immunoglobulin-containing cells from the onset. These results for the larval amphibian thymus disagree with a number of attempts to detect immunoglobulin on the surface of mouse thymus lymphocytes by use of fluorescent (Rabellino et al. 1971) or radioactively-labelled antisera to immunoglobulins (Nossal et al. 1972). Although it has been shown that the majority of mouse thymus lymphocytes possess immunoglobulin light-chain determinants by use of $^{125}$I-labelled antiglobulins, more rigorous conditions were required than those necessary to demonstrate immunoglobulins on spleen lymphocytes, which are presumably B cells (Bankhurst, Warner, and Sprent 1971). Since it appears that the thymus in amphibians is required for the induction of cell-mediated immune phenomena such as allograft rejection (Curtis and Volpe 1971; Horton and Manning 1972; see also Chapter 11), thymus lymphocytes probably function in a manner similar to their role in mammals. The surprising ease of detectability of immunoglobulin on thymus lymphocytes of *Xenopus* suggests that the immunoglobulins are relatively more exposed than are the corresponding proteins on mouse T cells. Two final points of proof are required in attempts to establish that IgM immunoglobulin is the thymus-lymphocyte receptor for antigen in *Xenopus*. First, it must be demonstrated that this protein is synthesized by the thymus cells themselves. It is also imperative to show that this immunoglobulin possesses binding specificity for antigen. In general, firm evidence is lacking that antibody formation in amphibians requires T cell and B cell collaboration, although some data which are consistent with this point are now available.

The development of the thymus of larval amphibians provides an important model for the generation of both thymic and peripheral lymphocytes. Turpen et al. (1973) have recently shown by reciprocal transplantation of undifferentiated thymic primordia between diploid and triploid chromosomally marked *Rana pipiens* embryos that thy-

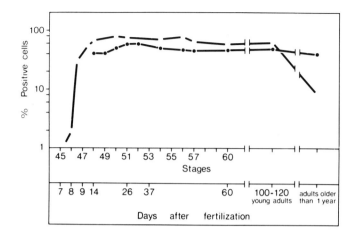

Fig. 50. Kinetics of appearance of immunoglobulin on the surfaces of thymus (o———o) and spleen (●———●) lymphocytes of developing *Xenopus* tadpoles. The presence of immunoglobulin on the lymphocytes was ascertained by binding of fluorescein-labeled antiserum to *Xenopus* IgM immunoglobulin. (From Du Pasquier, Weiss, and Loor 1972. Figure courtesy of Dr. L. Du Pasquier).

mic lymphocytes are ontogenetically derived from cells of the thymic primordium, rather than blood-bone stem cells which migrated into the developing organ. The authors stress that the techniques required to perform this demonstration cannot be carried out using murine embryos. Moreover, Turpen, Volpe, and Cohen (1973) show that virtually all the lymphocytes in the spleen, kidneys, and bone marrow of adult frogs are descendants of the original thymic stem-lymphocytes. These results illustrate the concept of thymus function stated by Beard (1894, 1900) at the beginning of this chapter.

## Development of the lymphatic system in pouch young marsupials

The preceding sections have discussed the utility of developing amphibians and avians as model systems for the study of the phylogenetic and ontogenetic development of the immune system. Although the nature of marsupial immunoglobulins remains a relatively obscure area, detailed studies have been performed on the ontogenetic development of immunity in certain species (Miller et al. 1965; Rowlands, La Via, and Block 1964; Rowlands, Blakeslee, and Lin 1972). Marsupials are particularly valuable for ontogenetic studies because birth follows an extremely short gestation period and the newborn pouch young of an opossum, for example, correspond in state of development of blood-forming tissues to an 8-week human fetus or a 10-day rat or mouse embryo (Block 1964, 1967). The opossum at birth lacks lymphatic tissues, bone marrow, plasma cells, and medium or small lymphocytes. Large lymphocytes, presumably stem cells, are present in the liver. The thymus is an unvascularized epithelial sheet, which may contain a small number of mesenchymal cells. Block (1967) has described the development of blood-forming tissues and blood of the newborn opossum in great detail, providing a foundation for investigations of the roles played by various lymphoid organs. As in other species described, the thymus first becomes lymphoid while the lymph nodes and spleen (secondary lymphoid organs) lag behind (Block 1967; Yadav, Stanley, and Waring 1972a, b). Plasma cells have not been found in opossums less than 40 days old; but the animals can form certain antibodies to bacteriophage prior to this time, when secondary lymphoid nodules are not present (Kalmutz 1962). Although the early antibodies produced by pouch young appear to be primarily IgM type, Rowlands, Blakeslee, and Lin (1972) provided evidence for the presence of a fetal-type immunoglobulin of intermediate size between 19S and 7S. The present data on marsupials, therefore, raises a number of tantalizing issues relevant to the phylogeny and ontogeny of immunity.

## Synopsis

All vertebrates are capable of producing antibodies to a variety of antigens, even though the lower species lack the sophisticated lymphoid cytoarchitecture characteristic of higher mammals. This generalization applies most clearly to the phylogenetic distribution of lymph nodes, because only metatherian and eutherian mammals possess the complex multifollicular nodes generally taken as characteristic of those mammals. The lymphoid tissues of all vertebrates of placoderm lineage consist of discrete thymuses and spleens, with accessory collections of lymphocytes associated with the kidneys, gut, and possibly other structures. Lymphocytes are present in all vertebrate species, but plasma cells have not been observed in cyclostomes and lower sharks. It is striking that adult cyclostomes can mount both cellular and humoral immune responses in the complete absence of organized lymphoid structures. This point illustrates that the basic attributes required for immunity are the presence of circulating lymphocytes and local aggregations of lymphocytes in the gills or gut regions. The development of discrete primary and secondary lymphoid organs reflects elaborations on the basic pattern, which are probably correlated with increased efficiency in antigen processing and antibody production.

# 11 Phylogenetic Distribution of Cell Mediated Immune Functions

## *Nature of cell mediated immune phenomena*

The previous chapter treated primarily the distribution of functional cells of the antibody-producing series (B cells) throughout the vertebrate phylum. Antibody-forming cells of cyclostomes and primitive elasmobranchs possess the morphology of lymphocytes and lymphoblasts, whereas those of the more advanced classes consist of plasma cells as well. This section is devoted to investigations designed to ascertain the presence in lower vertebrates of cells exhibiting functions ascribed to T lymphocytes of mammals. These functions are those involved in the mediation of so-called cellular immune reactions which are as follows:

Allograft rejection (AGR)
Graft versus host reaction (GVH)
Mixed lymphocyte reaction (MLR)
Helper function in antibody formation to thymus-dependent antigens (specific or nonspecific)
Delayed type hypersensitivity (DTH)
Production of macrophage inhibiting factor (MIF)

The list provided is not an exhaustive one, but examples of the reactions cited have been found among a variety of nonmammalian vertebrates. The phenomena tabulated here involve specific recognition and differentiation of lymphocytes that have been identified as T lymphocytes in mice (WHO 1969a).

It merits comment here that all immunological responses are carried out by lymphocytes or their progeny. The converse of this statement is not true, however; all activities of lymphocytes cannot be considered immunological in nature. Because of this fact, certain of the reactions cited here are not thought to be true immune responses by all workers (Hildemann and Reddy 1973; Lafferty 1973). Proof of the role of T cells in the reactions listed above was originally obtained by extirpation of thymus of neonatal animals and subsequently showing that certain immune responses were markedly depressed as a consequence of this treatment (Miller 1961). For example, neonatally thymectomized mice of one strain neither reject skin grafts from mice of another strain nor do they form antibodies to sheep erythrocytes (a thymus-dependent antigen). The identification of T cells in these murine responses was further aided in studies using antiserum directed against the Θ, or Thy 1, antigen which is present on T lymphocytes but not on B lymphocytes or their progeny (Raff 1971).

Use of anti Θ (Thy 1) serum has provided the most straightforward approach to the identification of T cells in mice. Although other rodents such as rats possess a T-cell surface marker antigenically cross-reacting with Θ (Douglas 1972), identification of T cells is more difficult in the majority of mammals which lack a clear-cut surface marker of this type. Arguments for the presence of T cells in guinea pigs and man have been made on the basis of functional analogy and the presence of cell surface properties generally associated with murine T cells, such as a lack of readily detectable surface immunoglobulin. B cells, in contrast, possess surface immunoglobulins which are easily detected by a wide variety of immunological techniques (Marchalonis 1974b; Warner 1974). An additional means of discriminating between murine T and B lymphocytes merits mention here, although it is not an immunological property. T lymphocytes are stimulated to synthesize DNA and divide following binding to the purified protein mitogen, concanavalin A, in soluble form, but B lymphocytes are not (Möller et al. 1973). In contrast, bacterial endotoxin (lipopolysaccharide) activates murine B cells but not T lymphocytes. Proof of the existence of distinct populations of lymphocytes in man was obtained through study of clinical situations which mimicked experimental thymectomy or bursectomy in lower species. Patients suffering from Bruton-type agammaglobulinemia have low levels of circulating immunoglobulins and are poor antibody producers. However, they give normal cellular immune responses and are considered to lack B cells but to possess T cells (Cooper, Gabrielsen, and Good 1967). Children suffering from Di George syndrome, in contrast, possess immunoglobulins and can form antibodies to certain antigens, whereas their capacity to mount cellular immune responses is lacking. These patients have been considered analogues of neonatally thymectomized mice or of congenitally athymic mice (Cooper, Gabrielsen, and Good

1967). I would emphasize that the search for clear-cut indicators of T cells and their functions is a challenging area of human and mammalian biology, and many of the problems involved are far from solution. Attempts to establish the presence of T cells in lower species must be based primarily upon functional analyses, because no one has yet separated populations of T and B lymphocytes in lower species. Moreover, studies of cellular immunity and cell collaboration in immune responses of lower vertebrates stand to contribute a good deal of basic information toward the elucidation of the mechanisms of recognition, differentiation, and collaboration by T lymphocytes.

Table 23 gives a general summary of the phylogenetic distribution of responses attributable to murine T cells. Certain responses such as the rejection of allografts occur within representatives of all vertebrate classes (Hildemann 1972) and possibly even the protochordates (Oka and Watanabe 1960). Others, such as the *in vitro* mixed lymphocyte reaction, were not found among lower vertebrates but definitely could occur in anuran amphibians and birds (Cohen 1975). Unfortunately, the species distribution of certain reactions, such as production of macrophage inhibiting factor (MIF) by lymphocytes sensitized by antigen, has not been examined in sufficient detail to allow phylogenetic generalizations to be made at this time. In the remainder of this chapter, I shall discuss the functional properties of putative T cells of lower vertebrates in more detail and will also consider evidence implicating the thymus in antibody formation to thymus-dependent antigens and in the capacity to reject allografts.

## *Phylogeny of transplantation reactions and antigens*

A transplantation reaction (allograft) is one in which a piece of skin or an organ of one individual of a species is transferred to another of the same species. If, for example, the scale of a fish (Hildemann 1957) or the kidney of a man is grafted onto a second member of the same respective species, the grafts usually become vascularized but then are infiltrated by host lymphocytes and are subsequently destroyed in an inflammatory reaction. Although circulating antibody may play a role in increasing the rate of graft rejection, lymphocytes themselves are capable of initiating and mediating the process of allograft destruction. Extensive studies of allograft reactions in man, mice, and chickens have allowed certain generalizations, such as the so-called genetic law of tissue transplantation, to be made regarding the relationships between transplantation antigens and histocompatibility (H) genes. This law states that a one-to-one relationship exists between histocompatibility genes and transplantation antigens, and it applies in the case of strong histo-

TABLE 23: Phylogenetic distribution of responses attributable to murine T cells.

| Murine T-cell response | Cyclostomes | Elasmobranchs | Teleosts | Urodele amphibians | Anuran amphibians | Reptiles | Birds |
|---|---|---|---|---|---|---|---|
| Allograft rejection | +[a] (Chronic) | + (Chronic) | + (Acute) | + (Chronic) | + (Acute) | + (Chronic) | + (Acute) |
| Mixed Lymphocyte reaction | -[b] | - | - | NT[c] | + | NT | + |
| Graft versus host reaction | NT | NT | NT | + | NT | + | + |
| Helper function | "Carrier specificity" in *in vitro* secondary response | "Carrier specificity" in AB production | NT | "Carrier effect" | "Carrier effect" | NT | T-B cooperation |
| Delayed type hypersensitivity | + | + | + | NT | NT | NT | + |
| MIF production | NT | NT | NT[d] | NT | + | NT | NT |

a. + = Effect has been observed.
b. - = Response has been sought, but not found.
c. NT = Not tested.
d. A representative holostean fish, the gar, gave positive MIF production (McKinney and Sigel 1974).

compatibility antigens, is illustrated in Table 24, which depicts genetic crosses between mice of strain A (genes AA) and B (genes BB). It is observed that rejection of a graft occurs if the graft possesses one gene (hence H antigen) additional to that of the genotype (and H pheno-type) of the recipient. The $F_1$, for example, accepts grafts from both parents but $F_1$ tissue is rejected by both parents. Exceptions to this pattern have been observed in the case of weak histocompatibility antigens, where a variety of quantitative effects have been reported (Hildemann 1972). Nevertheless, in vertebrate species which have been studied carefully, such as platyfish (Kallman 1970) and chickens (Hilde-mann 1972), where inbred or isogenic strains of animals occur, these rules are operative.

Table 25 presents a simplified summary of the rules describing rejec-tion (lack of formation of a common vascular system) among colonial tunicates of the species *Botryllus primigenus* (Oka and Watanabe 1960). This scheme contrasts with that of H antigens and genes in mice, because the graft is accepted if the recipient and the donor share one gene (and hence one antigen). Thus, presently available data suggest that tunicates possess multiple histocompatibility genes, but the mechanisms under-lying recognition and rejection differ from those of vertebrates. As pointed out by Oka (1970) and Burnet (1971), the general pattern present in this system resembles that involved in the prevention of self-fertiliza-tion among flowering plants. Hildemann and Reddy (1973) consider the allogeneic incompatibility of tunicates to represent a process of quasi-immunorecognition because a considerable lag period, involving

TABLE 24: Genetic rules governing rejection of histocompatible (H-2 differences) grafts in mice.

| Donor | | Recipient | | Accept or Reject |
|---|---|---|---|---|
| AA | | AA | | accept |
| AA | | BB | | reject |
| BB | | AA | | reject |
| AA | | $F_1$ { AB | | accept |
| BB | | { AB | | accept |
| $F_1$ { AB | | AA | | reject |
| { AB | | BB | | reject |
| $F_2$ { AA | | $F_1$ { AB | | accept |
| { AB | | { AB | | accept |
| { BB | | { AB | | accept |

TABLE 25: Genetic basis of allogeneic recognition in tunicates.

| AB x CD AC, AD, BC, BD | | |
|---|---|---|
| Donor | Recipient | Accept or Reject |
| AB | AB | accept |
| AD | AB | accept |
| BC | AB | reject |
| BD | AB | accept |
| AB | CD | reject |
| AD | CD | accept |
| BC | CD | accept |
| BD | BD | accept |

healing of the graft, precedes the initiation of cell destruction. More-over, no specific memory is demonstrated in a second challenge with tissues of the same animal used in the initial sensitization. The lack of an enhanced response to a second challenge is also characteristic of cer-tain cellular responses of true vertebrates; in particular, this property obtains for the mixed lymphocyte reaction and the graft versus host reaction. The former of these has been placed in the category of quasi-immunorecognition by Hildemann and Reddy (1973), and Klein and Park (1973) have shown that both responses have a common genetic basis.

All vertebrate species are capable of recognizing and eliminating grafts from members of the same species (Hildemann 1972). However, as illustrated in Figure 51, the rejection response is not exactly the same in all vertebrate classes and species (Cohen and Borysenko 1970). Al-though the graft is invaded by lymphocytes and eventually destroyed in all examples, the rate of rejection can vary markedly even if ecto-thermic species are maintained at the same temperature. Cohen and Borysenko (1970) have found that the rate of rejection can be divided into acute rejection, in which a graft survives for less than 14 days; subacute, in which survival continues for approximately 20 days; and chronic, where grafts survive for over 30 days. They observed that cyclostomes, elasmobranchs and lower teleosts show only chronic graft rejection. Higher teleosts, however, can reject allografts in an acute manner. Progressing up the phylogenetic ladder, caudate amphibians show chronic rejection rates whereas anurans exhibit acute reactions. All of the reptiles studied reject grafts in a chronic fashion, while birds and mammals carry out acute reactions. Birds and mammals (except possibly for monotremes, which show considerable variation in body

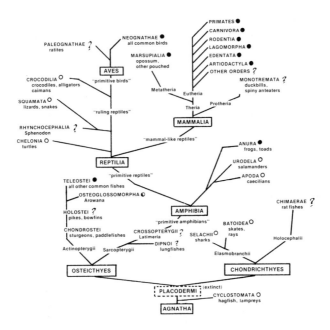

Fig. 51. Scheme for the phylogeny of allograft rejection. Open circles indicate chronic rejection; dark circles indicate acute rejection; black and white circles indicate intermediate rejection times. (From Cohen and Borysenko 1970).

temperature) may not quite represent a fair comparison, however, because they function at constant temperatures of 37°C or higher in contrast to the ectothermic vertebrates which were maintained at 20-25°C. Temperature, as illustrated in Figure 21, Chapter 3, plays a major role in antibody production. Moreover, the rate of graft rejection in species that cannot regulate their body temperature is markedly affected by ambient temperature. Macela and Romanovsky (1969) have observed that the initial phases of allograft rejection in frogs occur at relatively low temperatures, but the effector phase of destruction proceeds more rapidly at higher temperatures. Possible mechanisms involved in the temperature effect will be considered below.

The observations within the lower vertebrate classes pose an interesting evolutionary problem. Two explanations have been proposed to account for the fact that lower fishes and amphibians reject grafts in a chronic fashion, whereas higher members of the same classes routinely demonstrate acute rates of rejection. One possibility is that the so-called lower species are immunologically deficient relative to the teleosts or anurans. Cohen and Borysenko (1970) find this argument unlikely because instances of acute or subacute rejections have been observed in urodeles (Cohen and Hildemann 1968; Hildemann and Haas 1959) and caymans (Cohen 1971). Since the potential for a vigorous cellular immune response is present in these species, which do not usually mount an acute response, Cohen and Borysenko (1970) propose that the failure to give an acute reaction may stem from lack of strong histocompatibility antigens. Histocompatibility antigens are particular proteins which are present on the surfaces of cells. They are glycoproteins, characterized by a mass of about 50,000 daltons (Koene et al. 1971). A large number of antigenically distinct components occur in mammals. The major histocompatibility system in man is the HLA system (human leucocyte antigens), which consists of proteins encoded by two linked structural genes, each of which possesses multiple alleles (Reisfeld and Kahan 1971). The well studied H-2 system, which is the major histocompatibility series of mice, shows basic similarities to the human system (Thorsby 1971; Shreffler et al. 1971; Snell, Cherry and Demant 1971). Certain histocompatibility antigens have been classed as strong because congenic individuals that share all antigens except for a particular one reject grafts in an acute manner. Weak histocompatibility antigens induce chronic rejection. To substantiate their argument, Cohen and Borysenko (1970) have pointed out that the transplantation rejection phenomena observed with outbreeding urodele amphibians are very similar to those associated with interaction across weak histocompatibility barriers in mammals (Graff et al. 1966).

The observed scheme for the phylogenetic emergence of rapidity of allograft rejection, coupled with the above hypothesis, raises another important evolutionary question. Since the vertebrate classes that exhibit acute rejection occupy advanced places on evolutionary lines separated from each other by common ancestors characterized by chronic rejection, strong histocompatibility systems of teleosts, anurans, birds, and mammals most probably represent examples of parallel evolution. This is equivalent to stating that the strong histocompatibility antigens of teleosts may show little direct homology to those of frogs or man. Furthermore, strong histocompatibility antigens are not the only means by which acute rejection is initiated. In certain cases, multiple weak antigens can act in a synergistic manner, thereby mimicking the presence of a single strong antigen (Graff et al. 1966). Histocompatibility antigen systems of lower species obviously deserve further attention both at the functional level and in terms of isolation and chemical analysis. As I shall discuss in Chapter 17, such a system, which generates strong transplantation antigens and extreme antigenic polymorphism, may serve as the driving force in the establishment and maintenance of an essential surveillance system that polices against the spontaneous appearance of neoplasms.

A direct approach to the question of the involvement of thymus-derived lymphocytes in graft rejection observed in lower vertebrates is to remove the thymus early in life and determine if the response is impaired. Miller (1961) made such observations in mice and a number of workers have shown that thymectomy suppresses to some extent the capacity of fish (Sailendri 1973), urodele amphibians (Charlemagne 1972; Tournefier 1972; Cohen 1975) and anuran amphibians (Cooper and Hildemann 1965; Du Pasquier 1973) to reject allografts. Du Pasquier (1973) and Cohen (1975) emphasize that care must be taken to remove the thymus completely; otherwise inconclusive results may be obtained.

The preceding discussion raised the possibility that strong histocompatibility antigens do not occur in primitive vertebrate species. Hildemann (1972) emphasizes that strong or major histocompatibility loci have not been found below the avian level of phylogeny. Chickens possess the so-called B blood group system, which functions both as a blood group locus and as a major histocompatibility locus (Schierman and Nordskog 1963; Hala et al. 1966). At least 21 alleles are detectable by serological testing (Allen and Gilmour 1962). The major histocompatibility loci of man (HL-A) and mouse (H-2) are likewise complex, but recent genetic studies have enabled the construction of linkage maps showing the relationships of genes encoding H antigens to each other and to other markers (Shreffler et al. 1971; Klein and Park 1973). The diagram given in Figure 52 depicts the histocompatibility antigen (H-2) gene region of the ninth linkage group of the mouse genome. A similar arrangement is thought to obtain for genes specifying HL-A

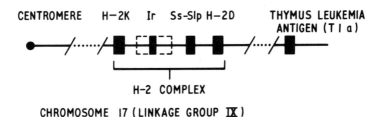

CHROMOSOME 17 (LINKAGE GROUP IX)
OF THE MOUSE

Fig. 52. Simplified diagram of the genetic region of chromosome 17 (linkage group IX) encoding murine strong histocompatibility antigens (H-2). The dotted box enclosing the Ir region indicates that the region is complex. It includes Ir-1, Ir-IgG and Ir-IgA loci which are closely linked, but not identical. The genes responsible for the graft versus host reaction (GVHR) and the mixed lymphocyte reaction (MLR) lie within this Ir region. (Based upon McDevitt and Landy 1972; Klein and Park 1973).

antigens (Thorsby 1971), which are homologous to murine H-2 antigens as demonstrated by antigenic cross-reactivity (Pellegrino et al. 1974) and similarity in amino acid compositions (Mann, Fahey, and Nathenson 1970; Robert et al. 1972). A number of alleles occur at each of the H-2K and H-2D regions which differ by one map unit, a stretch of chromosome of sufficient length to include at least 100 genes of average size (Cohn 1972). It is worth considering the H-2 gene linkage group in some detail, because genes regulating certain immune responses (Ir genes), the mixed lymphocyte reaction (MLR), and a serum protein (Ss-Slp) related to complement activity (Demant et al. 1973), have been shown to map between the H-2K and H-2D regions. Moreover, evidence to be discussed in Chapter 16 suggests that histocompatibility antigens and immunoglobulins may share a common ancestry, albeit distant, and may interact in some functional sense in the elaboration of antibody diversity (Burnet 1970a; Jerne 1971). Nevertheless, it would be premature to propose that the putative major histocompatibility loci of teleosts and anuran amphibians, which must represent parallel or analogous systems to the H-2 or HL-A systems of mouse and man, are arranged in a similar fashion in association with immune response genes. The presence of immune response genes in nonmammalian vertebrates remains to be ascertained. However, as I will illustrate below, mixed lymphocyte reactions clearly occur in species as primitive as anuran amphibians. Genetic linkage relationships between the putative H antigen and MLR loci of anurans must be determined. This point merits further study, because the MLR locus of mice is genetically indistinguishable from the Ir-1 immune response locus of mice, and the latter locus has been hypothesized to code for the antigen receptors on cells involved in cell mediated immunity (Benacerraf and McDevitt 1972). Evolutionary implications of the association of genes encoding H antigens, the mixed lymphocyte reaction, and immune response genes will be followed throughout the remainder of this text.

All true vertebrates, from the hagfish to man, are capable of rejecting grafts from members of the same species. The genetic law of transplantation, of "one gene—one transplantation antigen," definitely applies to vertebrates as primitive as teleosts. In order for rejection to occur the donor must possess at least one histocompatibility gene, therefore one H antigen, distinct from that of the recipient, while the recipient must possess on its lymphocytes a receptor capable of recognizing this antigen which is distinct from its own H antigens specified by its individual constellation of H genes. The histocompatibility antigen pattern is complex even within lower vertebrates, such as the teleost *Xiphophorus maculatus* (platyfish), where 10-15 independent H loci have been reported (Kallman 1970; Hildemann 1972). Evidence has been presented that thymus-derived lymphocytes play a definite role in allograft rejection in lower vertebrates.

Although no evidence presently exists regarding the arrangement of H loci or the presence of immune response genes in lower vertebrates, I have introduced these topics here because of the possible relevance of these genes to the question of the recognition of antigens by thymus-derived lymphocytes. This issue will be explored throughout the remainder of this book.

## *Phylogenetic aspects of the graft versus host reaction (GVH)*

As discussed in the introduction to this chapter, a GVH response in mammals occurs when T cells bearing given H antigens are transferred to an immunologically incompetent allogeneic host possessing different H antigens. Since this reaction has been studied only in birds, reptiles, and amphibians, it is not possible to assess the phylogenetic distribution of this reaction. Murakawa (1968) performed a series of experiments using the Japanese newt (a urodele), which demonstrated that the spleen is the richest source of cells required for transplantation immunity and, in addition, obtained evidence for the presence of a GVH reaction in this species. In the first experiments he removed either the thymus, spleen, or a small fragment of liver from the animals. He then exposed the newts to a dose of whole body irradiation calibrated to abolish the capacity of control animals to reject allografts. Following irradiation he grafted certain of the removed organs to the same animals from which the organs had been taken. In all cases the capacity of the irradiated autografted animals to reject allografts was restored. Spleen autografts brought about the most rapid graft rejection and thymus grafts the slowest. This procedure of replacing the animal's own organs is cumbersome, but had to be carried out because isogeneic strains were not available. In contrast to these results, if Murakawa gave allogeneic spleens or spleen cell suspensions to irradiated newts, lethal GVH reactions developed. Liver, which might contain hematopoietic and lymphopoietic stem cells, prolonged the lives of the x-irradiated animals but did not promote a GVH reaction. Further study is required to delineate precisely which cell lines within the spleen actually mediated the allograft rejection and GVH responses. The fact that thymus autografts gave a poor reconstitution of transplantation immunity does not necessarily imply that thymus-derived cells are not involved in this response. The vast bulk of lymphocytes present in murine thymus are not readily stimulated to cellular immune activity, whereas the so-called peripheral T cells, such as those of the spleen, are highly reactive (Miller 1972). The distribution of thymus-derived lymphocytes within the thymus and spleen of newts may parallel this situation. A more likely explanation however, is that the spleen contains stem cells capable of developing into cells which respond in the GVH reaction.

Studies of snapping turtles indicate that the spleens of these reptiles contain cells which can carry out GVH reactions both *in vitro* (Sidky and Auerbach 1968) and *in vivo* (Borysenko and Tulipan 1973). *In vivo* studies have been aided by the surprising finding that snapping turtles are immunologically incompetent during the first few months of life (Cohen 1971). It was therefore a simple matter to give young turtles intraperitoneal injections of adult allogeneic spleen cells and observe the development of GVH disease (Borysenko and Tulipan 1973). GVH disease developed rapidly when the animals were maintained at 30°C but much more slowly when they were kept at 20°C.

Graft versus host reactions of chickens have been studied in some detail and serve as an important model for the investigation of this phenomenon (Simonsen 1957; Warner 1964; Lafferty 1973). Such studies can be performed quite easily by injecting embryos with blood leucocytes or spleen cells of immunologically competent allogeneic adult chickens. The degree of reaction can be assessed merely by weighing the spleens of the embryos, because the transferred leucocytes localize and proliferate there. Although the intuitive approach to GVH reactions would be that such responses are merely the reverse of the usual host versus graft situation and the principles of transplantation immunity described above should be generally applicable, certain differences have been found between the two types of response. In allograft rejection, the rate of rejection is accelerated by preimmunization of the host with the tissue to be tested, i. e., a clear-cut secondary response occurs. However, in the GVH reaction, if the donor and recipient differ in major histocompatibility antigens (B locus in chickens), preimmunization of the donor does not increase the efficiency of the response. In contrast, if the donor and host differ only in minor histocompatibility antigens, preimmunization of the donor increases the number of active cells. Another difference between host versus graft rejection and the graft versus host reaction is that grafts from other species are rapidly rejected in the former response but virtually no reaction occurs in the latter. This effect is illustrated by data of Lafferty (1973), who reported a mean spleen weight of 134 mg for chicken embryos injected with $10^6$ adult chicken cells from outbred chickens, whereas the spleen weights resulting from injections of duck and sheep leucocytes was only 8 and 10 mg respectively. The particular properties of the GVH reaction have yet to be fully explained (WHO 1969a). Workers such as Lafferty propose that this reaction in the chicken, although it involves immunologically competent lymphocytes to some extent, is a phenomenon which is distinct from the classical cellular and humoral immune responses to antigens. Recent studies (Klein and Park 1973) have shown that the genes which condition this response in mice are closely linked, if not identical, to those which influence the mixed leucocyte reaction (Bach et al. 1972). Moreover, as will be discussed below, both genetic loci map with the so-called immune response (Ir-1) locus.

## The mixed lymphocyte reaction (MLR)

In the section describing allograft immunity, I discussed the presence of strong histocompatibility antigens, which are linked genetically in mice and guinea pigs to other markers involved to some degree with the immune system. These included, in the mouse, the Ir-1 locus, which regulates the capacity to form antibodies to certain antigens; the Ss-Slp marker, which is related to serum complement levels; and the MLR locus, which governs the capacity to give a mixed lymphocyte reaction. Mixed leucocyte (lymphocyte) reactions were attempted initially in man in the hope of obtaining a simple *in vitro* measure of the degree to which cells of certain individuals were compatible with those of others. Such information would, of course, be of tremendous importance in terms of potential organ transplantation. Peripheral blood cells (buffy-coat-containing lymphocytes, polymorphonuclear leucocytes and mononuclear cells) of individual A are mixed in culture with those of individual B, and tritium labelled thymidine (a DNA precursor) is added. If the cells recognize the right sorts of antigenic differences, they are stimulated to proliferate, a process which requires DNA synthesis. It was thought originally that observed proliferation reflected the presence of distinct strong histocompatibility antigens. More recent analysis, however, indicates that the reaction is determined not by H loci but by the MLR locus linked to the H genetic system. Furthermore, the MLR locus (Bach et al. 1972) is extremely closely linked to, if not genetically indistinguishable from, both the locus conditioning the GVH response and the Ir-1 locus (Klein and Park 1973). This last genetic locus has been proposed to be a structural gene encoding the receptor for antigen on thymus-derived lymphocytes (Benacerraf and McDevitt 1972; McDevitt and Landy 1972). McDevitt (1972) suggested that this T-cell system for the recognition of antigen was not immunoglobulin but some more primitive molecule. The phylogenetic distribution of the capacity to generate the mixed lymphocyte response, therefore, can provide some assessment of the above hypothesis.

Goldshein and Cohen (1972) have obtained clear-cut evidence show-ing that cells of an anuran amphibian, the marine toad (*B. marinus*), can give mixed lymphocyte reactions *in vitro*. These workers used spleen cells rather than peripheral blood in their assay. The maximum uptake of ³H-thymidine occurred in the allogeneic cultures after 7 days of incubation at 26°C. Sigel et al. (1972) have performed experiments designed to determine the properties of MLR in teleosts and sharks and were unable to demonstrate this response. Consistent with these results were those of Cooper (1971), who did not obtain an allogeneic response when leucocytes of larval lampreys were incubated in mixed cultures. Furthermore, E. C. McKinney and M. M. Sigel (personal communication) emphasize that significant results have never been obtained in their laboratory with mixed lymphocyte cultures involving any species of fish. Although negative results must always be interpreted with caution, especially in systems where optimum conditions might not have been tried, a tentative pattern emerges that vertebrates more primitive than anuran amphibians cannot mount an *in vitro* mixed lymphocyte reaction. The reasons for this situation remain to be determined. If the MLR locus encodes an MLR antigen expressed on cells, a receptor molecule should also exist on cells of an animal bearing distinct MLR antigens. The combination of MLR antigen on cell A with the receptor for it on cell B would initiate the process of division and differentiation. Cyclostomes, elasmobranchs, and teleosts would give negative responses if either or both antigen and receptor were absent. Another interesting explanation for the lack of an MLR is the presence of suppressor cells in the test population which act in some fashion to prevent recognition or proliferation. Precedent exists in mammals for the existence of T cells capable of suppressing responses to antigen (Gershon and Kondo 1971; McCullagh 1973) and mitogens (Gershon, Gery, and Waksman 1974). These possibilities are considered in detail by Cohen (1975). The issue of mixed leucocyte reactions in primitive vertebrates deserves further consideration because it is technically not difficult to assess and provides a test of certain current theories regarding the nature of the antigen receptor on thymus-derived lymphocytes.

## Collaboration between thymus-derived lymphocytes and bone-marrow-derived lymphocytes

Comparisons of cellular immune response of lower animals and those of the mouse represent an uncertain area in which we must deal with functional analogies. When responses formally similar to those of the standard system are observed, we must be judicious in interpreting them in terms of our standard model. The problem is complicated because the exact mechanisms of cooperation between T cells and B cells are presently objects of contention among differing schools of thought (see Chapter 1). In this section I will present the experimental approaches used to ascertain the presence of helper T cells in lower animals. Very few studies designed to provide direct information on cell-cell collaboration in lower animals have been performed at this time (Ruben 1975). Ruben, van der Hoven, and Dutton (1973) used the so-called carrier effect to obtain data consistent with cell-cell collaboration in the immune response of urodele amphibians to haptens, a simple means of verifying the possible properties of responses to antigens that are known to be thymus dependent in mammals.

It is now well established that the formation of antibodies to foreign erythrocytes and serum proteins requires collaboration between T cells and B cells in mice. Initial evidence establishing the key role of thymus-derived lymphocytes was obtained by removal of thymuses of neonatal animals (Miller 1961) and testing such animals for their responses to certain antigens. In an experiment of this nature, it is crucial that the thymus be removed prior to the export of large amounts of thymus-derived lymphocytes to the periphery, and care should be taken to determine whether the thymus regenerates. Furthermore, even if thymectomy is performed with utmost care in early development, certain cellular immune responses might be unimpaired. For example, in mammals such as the sheep, removal of the thymus rudiment *in utero* does not result in pronounced suppression of immune responsiveness (Cole and Morris 1971). Sailendri (1973) studied the effect of thymectomy on the response of the mouth breeder (a teleost common among tropical fish collections) to sheep erythrocytes (SRBC), a classical thymus-dependent antigen in the mouse. Thymectomy of fish less than 2 months old totally suppressed the immune response to SRBC. Du Pasquier (1970) showed that thymectomy of larvae of the midwife toad (*Alytes obstetricans*) within 40 days of hatching virtually eliminated the appearance of hemolytic plaque-forming cells or immunocytoadherent cells to SRBC. Turner and Manning (1974) performed thymectomies on larval *Xenopus* during the first two weeks of life and found that the capacity of the treated animals to respond to SRBC and to another T-dependent antigen, human gamma globulin, was impaired. Thymectomy suppressed both the IgM and the IgG(?) response to human gamma globulin. Under the experimental conditions used, *Xenopus* made only IgM antibodies to SRBC. The capacity to form these antibodies was totally destroyed by thymectomy. It is tempting to interpret these findings to signify that thymectomy removed T lymphocytes capable of acting as helper cells in cooperation with antibody-forming B cells. Further studies are required to establish this hypothesis. Mitchison, Rajewsky, and Taylor (1970) performed extensive studies on the cellular basis of the response of mice to hapten-carrier conjugates. The basic phenomenon is that the secondary antibody response to a particular hapten is better if the carrier used in the booster injection is the same one that was used in the priming injection. It is now generally considered that priming with the carrier-hapten complex induces the production of a large number of T cells reactive to the carrier, which cooperate with B cells bearing receptors for the hapten. A secondary injection of hapten on the same carrier ensures that the enlarged pool of reactive T cells is available for collaboration with B cells. A simplified scheme illustrating collaboration between a T cell which recognizes a carrier determinant and a B cell specific for the haptenic group is given in Figure 53. Weinbaum, Gilmour and Thorbecke (1973) have obtained a direct demonstration of T-B collaboration in the chicken which illustrates the carrier effect. The result is made more elegant because of the clear-cut demarcation between the source of T cells (thymus) and B cells (bursa) in the chicken (see Chapter 10).

T cells acting as helper cells were obtained by immunizing bursectomized (lacking in B cells) chickens with bovine serum albumin (BSA). Normal chickens were immunized with the 2, 4, 6-trinitrophenyl (TNP) hapten coupled to keyhole limpet hemocyanin (KLH). After the proper immunization period, spleen cells were removed from these animals and injected either singly or mixed into lethally irradiated syngeneic animals.

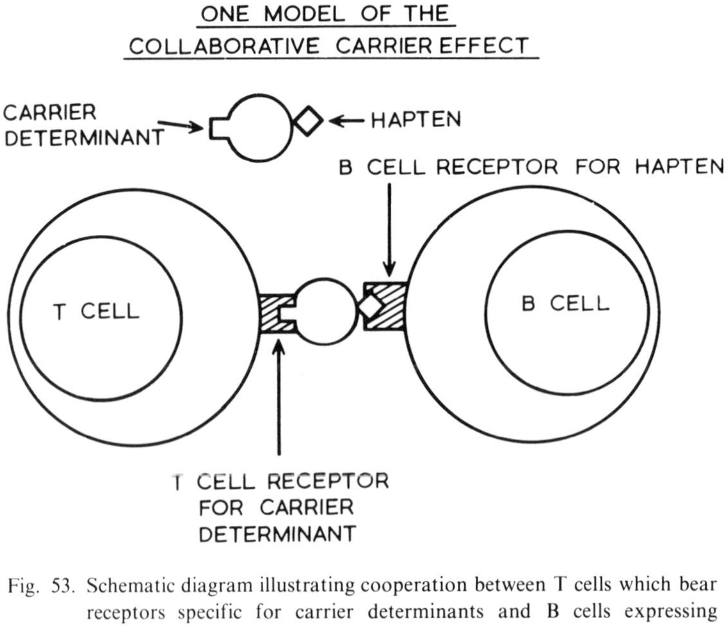

Fig. 53. Schematic diagram illustrating cooperation between T cells which bear receptors specific for carrier determinants and B cells expressing receptors which recognize haptenic determinants.

This adoptive transfer to irradiated hosts serves under ideal conditions merely as a culture vessel; similar experiments in mice are routinely carried out *in vitro*. After the cell transfer, the recipient chickens were given an injection of TNP coupled to BSA. In the system there are B cells from animals immunized with TNP-KLH, therefore some should be reactive to TNP; and there are T cells activated to the protein carrier BSA. If T-B cooperation occurs, the recipient animals having both primed B cells and T cells and immunized with the TNP-BSA complex, should give a definite response to the hapten. In a typical experiment that measured passive hemolytic plaque forming cells (PFC) to TNP, chickens given B cells plus TNP-BSA had 135 PFC per spleen, those receiving T cells plus TNP-BSA had 14 PFC per spleen, whereas those receiving TNP-BSA plus both T and B cells had 4,441 PFC per spleen. It is worth pointing out that this carrier effect operates only when the hapten and carrier are on the same molecule.

The above data demonstrate that cooperation between T and B cells occurs in an avian species, the chicken. Studies involving carrier specificity as a probe for the presence of cells analogous (or homologous) to T cells have been performed in cyclostomes (Cooper 1971), teleosts (Yocum, Cuchens, and Clem 1975), and anuran amphibians (Ruben, van der Hoven, and Dutton 1973). Although it is tempting to interpret carrier reactivity as a T-cell function and hapten recognition as a B-cell property, this cannot be done uniquely, because murine T cells have recently been shown to bind haptens (Lawrence, Spiegelburg, and Weigle 1973; Rutishauser and Edelman 1972; Rolley and Marchalonis 1972; Möller, Bullock, and Mäkelä 1973), and hapten-specific T-helper cells have also been generated under certain conditions (Rubin and Wigzell 1973; Alkan et al. 1972; Hanna, Bhan, and Leskowitz 1973). Cooper's work (1971) is particularly interesting because his studies were performed on the ammocoete larvae of lampreys, which exemplify the most primitive living vertebrates. His study consisted of immunization of the animals *in vivo* with hapten carrier conjugates, followed by *in vitro* assay of the ability of blood lymphocytes to undergo blast transformation in a secondary challenge with heterologous and homologous haptens and carriers. *In vitro* secondary challenge with the homologous hapten clearly initiated specific blast transformation. This was the first illustration that animals as primitive as cyclostomes could respond immunologically to small molecules such as haptens. The carrier-specific effect was not as pronounced, but in one experiment, in which the larvae were primed *in vivo* with orthanilic acid coupled to guinea pig serum, the best *in vitro* blastogenic response was observed when the secondary stimulus consisted of the same complex.

Yocum, Cuchens, and Clem (1975) obtained presumptive evidence for the presence of carrier-reactive cells in the secondary anti-hapten response of sea robbins. The primary challenges consisted of either BSA, bovine gamma globulin (BGG), DNP-BSA, or DNP-BGG. Approximately one month later the fish were reinjected with either the homologous or heterologous DNP conjugate. At suitable times thereafter the fish were bled and the sera assayed for antibodies to the DNP hapten by the technique of neutralization of bacteriophage to which hapten had been coupled. The maximal neutralization constants obtained for the primary response to hapten was about 20; those of the secondary response in animals given two injections of the same hapten-carrier complex were ten times higher than the primary level. Fish given carrier in the primary injection and DNP coupled to the same carrier in the booster injection possessed neutralization constant values comparable to those of the secondary level. In contrast, fish given DNP on the heterologous carrier gave only a primary response. These results parallel those cited above for the chicken, although no information is presently available on the nature of collaborating cells in the sea robbin.

Ruben, van der Hoven, and Dutton (1973) have approached the problem of cell collaboration in the immune response of the newt, *Triturus viridescens*, to the hapten trinitrophenol (TNP) by studying the carrier effect. Analysis of this response by cell transfer experiments (Mitchison 1971a) and by *in vitro* systems (Kettman and Dutton 1971) established that thymus-derived lymphocytes recognize the carrier antigenic determinants and bone-marrow-derived cells respond to the hapten by producing antibodies specific for this moiety. Ruben, van der Hoven, and Dutton (1973) used chicken or toad erythrocytes as carriers for the TNP hapten. Preimmunization with the carrier immunogen caused the development of immunocytoadherent cells (rosettes) specific for the hapten in the spleen, kidney, and liver of the newt after challenge with TNP on the same carrier. In accordance with results described above for teleosts and birds, preimmunization with TNP coupled to another carrier or injection of the carrier alone did not induce a response of TNP- specific rosette-forming cells. These workers proposed that the response of the newt to hapten carrier systems, analogous to that of the mouse, requires the interaction of two cooperating cellular populations. Ruben (1975) has further attempted to distinguish between helper rosettes and rosettes representing antibody-producing cells on morphological and functional grounds. Greaves, Möller and Möller (1970) previously defined nonsecretory rosette-forming cells (RFC) as those which possess only a single layer of RBC bound to their surfaces, whereas secretory RFC have at least two layers of bound RBC. The multiple layers would result from diffusion of secreted immunoglobulin outward from the surface of the lymphocyte. Applying these criteria, Ruben observed that the vast majority (approximately 90%) of RFC at 2 days following immunization were the nonsecretory type, while secretory RFC comprised about 30% of the RFC population at day 8. On the basis of this difference in time of appearance between secretory and nonsecretory RFC, he was able to test for the correlation of helper function with one or the other. This was done by priming the newts with carrier erythrocytes and then giving the booster injection of TNP bound to the carrier erythrocyte at various times following the primary injection. He found that the maximum production of RFC specific for TNP occurred when the second injection was given 2-4 days following the first, that is, when the number of nonsecretory rosettes was maximal. This was well before circulating antibodies appear in amphibians maintained at room temperature. Therefore, the likelihood that this carrier effect was mediated by carrier antibody was small, and it was feasible to make a tentative identification of nonsecretory RFC as helper cells. It was interesting also that maximal helper activity was correlated with priming by use of low doses of carrier RBC. This observation was consistent with effects observed in mice that helper T-cell phenomena are usually most pronounced at low doses of antigen (Mitchison 1971a; Benacerraf and McDevitt 1972).

The existence of antigen-binding helper cells contrasts with data of Hunter, Munro and McConnell (1972), who failed to find RFC in populations of murine T lymphocytes activated as helper cells to erythrocytes. However, the results of Basten et al (1971) established that helper T cells directed against protein antigens can be specifically inactivated by incubation with heavily radioiodinated antigen (so-called hot antigen suicide). The helper T cells must bind the radioiodinated antigen in order for killing to occur. The finding of immunocytoadherent helper cells in the newt is not surprising in the context of amphibian cellular immunity, because immunization of the leopard frog (*Rana pipiens*) with erythrocytes induces the appearance of nonsecretory RFC within the thymus (Ruben 1975). This response peaks at days 2-4. Furthermore, Du Pasquier, Weiss, and Loor (1972) have shown that thymus lymphocytes of larval anuran amphibians possess sufficient surface immunoglobulin for it to be detected by the relatively insensitive technique of immunofluorescence (see Chapter 10). In a discussion of the temperature dependence of immunity in amphibians immunocytoadherence data will be presented which parallel those of Ruben and his colleagues and suggest the existence of two populations of lymphocytes in these species. One population apparently represents cells which recognize antigen, but do not form antibodies, whereas the other consists of antibody-producing cells. It would be premature to identify these populations as directly homologous to T and B lymphocytes of mice. Such a result is extremely significant to the phylogeny of immunity, because urodeles are the most advanced vertebrate group which do not, either as larvae or adults, possess a bursa of Fabricius or use bone marrow as a source of immunologically competent cells.

## Other cell mediated immune responses in nonmammalian vertebrates

The phylogenetic distribution of various cell mediated immune responses given in Table 23 resembled a jigsaw puzzle in which most of the pieces were missing. This simile applies also to studies of delayed type hypersensitivity and the production of macrophage inhibitory factor. Finstad and Good (1966) found that cyclostomes and lower fishes develop pronounced necrotic lesions at the site of subcutaneous injections of complete Freund's adjuvant (a substance containing oils and *Mycobacterium tuberculosis*). I observed a similar result in dogfish sharks (unpublished observation). These observations might indicate that a phenomenon resembling delayed type hypersensitivity occurs in primitive vertebrates because tuberculin sensitivity is taken as a classical example of the response in mammals. Theis and Thorbecke (1973) obtained unequivocal evidence of delayed type hypersensitivity to horse spleen ferritin in bursectomized chickens, thereby demonstrating the pure T-cell nature of the cells specifically initiating the response. Furthermore, they were able to inhibit the response by rabbit antiserum to chicken immunoglobulin. In particular, they concluded that antibodies directed against light chains were effective inhibitors of specific delayed type hypersensitivity responses. Because of the lack of B cells in the bursectomized chickens, this study provided clear-cut evidence for the existence of immunoglobulin on chicken T cells and suggested a receptor function for this immunoglobulin.

The tubercle bacillus and its attenuated strain, bacillus *Calmette guerin* (BCG), are antigens which have been extensively used in the study of delayed type hypersensitivity and production of macrophage inhibitory factor (MIF) in man. MIF production to some degree reflects delayed hypersensitivity but can be carried out *in vitro*. Studies of Ambrosius and Drössler (1972), using the anurans *B. bufo, R. esculenta*, and *R. temporaria*, serve to illustrate both the phenomenon and the extent of our limited knowledge regarding its phylogenetic distribution. Animals of these species were immunized with BCG. Subsequently peritoneal exudate cells or spleen cells, both mixtures containing lymphocytes and macrophages, were cultured *in vitro* in the presence of purified protein derivative of the tubercle bacillus (PPD) or heterologous antigens, and inhibition of macrophage migration was observed.

MIF production occurred in the marine toad *B. marinus*, although attempts to demonstrate delayed type hypersensitivity in anurans by the usual mammalian technique of skin testing have been unsuccessful (Cohen 1975). The MIF production response was also demonstrated in species more primitive than amphibians. McKinney and Sigel (1974) observed antigen-specific inhibition of cell migration in a holostean fish, the gar (*L. platyrhincus*).

## Synopsis

All true vertebrates possess lymphocytes which can recognize allo- and xenoantigens and initiate a process of specific graft destruction. Extirpation of the thymus of larval or neonatal forms of such diverse species as teleost fish, anuran amphibians, chickens, and mice substantially eliminates transplantation immunity and suggests that thymus-derived lymphocytes play a major role in the process. Leucocytes of anuran amphibians give a definite MLR response when grown in culture, but leucocytes of cyclostomes, teleosts, and elasmobranchs failed to exhibit this response. These observations are consistent with recent data establishing that the MLR gene, which governs the MLR response, is distinct from the genes encoding the strong histocompatibility antigens of mammals. If the lack of MLR reactions in pretetrapod vertebrates proves to be real, it raises objections to the hypothesis of McDevitt (1972) that the Ir-1 gene of mice, which is genetically indistinguishable from the MLR locus and GVHR locus, encodes a primitive T-cell receptor which preceded immunoglobulin.

Studies based upon the carrier effect provided preliminary evidence suggesting that cell-cell cooperation occurs in cyclostomes, teleosts, urodele amphibians, and birds. Although it is tempting to propose that such results establish the presence of T cells and B cells in all vertebrate species, we must temper our enthusiasm and emphasize that cooperation experiments involving separated populations of cells have not been carried out. Moreover, the elaboration of carrier-specific antibodies which enhance antigen localization or presentation might account for the observed results in lower species. Circumstantial evidence on the effect of neonatal or larval thymectomy on the immune response of teleosts and amphibians to antigens known to be thymus dependent in mice, suggests that lymphocytes arising in the thymus are required for antibody production. In addition, tentative evidence has been provided by Ruben (1975) for the morphological and functional separation of lymphocytes serving as helper cells (nonsecretory immuno-cytoadherent cells) and antibody-producing cells (secretory immuno-cytoadherent cells). I shall pursue these results and their implications in analyses of the temperature dependence of immunity in ectotherms (Chapter 14) and the nature of recognition molecules in immunity (Chapter 17).

Taking the overall results of this chapter in conjunction with those of the phylogenetic distribution of antibodies, it is clear that cell mediated immunity (T-cell function) and antibody production (B-cell function) are general properties of vertebrate species. The cell mediated immune system, like the antibody-producing system, underwent substantial evolution with the evolution of vertebrate classes. The emergence of strong histocompatibility antigens in different vertebrate classes as outlined by Cohen and Borysenko (1970) serves as a key illustration of this evolution. I would emphasize that workers intending to study the mechanisms of lymphocyte function in diverse vertebrates should proceed with caution and guard against the pitfalls of interpreting their findings directly in terms of the dogmas of murine immunology. Two points in particular warrant reiteration here. In the first place, not all reactions carried out by T cells are immunological (Hildemann and Reddy 1973; Lafferty 1973; Segall, Schendel, and Zur 1973). In the second place, criteria for distinguishing T cells and B cells in the mouse may not apply even in such closely related species as the rabbit (Sell and Sheppard 1973). Peripheral blood lymphocytes of the rabbit, for example, possess readily detectable surface immunoglobulin, but exhibit a panel of response to various mitogens which is characteristic of murine T lymphocytes.

# 12 Immunological Memory and Tolerance in Nonmammalian Vertebrates

The two major characteristics of an antibody response to a particular antigen are inducibility and specificity for the challenged antigen. Another important property of antibody production is that a second challenge with the same antigen can lead to a quantitative, and sometimes qualitative, change in the response to that antigen. In Chapter 1, I described positive immunological memory, in which a second injection of antigen induced a significantly greater amount of antibody than did the initial immunization to that antigen. Nossal, Austin, and Ada (1965) have defined a set of criteria to assess immunological memory of mammals. These are as follows: (a) a rapid increase in antibody formation, (b) the attainment of higher levels of antibody, and (c) the predominance of 7S (IgG) antibodies in the secondary response. Another criterion which was considered in Chapter 9 is an increase in the binding affinity of antibody for the antigen. The criteria cited here are useful, but they cannot be applied dogmatically to certain lower animals such as stingrays, for example, because these animals do not possess IgG immunoglobulins. The criteria of increased rate of antibody production and higher levels of antibody are generally applicable, however.

Thus far I have considered only positive immunological memory. Mammals can be rendered unresponsive or tolerant by prior treatment with antigen. This specifically unresponsive condition can be induced by injection of embryonic or larval animals with antigen (Smith and Bridges 1958) or by injection of adult animals with excessive (Felton 1949) or minimal (Mitchison 1964) doses of antigen. The former case of adult tolerance has been termed "immunological paralysis" or "high-zone tolerance," and the latter has been named "low-zone tolerance." The exact mechanisms responsible for tolerance in mammals have not yet been completely elucidated (Weigle 1973; Nossal 1974). However, it is clear that interpretation of immune responses in lower animals might be difficult because the dose of antigen given or its manner of presentation could be tolerogenic rather than immunogenic. This possible complication must be taken into account in the analysis of animal systems in which it is often not practical to carry out studies using large groups of animals and a wide range of antigen doses. Evidence of positive memory in antibody formation has been reported for representatives of all of the vertebrate classes, including cyclostomes (Hildemann 1972). Studies designed to demonstrate specific tolerance in adult animals of lower vertebrate classes are much fewer in number, and only one such report is presently in the literature (Marchalonis and Germain 1971).

## *Immunological memory in antibody formation*

Thoenes and Hildemann (1970) found evidence of immunological memory in the antibody response of the Pacific hagfish (*E. stoutii*) to keyhole limpet hemocyanin, a protein antigen. Sigel and Clem (1966) attempted to demonstrate secondary responses of the lemon shark (*Negaprion brevivostris*) to influenza virus and to bovine serum albumin. Under certain conditions of immunization, these cartilaginous fishes manifest an increased immune response to the virus. However, immunological memory to the protein was not found under the conditions used. In another series of experiments by the same workers (Clem and Sigel 1966), two teleost fish, the gray snapper (*Lutjanus griseus*) and the margate (*Haemulon album*) showed specific immunological memory to bovine serum albumin (BSA), whereas a holostean fish, the gar (*Lepisosteus platyrhincus*) did not. Three findings in the studies of Sigel and Clem warrant comment. (1) The increased responses, when present, were not impressive by mammalian standards. (2) The shark and the gar were not made tolerant by high doses of BSA. (3) Only high molecular weight antibody (IgM-like) was present in the snapper and margate. Although technical alterations might provide more dramatic results, it is clear that the capacity of the most primitive vertebrates to exhibit immunological memory is not well developed.

Amphibians, in particular the marine toad (*Bufo marinus*), exhibit increased antibody responses to booster injections of bacteriophage antigens (Evans et al. 1966; Lin, Caywood, and Rowlands 1971) and to BSA (Evans et al. 1966). Once again, these secondary responses are not impressive when compared to the increased reactivity which a rabbit or other mammal elaborates to the same antigens. Certain reptiles, in contrast, exhibit definite secondary responses to booster injections of antigen (Ambrosius et al. 1970; Ambrosius and Fiebig 1972). Turtles and lizards showed clear-cut memory to protein antigens. The ability of lizards to give a secondary response to protein antigens is illustrated in Figure 54, which shows data obtained by Ambrosius et al. (1970). These workers gave the legless European lizard (*Ophisaurus*) a series of injections of pig serum. A plateau of titer was reached about forty days after the initial injection of antigen. When the animals were given a second injection, there was a rapid increase in titer to a level three times higher than that achieved following the primary injection. Furthermore, the antibody present following the second injection was primarily low molecular weight as judged by gel filtration on Sephadex G-200 (Figure 55). The response of *Ophisaurus* to pig serum proteins met the criteria for immunological memory as defined above by Nossal, Austin, and Ada (1966).

Data providing unambiguous evidence of secondary immune responses and the existence of low molecular weight (7S or smaller) antibodies in reptiles have not always been obtained, however. Maung (1963), in his pioneering study of the characteristics of turtle antibodies, found that

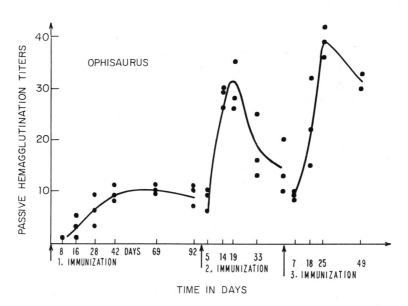

Fig. 54. Primary response and immunological memory in the legless lizard, *Ophisaurus*. Animals were immunized with pig serum and titers measured by passive hemagglutination. (From Ambrosius et al. 1970).

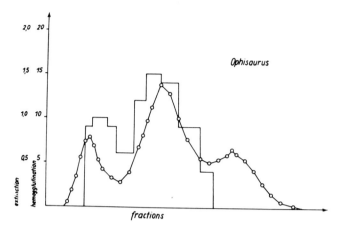

Fig. 55. Presence of high and low molecular weight antibodies in the serum of legless lizard, *Ophisaurus*. Antiserum was obtained 20 days after immunization with pig serum fractionated by gel filtration on Sephadex G-200. o——o, protein concentration as judged by extinction at 280nm; ——, hemagglutinin titer. (From Ambrosius et al. 1970).

only high molecular weight antibodies were produced to the bacterium *Brucella abortus*. Furthermore, studies involving immunization of the tuatara (*Sphenodon punctatum*) with *Salmonella adelaide* flagella failed to disclose immunological memory. I shall consider these results in parallel with similar observations with amphibians in detail following consideration of immunological memory in avian species.

It is well established that birds, as represented by chickens and ducks, possess more than one class of immunoglobulins (Dreesman et al. 1965; Szenberg, Lind, and Clarke 1965; Grey 1967). Moreover, there is a general tendency for antibodies that appear early following immunization to consist of 19S IgM molecules, while those that appear later tend to be considerably smaller, approximately 7S in the chicken and 5.7S in the duck. However, although workers in the field routinely discuss memory in avian species, there are few clear-cut criteria to distinguish between the the primary and secondary response in chickens (Orlans 1967). A number of workers found that in order to detect marked differences in the rate or magnitude of an antibody-forming response, it was necessary to give several injections of antigen spaced at rather long time intervals, e.g., 5-10 weeks in adult chickens (Hofstad 1953; Wolfe et al. 1960). One interesting difference between primary and secondary responses to BSA was noted by Orlans (1967) and Gildin and Rosenquist (1963); the antibody induced by a single injection of BSA reaches a peak at 7 days and is usually undetectable by three weeks following primary stimulation. In contrast, high antibody titers continue for over a year if multiple injections of this antigen are given.

The data reviewed here indicate that some degree of immunological memory was observed in antibody production by representatives of each of the living classes of vertebrates. Moreover, a number of cases were considered where the investigators did not obtain evidence of secondary responses. A variety of technical reasons might account for these results; for example, lack of sensitivity of assay methods, environmental temperature at which the animals were maintained, improper feeding, and dosages and routes of antigen administration. The consequences of these technical difficulties may be far from trivial and might provide some insight into the cellular regulatory mechanisms involved in the induction of immunity and tolerance.

## Induction of tolerance in amphibians and reptiles

Results have been described that showed that amphibians and reptiles produce low molecular weight antibody and mount secondary responses to serum protein and bacteriophage antigens. In contrast, they form only high molecular weight IgM and do not exhibit immunological memory to bacterial antigens such as intact *Brucella* or *Salmonella* flagella. My colleagues and I have found the marine toad, *Bufo marinus,* to be a useful model for studying antibody formation and tolerance induction in anura. This tropical species can be maintained successfully at 37°C and exhibits a vigorous response to *Salmonella adelaide* flagellar antigens, mammalian serum proteins, and erythrocytes. Our investigations of this system have been directed toward gaining detailed information on the following aspects of antibody formation: (1) kinetics of appearance of antibodies and antibody-forming cells, (2) retention and processing of antigen, (3) induction of tolerance to protein antigens, and (4) nature of the antibodies produced. I will use the primary immune response of *B. marinus* to *Salmonella adelaide* flagella to illustrate the kinetics of appearance of antibody-forming cells and serum antibodies in amphibians and the induction of high-zone tolerance to BSA.

Injection of *B. marinus* with *S. adelaide* flagella induced the immune response depicted in Figure 56. The animals were each given a single intraperitoneal injection consisting of 100μg of antigen and were maintained at 37°C. Serum antibody titers were measured by the bacterial immobilization technique of Ada et al. (1964); antibody-forming cells (AFC) obtained from the spleens were quantified according to the immunocytoadherence colony method of Diener (1968). A significant number of antibody-forming cells was first observed at 3 days following immunization, and a plateau was reached at 7 days.

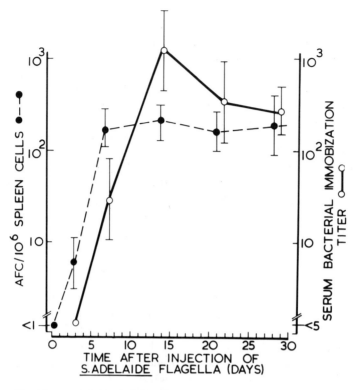

Fig. 56. Appearance of antibody-forming cells (AFC) and circulating antibodies in the toad (*B. marinus*), following injection of *Salmonella adelaide* flagella. AFC were enumerated by the immunocytoadherent method of Diener (1968). The animals were maintained at 37°C. (From Marchalonis 1971b).

Antibody titers lagged slightly behind, but the peak was reached by 14 days. The titers and numbers of antibody-forming cells observed here were comparable in magnitude to those obtained in mice or rats following primary stimulation with this antigen.

In most respects the toad primary response to flagella was very similar to that given by mammals to the same antigen. However, we observed a number of differences that merit comment; namely, (1) the toads did not show an enhanced response resembling immunological memory upon secondary stimulation with flagella, (2) multiple injections of adult toads with large doses of the antigen failed to induce a specific lack of response of the type termed high-zone tolerance in mammals, and (3) only antibodies of the IgM type were produced. The fact that only IgM antibodies are produced to flagella has proved to be a useful means of producing IgM antibodies to haptens in bullfrogs (Rosenquist and Hoffman 1972). If DNP is coupled to flagella, only antibodies of this class are produced. These results contrast markedly with mammalian responses to this antigen where memory, tolerance, and IgG antibodies are readily induced (Nossal, Austin, and Ada 1965). We were confronted by two possibilities; either this species possesses a very restricted immune capacity, or the nature of the antigen used determined the properties of the response. Experimental studies from other laboratories independently confirmed our findings with *Salmonella* antigens (Evans et al. 1966; Acton et al. 1972a) and also showed that the toad could mount an enhanced secondary response to bacteriophage antigens (Lin, Caywood, and Rowlands 1971). We observed that toads immunized with bovine serum albumin formed 7S antibodies as well as IgM antibodies. Similar results were obtained when the tuatara was immunized with *S. adelaide* flagella (Marchalonis, Ealey, and Diener 1969). High-titer antibodies were produced after a single injection and this titer could not be increased by secondary injections. Moreover, antibody activity was localized completely in the IgM immunoglobulins for the entire period of the experiment (six months). The tuatara, like the toad, possesses multiple classes of immunoglobulin. Unfortunately, we were unable to obtain more tuataras so that other antigens could be tested. This restriction did not apply to the common turtles, and Ambrosius et al. (1970) performed extensive studies on the tortoise (*Testudo*). Tortoises immunized with pig serum show a definite capacity to give a secondary response meeting the rigorous criteria defined above. I will propose a tentative explanation for the divergent responses of amphibia to certain antigens below. The parallel developed between responses of mice and amphibians to certain antigens probably holds true for responses of reptiles to the same or comparable antigens. *Brucella* and *S. adelaide* flagella are thymus-independent antigens under certain conditions, whereas serum proteins require the presence of thymus-derived helper cells to initiate an immune response. The response of the tortoise to *Brucella* is quite similar to the response which congenitally athymic mice (homozygous for the *nu* gene) give to the same antigen (Crewther and Warner 1972). The athymic mice, however, are unable to respond to serum proteins.

Thus, profound qualitative differences were observed when amphibians and reptiles were immunized with polymerized bacterial antigens rather than with serum proteins or bacteriophage. These differences were also apparent at the level of induction of tolerance.

Since serum proteins such as bovine serum albumin are effective in inducing tolerance in adult mice, we (Marchalonis and Germain 1971) used this antigen to continue our studies of immunological tolerance in anura. We found that high doses of soluble BSA did not induce antibodies in toads. Furthermore, such treatment specifically suppressed the capacity of the animals to form antibodies to this antigen, even when immunization was carried out in adjuvant to provide maximal stimulation of the immune system. This pretreatment with BSA did not affect the ability of toads to mount an immune response to the unrelated antigen human IgG immunoglobulin.

The immunosuppressive effect of large doses of BSA is illustrated in Figure 57, which compares the kinetics of clearance of radioactively labeled BSA from control and immunologically suppressed animals. These data show that toads given high doses of BSA exhibit only non-immune metabolic clearance when challenged with an immunogenic dose. Control animals, in contrast, eliminate the antigen in a nonimmune fashion for approximately 8 days, then switch to a rapid immune clearance of the protein. Furthermore, control toads possessed no detectable $^{125}$I-labeled BSA by day 22 after injection and exhibited passive hemagglutination titers ranging from 64 to 256 at that time.

Our findings on the induction of high-zone tolerance to BSA in toads paralleled Mitchison's observations with mice (1964). The doses required to induce either antibody production or tolerance, when normalized to a dosage per gram of body weight, were quantitatively very similar in these phylogenetically diverse species. Furthermore, the ease by which tolerance was induced in *B. marinus* may account in part for many of the early difficulties incurred in immunizing amphibian species.

The above observations demonstrate that adult anuran amphibians, as represented by the marine toad, can carry out a range of immune responses comparable to that of mammals in complexity. The pure IgM response to bacterial antigens is not restricted to amphibians, however. Reptiles as diverse as the tuatara and turtle, form only IgM antibodies when injected with flagella or whole bacteria (Marchalonis, Ealey, and Diener 1969; Maung 1963), although these species possess multiple classes of immunoglobulins and may respond with recognizable secondary responses to certain antigens (Marchalonis, Ealey, and Diener 1969; Chartrand et al. 1971; Benedict and Pollard 1972). An analogous situation has been observed in mammals given highly polymerized antigens. Mice respond to pneumococcal polysaccharide (Howard et al. 1971) or the synthetic polyvinylpyrrolidone (PVP) (Andersson and Blomgren 1971) by producing only IgM antibodies. Furthermore, immunological memory is weak or absent. These antigens are considered to be thymus independent because bone-marrow-derived lymphocytes alone are capable of giving the full response. The hypothesis has been presented that polymeric antigens consisting of multiple identical antigenic determinants are capable of binding directly to B lymphocytes and inducing an antibody-forming response (Feldmann and Nossal 1973). Monomeric antigens in contrast, require thymus derived helper cells to

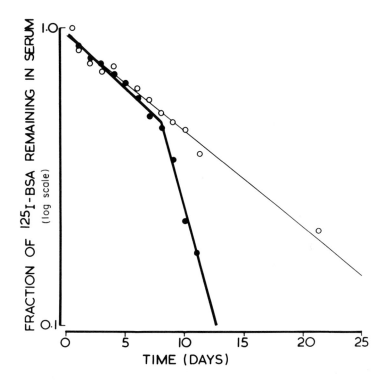

Fig. 57. Kinetics of elimination of 125I-labeled bovine serum albumin (BSA) from the blood of normal (●——●) toads and toads rendered tolerant (o——o) by injection of 250mg BSA in saline. Normals were given only diluent. Both groups were challenged with an immunogenic dose of soluble BSA containing a trace amount of the radioiodinated antigen. Tolerant animals cleared the BSA at a uniform rate throughout the course of the experiment. Normal animals eliminated BSA at the same rate as tolerant animals for the first week. The break in slope and rapid elimination denoted the onset of immune clearance.

mediate in the induction of antibody formation. Polymerized *S. adelaide* flagellin behaves as a completely thymus-independent antigen in tissue culture (Diener, O'Callaghan, and Kraft 1971), but appears to be thymus dependent *in vivo* (Crewther and Warner 1972). This discrepancy probably results from the fact that polymers of flagellin are not covalently linked and may dissociate into monomers of molecular weight 40,000 within the animal. Barring the presence of other complicating effects, these monomers would then be processed in a manner similar to that occurring for serum albumin (M. W. 65,000) which is a classic thymus-dependent antigen (Mitchison, Rajewsky and Taylor 1970). The polysaccharide and PVP antigens are covalently linked and are difficult to digest *in vivo*. Consequently, deposits of these antigens persist for long periods. The fact that anuran amphibians and reptiles respond to polymerized flagellin in a manner that mimics that given by mice to thymus-independent antigens raises the possibility that divergence of lymphocytes into T and B cells might have occurred in the ancestors of amphibians. It also suggests that amphibians and reptiles might serve as useful models for the delineation of basic events in immune reactions of higher vertebrates.

## Suppression of antibody formation to protein antigens in chickens

Detailed studies of the effects of dosage of human serum albumin (HSA) on the synthesis of antibodies to this antigen were performed by Cerny and his colleagues (Cerny et al. 1965; Cerny and Ivanyi 1966; Ivanyi and Cerny 1965). These workers observed that the onset of antibody formation to this antigen was delayed as the antigen dosage within the optimum range was increased, and that IgM antibody production was totally inhibited by suboptimal doses. It was interesting that formation of IgG-like antibodies. IgG(Y), continued even though the IgM response had been suppressed. The investigations were continued into dosage effects upon cells which bound fluorescein-labelled HSA (Cerny and Ivanyi 1966). It was found that a delayed onset of appearance of antigen-binding cells correlated with the previously observed suppression of antibody production. The cells which predominated in the spleen when IgM synthesis prevailed resembled lymphocytes, whereas cells resembling plasmocytes comprised the majority during the phase of IgG (Y) production. The authors proposed that, at a given antigen dose, the response of the lymphoid cell population was diverse; e.g., some cells were stimulated to produce antibodies, whereas others were inhibited or not stimulated at all. A similar notion was expanded mathematically by Marchalonis and Gledhill (1968), who argued that the incomplete nature of certain tolerance phenomena in mice and rats, particularly low-zone tolerance, resulted from the fact that a so-called tolerant population of cells contained subpopulations of stimulated, suppressed, or unimpressed cells.

## Induction of tolerance in larval amphibians

The preceding treatments of the induction of tolerance or immunological paralysis dealt with suppression of antibody formation and immune clearance of antigen. Tolerance can also be manifested at the level of cellular immunity. Specific failure to reject allografts due to pretreatment of embryonic animals with cells bearing the antigen in question has long been considered a form of immunological tolerance (Billingham, Brent, and Medawar 1956). The use of embryonic animals for the induction of tolerance stems from the classical observation in which Owen (1945) found that twin cattle do not form antibodies to blood groups of their twin. Cows bear twin calves which may be derived from two separate fertilized ova and are thus nonidentical. Therefore, they are neither more nor less closely related to each other than are any two siblings. However, twin fetal calves share a single placenta, and extensive mingling of the blood of the developing embryos occurs. Even though the red cells of the twin cattle may be antigenically distinct, their presence within the other's blood stream throughout embryonic life and after birth acts as a stimulus for the induction of tolerance to the distinct antigens of the twin.

In the type of tolerance illustrated for the cattle twins, an accidental mixing of blood and blood-forming stem cells produces blood group chimeras which are immunologically stable. Billingham, Brent, and Medawar (1956) showed by deliberate experiment that allograft rejection in mice could be specifically suppressed by injecting mouse embryos *in utero* with living spleen cells of mice bearing foreign histocompatibility antigens. Larval amphibians lend themselves quite readily to the type of experimental manipulation required for studying the establishment and breaking of tolerance. Volpe and his colleagues (Volpe and McKinnell 1966; Volpe and Earley 1970; Volpe 1971) have applied classical embryological techniques to prepare chimeric tadpoles and frogs for use in analysis of tolerance to skin allografts. Pairs of nonsibling embryos can be joined side to side in the area of the gill primordium to insure vascular communication (Figure 58). Throughout embryonic and larval development there is a complete intermingling of blood through the artificially established vascular connections. The animals are viable and will undergo metamorphosis into parabiosed frogs. The copartners are highly tolerant of grafts of skin from each other, but are intolerant of skin allografts from unrelated frogs. It was possible to show that the copartners exhibit blood cell chimerism by performing experiments in which one partner possessed the normal diploid complement of chromosomes and the other had a triploid set. The proportions of donor and host blood cells vary with the age of the parabionts. In premetamorphic larvae the percentage of donor cells is comparable to that of the host cells. However, with transition from juvenile life (postmetamorphic) the donor-type cells are at a selective disadvantage. The number of donor cells decreases to a point where new stem cells of the host that can recognize the donor cells as foreign arise, and tolerance is broken. Volpe's data on parabiotic frogs allow the interpretation that the tolerant state results from the

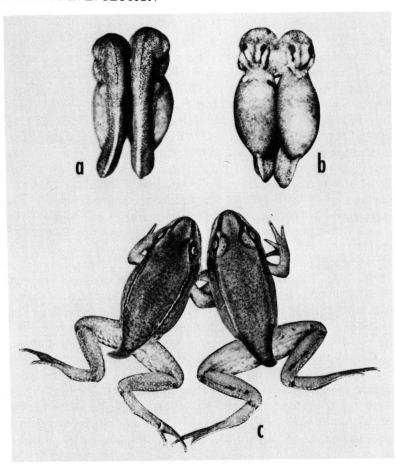

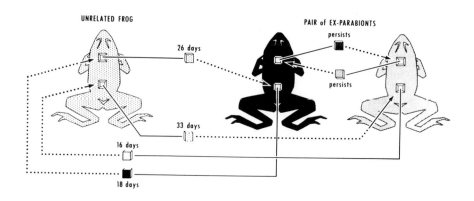

depletion of lymphocytes capable of reacting immunologically to donor cells, and breaking of tolerance is the result of the reappearance of such reactive clones of lymphocytes (Figure 59). Volpe and McKinnell (1966) have also explored another technique of amphibian embryology that possesses great potential in immunological studies. It is possible to remove nuclei from cells of *R. pipiens* embryos at the blastula stage and implant these in enucleated eggs using the elegant approach of Briggs and King (1952) and Gurdon and Brown (1965). Such embryos produced by nuclear transfer develop into a clone, or isogenic group of frogs. Volpe and McKinnell (1966) have shown that such groups do, in fact, behave as a clone in terms of allograft transplantation. The production of isogenic frogs is a step required for detailed elucidation of the cellular basis of immunity in nonmammalian species. For example, experiments designed to ascertain whether cells carrying out functions analogous to those of murine T cells are separate from those carrying out B-cell functions can be carried out by transfer of presumptive cells into lethally irradiated hosts. If the transferred cells are antigenically identical to those of the host, no complicating side effects such as a graft versus host reaction will occur.

The model for induction of tolerance derived by Volpe from his frog data represents the most straightforward interpretation of the observed suppression of specific reactivity. Other models for the induction of tolerance are presently under consideration based upon data from studies involving mice. One of these is the suppressor T-cell concept (Gershon and Kondo 1971). Rather than having a tolerant state result from the elimination of cells that are specific for the test antigen, Gershon and Kondo proposed that tolerance results from the function of a suppressor cell that is activated by antigen and subsequently prevents antibody formation to that antigen. It is an interesting concept, because it requires that at least some forms of tolerance act in an infectious manner, i.e., a population of tolerant cells when transferred to a normal recipient animal should render the recipient tolerant. In contrast, a population of cells that had been specifically depleted of reactive cells would not diminish the immune capacity of the host. At present there is

---

Fig. 58. Induction of allograft tolerance in *Rana pipiens* by parabiotic union of larvae. (A) Tadpoles (68 hrs post fertilization) joined in parabiosis in the region of the gill epithelium. a. dorsal view, b. ventral view, c. parabiotic frogs after metamorphosis. (B) After metamorphosis and surgical separation, members of ex-parabiotic pairs accept graft from each other but not from third party frogs. (From Volpe 1971. I thank Dr. E. P. Volpe for providing original illustrations.)

INDUCTION OF TOLERANCE

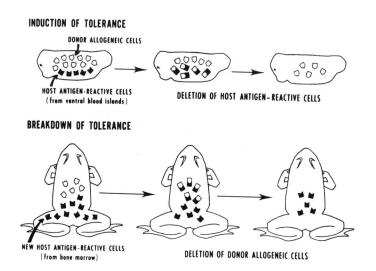

Fig. 59. Interpretation of the mechanism of acquired allograft tolerance in amphibians. Acquired immunological tolerance results from the specific depletion of reactive lymphocytes. The breakdown of tolerance occurs through the emergence of a new population of lymphocytes which are reactive to the antigen. (From Volpe 1971).

no evidence supporting this type of model in lower vertebrates, but too few studies of the establishment of tolerance in such animals have been carried out to exclude other mechanisms.

## Synopsis

It is possible, under proper experimental conditions, to demonstrate immunological memory in representatives of all the vertebrate classes. This memory is not very impressive in fishes or amphibians. Reptiles, in response to serum protein antigens, exhibit secondary responses that meet the rigid criteria applicable to mammalian species. Negative memory, or tolerance, can be demonstrated in lower species such as amphibians. However, tolerance in lower species has been largely ignored, and a great deal of work is required to elucidate humoral and cellular mechanisms involved in its induction and maintenance.

# 13 Phylogeny of Complement

## *The nature of complement function and structure*

Antibodies, although remarkably specific in their capacity to recognize and bind to foreign macromolecules, are not themselves usually effective in promoting significant biological effects. Combination of antibody with a bacterium, for example, can cause agglutination of the bacterial suspension but will do little, if any, harm to the invading organism. If this binding occurs in the presence of normal blood serum, however, a series of proteins termed complement components fix to the antigen-antibody complex and bring about lysis of the bacterial cell. This was one of the classical immunological observations which dates back to the work of Buchner (1889), von Fodor (1886), and Nuttal (1888), who also found that the capacity to lyse the cells was destroyed by heating serum to 56°C. Bordet and Gargov (1901) subsequently showed that the capacity of heated serum to lyse cells was restored when a small amount of fresh serum, inactive by itself, was added. The heat-labile substance was designated "complement." It is now well established that nine major complement components are present in the serum of man and the guinea pig (Müller-Eberhard 1968). These constitute a diffuse collection of proteins, including esterases and proteolytic enzymes, which cause a cascade effect that is initiated by combination of antibody to antigen and mediates secondary functions, such as augmented phagocytosis of antigen-antibody complexes and damage to the surfaces of cellular antigens.

The sequence of events involved in the lysis of erythrocytes (E) by combination with specific antibody (A) followed by addition of complement components is illustrated in Figure 60. This scheme of sequential reactions is known as the classical pathway of complement-mediated lysis. The antibody combines with the antigen via its combining site in the Fab piece and the first component of complement C1 attaches to the Fc portion of the bound antibody molecule. C1 consists of a complex of three components which are normally inactive unless bound to antibody. These C1 subunits are called C1q, C1r and C1s (Naff, Pensky, and Lepow 1964). C1q binds to the immunoglobulin, thereby activating C1r which subsequently activates C1s to the so-called C1̄s form (the bar symbolizes the activated state). The C1s subunit possesses esterase activity. The addition of the C1 complex to the sensitized erythrocyte requires the presence of Ca$^{++}$ ions. The fourth component then binds to the EAC1 complex and the subsequent order of addition of complement components is C2, C3, C5, C6, C7, C8 and finally C9. I have indicated that some biological functions, such as immune adherence and enhanced phagocytosis of antigen-antibody complexes, do not require the binding of all nine components but occur when components 1, 4, 2 and 3 have bound. Similar conclusions apply to neutralization of viral antigen by antibody plus complement (C1, 4) and chemotaxis of leucocytes towards antigen-antibody complexes. Complete lysis of cells in the classical pathway does, however, depend upon the combination of cells with antibody and all nine C components.

In addition to the classical pathway outlined here, a number of alternative pathways exist which utilize the terminal components of complement, e.g., C3-C9, and eliminate the requirement for erythro-

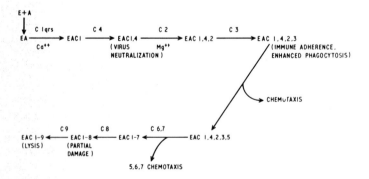

Fig. 60. Schematic diagram for the lysis of erythrocytes by the classical complement-fixation pathway. E. erythrocytes; A. antibody to erythrocytes. Complement components are designated by C followed by the number of the component (Based upon Gigli and Austen 1971).

cytes to bind specific antibodies in order to initiate the chain of lytic events. It was found, for example, that activation of C5 and C6 (in the presence of C7-C9) and unsensitized erythrocytes resulted in the lysis of such cells (Thompson and Rowe 1968; Lachmann and Thompson 1970). Moreover, the need for antibody and the first components of complement (C1, 2, 4) can be bypassed by the addition of substances which activate C3 (Pickering et al. 1969; Pillemer et al. 1955; Bitter-Suermann et al. 1972). One approach to the activation of C3 is to use a factor derived from cobra venom which combines with a cofactor present in serum and subsequently with C3 (Pickering et al. 1969; Bitter-Suermann et al. 1972). Experiments employing cobra venom factor (CVF) have been used to probe for the presence of components in the serum of lower vertebrates and invertebrates (Day et al. 1970, 1972) which are functionally similar to the later acting C1 components of higher mammals. These studies will be discussed below.

Some of the molecular properties of human complement components are listed in Table 26. The isolation of serum complement components and their physical and biological functions are reviewed by Müller-Eberhard (1968). The concentrations of C1 components found in serum are relatively low, maximum 1.5 mg/ml for C3, as opposed to 10-15 mg/ml for immunoglobulins such as IgG in mammals. This may account in part for the fact that very few attempts have been made to isolate complement components from the sera of nonmammalian vertebrates. The approach most commonly used by investigators studying the phylogenetic distribution of complement components has been a functional one designed to assess the presence of classical complement components by use of inhibitory conditions which are operative in man and the guinea pig (Gewurz et al. 1966; Gigli and Austin 1971).

## Phylogenetic distribution of classical complement activity

The general assay for classical complement is a hemolytic one. In essence, the investigator asks if the fresh serum of a lower vertebrate will lyse mammalian erythrocytes. Since most sera possess low concentrations of natural antibodies to sheep erythrocytes, these cells are usually used in the assay. Because mammalian complement is readily inhibited by simple procedures including removal of $Ca^{++}$ and $Mg^{++}$ ions by chelating agents or heating to 56°C, a putative complement-mediated lytic system is treated with these agents. If lytic activity is lost, one concludes that a classical lytic system is present. By this criterion, all vertebrates from cartilaginous fishes to man possess complement activity (Gewurz et al. 1966; Gigli and Austin 1971). Data available for representatives of the various vertebrate classes are summarized in Table 27.

TABLE 26: General properties of human complement components.

| Property | C1q | C1r | C1s | C4 | C2 | C3 | C5 | C6 | C7 | C8 | C9 |
|---|---|---|---|---|---|---|---|---|---|---|---|
| Electrophoretic mobility | $\gamma 2$ | $\beta$ | $\gamma 2$ | $\beta 1$ | $\beta 2$ | $\beta 1$ | $\beta 1$ | $\beta 2$ | $\beta 2$ | $\alpha 1$ | $\alpha$ |
| Sedimentation coefficient | 11S | 7S | 4S | 10S | 5.5S | 9S | 8.7S | 6.6S | 5.6S | 8S | 4.5S |
| Serum concentration ( μg/ml) | 100-200 | - | - | 600 | 12 | 1,500 | 99 | - | - | - | 2 |
| Thermolability (56° C. 30 min) | Yes | - | Yes | No | Yes | No | Yes | No | - | - | - |
| Inactivated by | - | - | DPF[a] | $N_2H_2$, NEM[b] | pCMB[c] | $N_2H_2$ | - | - | - | - | - |

*Sources*: Data compiled by Gigli and Austin 1971; Davis et al. 1969, p. 515.

a. DFP = disopropylphosphofluoridate (an esterase inhibitor); $N_2H_2$ - hydrazine.

b. NEM = N-ethylmaleinmide;

c. pCMB = p-chloromercuribenzoate (both NEM and pCMB react with free sulfhydryl groups).

Sera of the various species, excluding the lamprey, all possess a natural lytic complement that differs in temperature optima and species origin of antibodies that potentiate the lytic reaction. The differences in temperature optima would be expected in view of the fact that each ectothermic species is adapted to a characteristic range of environmental temperature. The observation that complement components are not functionally interchangeable from class to class within vertebrates is an important one, which has often led to technical difficulties in the development of reliable assays for immune hemolysis in lower species. The best-studied example is that of chicken antibodies, which do not fix mammalian complement (Rose and Orlans 1962). Since chicken IgG(Y) fixes chicken complement very efficiently, this finding suggests that the Fc of chicken IgG(Y) is sufficiently distinct structurally from mammalian IgG such that C1q cannot recognize it. Interestingly, Romano, Geczy, and Steiner (1973) have recently reported that bullfrog antibodies, both IgM and IgG(?), will fix guinea pig complement.

Gewurz et al. (1966) found that sera of cyclostomes such as lampreys lysed rabbit erythrocytes, but that this effect was not one of antibody plus classical complement proteins. Moreover, the lytic agent was characterized by an extremely low molecular weight as judged by gel filtration on Sephadex G-200 (Pollara, Finstad, and Good 1966). Further work is required to ascertain the molecular nature of this lytic agent and its possible relationship to the complement system of other vertebrates. At present we can conclude that classical lytic complement systems are present in all vertebrates derived from placoderm ancestors.

As indicated in Table 27, the first component of nurse shark complement was isolated recently by Ross and Jenson (1973b), who previously showed that the hemolytic properties of this component were very similar to those of mammalian C1. The intact molecule has a sedimentation coefficient of 13.9S and consists of three antigenically distinct precipitin arcs. More studies of this nature are required to obtain sufficient information to estimate the degree of structural homology among complement components of diverse vertebrate species.

## *Nonclassical lytic complement systems*

Although invertebrates and agnathans, including the hagfish and lamprey, do not possess hemolytic activity in their hemolymph or serum when tested by classical methods, Day et al. (1970, 1972) have demonstrated that certain of these species possess a lytic system that can be activated by cobra venom factor. This factor activates the third component of mammalian complement and leads to a lysis of erythrocytes in the absence of antibody and C1, C2, and C4. Day et al. (1970, 1972)

TABLE 27: Properties of naturally-occurring lytic systems of lower vertebrates.

| Species | Vertebrate class | Natural lysin to: | Temperature dependence | Heat labile |
|---------|-----------------|-------------------|------------------------|-------------|
| Lamprey[1] | Agnathan | Rabbit RBC | 4°C | Relatively stabile |
| Nurse shark[2,3] | Elasmobranch | Variety of RBC | 25°-30°C | Yes |
| Paddlefish[1] | Chondrostei | Rabbit RBC | 4°C | Yes |
| Number of species[4,5] | Teleost | Variety of RBC | 4°-28°C | Yes |
| Frogs, toads[6] Necturus | Amphibian | Rabbit RBC | 4°-28°C | Yes |
| Snakes, turtles | Reptile[5] | Human & sheep RBC | 15°-37°C | Yes |
| Chicken | Avian[5] | Variety of cells | 37°C | Yes |

reported that an arthropod (the horseshoe crab), a sipunculid worm (*Golfingia spp.*) and an agnathan (the hagfish) contained serum components which functionally resembled complement components acting in the terminal position of the lytic complement reaction sequence. An invertebrate phylogenetically related to the chordate stream, the starfish (*Asterias forbesi*) possessed the cofactor to which cobra venom factor binds but lacked hemolymph constituents functionally analogous to terminal complement components. Furthermore, these workers mention that lamprey serum possesses the 9th component of complement. Structural homology of these factors to those of mammals remains to be established. Further study, including chemical characterization of individual complement components of lower species, is warranted, because the complement system may play a role in inflammatory reactions under circumstances where the classical antibody-dependent pathway is not involved (Gigli and Austin 1971).

## Antibodies and complement: An evolutionary perspective

Two important generalizations can be drawn from the presently available data, albeit fragmentary, of the phylogenetic origins of the complement system; namely, (1) in species where classical antibodies

| Requires $Ca^{++}$, $Mg^{++}$ | Potentiated by antibody | Classical lytic components functionally present | Components isolated |
|---|---|---|---|
| Possibly (not all or none) | No (Mammalian AB) | None | None |
| Yes | Yes (shark & turtle AB) | C1, 4, 2, 3, 9[5] | C1 |
| Yes | Yes (mammalian AB) | C1, 4, 2, 3 | None |
| Yes | Variable | C1, 4, 2, 3 | None |
| Yes | Yes (rabbit & turtle AB) | C1, 4, 2, 3 | None |
| Yes | Yes (rabbit & snake AB) | Not known | None |
| Yes | Yes (Avian) | Probably all | None |

*Sources:* (1) Gewurz et al. 1966; (2) Ross and Jensen 1973a,b; (3) Jensen 1969; (4) Legler, Evans, and Dupree 1967; (5) Gigli and Austin 1971; (6) Legler and Evans 1966.

are present, classical complement occurs, and (2) the first complement components to emerge phylogenetically were the late-acting ones. The first statement requires some qualification because of exceptions like lamprey antibodies and non-complement-fixing antibodies of mammals. Although lampreys possess antibodies similar in polypeptide chain structure to IgM immunoglobulins, they are not linked via disulfide bonds and may possess aberrant properties in complement fixation. Furthermore, antibodies of mammals exist that do not fix complement via the classical pathway but will initiate bypass lysis (Gigli and Austin 1971). Neither of these qualifications nullifies the broad conclusions. They provide further questions in the consideration of the interrelationships between the complement system and the antibody-forming system.

The complement system, particularly the late-acting components (Figure 60), may be an extremely ancient one concerned with the non-specific destruction of pathogens (Day et al. 1970, 1972). The involvement of these complement components in defense reactions formally analogous to those that occur in protostomate invertebrates, such as enhanced phagocytosis, is consistent with this proposal. Antibodies, on the other hand, represent a specific recognition and response system that arose within the chordates or their immediate progenitors. Conceivably, during the evolution of chordates there emerged a structural and functional linkage between the ancient complement system and the more recent immune system. This connecting link might well be the early-acting complement components, C1, C4, and C2. The genetic basis of this event would have required the emergence of genes encoding a system of proteins that would recognize immunoglobulins and also interact with the C3-C9 complement components. Such a complex event might have required millions of years of evolutionary time, and primitive vertebrate forms, such as agnathans, which diverged early from the line leading to jawed vertebrates, could represent the condition where C3-C9 and antibodies resembling the original antibodies occur, but the C1, C4, and C2 molecules are lacking.

# 14 Temperature Dependence of Immunity in Ectothermic Vertebrates

## Temperature and immunity

One of the first studies of the immune response of lower vertebrates, that of Widal and Sicard in 1897, drew the important conclusion that the capacity of frogs to produce antibodies was dependent on the temperature at which the animals were maintained. It has been amply confirmed that immune responses of all ectothermic species are temperature dependent, and this conclusion applies to both cellular (Hildemann 1957; Macela and Romanovsky 1969) and humoral (Bisset 1948; Evans et al. 1965, 1966; Diener and Marchalonis 1970) immunity. The problem has both practical and theoretical aspects that merit investigation. Frog farming, for example, is carried out on a fairly large scale in some countries and fish farming will become increasingly important in future years. Epizootic diseases affecting populations of such ectotherms are related to environmental conditions including water temperature, water levels, and oxygen depletion (Avtalion et al. 1973). The theoretical aspects of the problem stem from observations that it is possible to separate the immune response into discrete steps by merely varying the temperature within a range which is physiological for the test animals in question (Cone and Marchalonis 1972). Data from a variety of systems suggest that the initiation of immunity is independent of temperature (within physiological limits, of course), but that the effector phases of immunity require a certain minimum temperature to occur (Bisset 1948; Elek, Rees and Gowing 1962; Macela and Romanovsky 1969; Ching and Wedgewood 1967). Krueger and Twedt (1963) obtained results consistent with this proposal by isolating frog cells that form antibodies to *Salmonella* by micromanipulation. They concluded that such cells contain antibody at 4°C but require higher temperatures to release it. Unfortunately the interpretation of this technically elegant experiment is not clear-cut, because the release at 37°C may not have represented active secretion but merely leaking out through the membrane. The interpretation of this temperature effect is further complicated by the possibility that the state of the cell membrane at 4°C represents a totally different phase state than that of the active cell at 37°C. Further data are required to establish that the immunoglobulin is released by an active process of secretion.

## Effect of temperature upon allograft rejection

Figure 61 depicts the effect of temperature on the rate of rejection of first- and second-set scale allografts in goldfish (Hildemann 1957). In the first place, a clear-cut secondary response (memory) obtained at all temperatures. In both the primary and secondary response, the rate of scale graft rejection was markedly temperature dependent. Graft rejection for the first-set grafts required 40 days at 10°C and less than 10 days at 20°C. Corresponding values for the second-set reaction were 20 days and about 5 days. In experiments designed to determine whether antigen recognition or effector function of activated lympho-cytes was the temperature-sensitive step in the reaction, Macela and Romanovsky (1969) grafted frogs (*R. temporaria*) at a low temperature (5°C) and then shifted the animals to 25°C. These workers found that grafted animals exposed to low temperature for a sufficient time after challenge will, after exposure to high temperature, reject the first-set graft in a rapid manner similar to that which occurs in second-set grafting at the elevated temperature. Therefore, antigen processing and recognition clearly take place at lower temperatures. The lesion in allo-graft immunity appears to be the reaction of the activated cells with the grafts. These results are consistent with Bisset's hypothesis for antibody formation (1948) and will be discussed below in the context of a scheme proposed by Avtalion et al. (1973) to describe the chronological determination of the temperature-sensitive effect in antibody formation.

## Induction of tolerance at low temperatures

I previously described the induction of high-zone tolerance to protein antigens in fish, amphibians, and chickens (Chapter 12). Avtalion and his colleagues (1973) observed that a low-zone tolerance phenomenon can be induced in carp by injection of the minimal antigen dose (0.04-

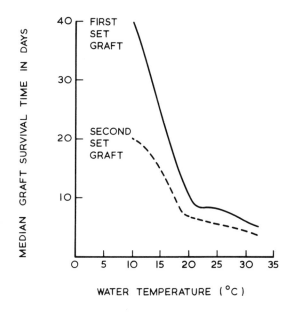

Fig. 61. Effect of temperature upon the rate of rejection of allografted scales in the primary (first set graft) and secondary (second set graft) responses of goldfish. (Based upon Hildemann 1957).

0.2 mg per kg body weight) of BSA into animals kept at 12°C. Such fish did not produce circulating antibodies even when acclimated to a temperature of 25°C and given a booster injection. These workers point out that if the second injection was given in complete Freund's adjuvant, the tolerance was less pronounced and had a tendency to break. Studies in mice (Mitchison 1964) previously indicated the fragile nature of low-zone tolerance relative to that obtained in high-zone paralysis. Since tolerance induction occurred at low temperatures, specific recognition of antigen by competent lymphocytes is not the sensitive step in the reaction. Furthermore, low-zone tolerance in mice is considered to involve inactivation of T cells (Mitchison 1971b), so this effect provides additional circumstantial evidence suggesting that cells analogous or homologous to both T cells and B cells are present at the level of teleost fish.

## The influence of temperature on antibody formation in the marine toad

Since the studies reported used the terminal stages of immune reactions such as graft rejection or the presence of circulating antibodies as their monitors of the effect of temperature on immunity, we decided to investigate the early cellular phase as well as the production of antibodies. Cone and I (1972) immunized *B. marinus* with horse erythrocytes and followed the reaction by assaying the appearance of immunocytoadherent lymphocytes (rosette-forming cells) in the spleen and the presence of hemagglutinating antibodies in the bloodstream. We observed the effects of variation of temperature within a range physiological for the toads (20°C-37°C) upon these parameters of immunity. The results obtained are illustrated in Figure 62. Toads given 600 million red blood cells and maintained at 37°C have a peak of specific rosette-forming cells in their spleens at day four after injection and possess serum titers of greater than 1000 by the eighth day. The response to the same dosage of antigen given by toads kept at 20°C contrasts markedly with this pattern. The peak of RFC is present and it is comparable quantitatively to that achieved at 37°C. However, only a negligible rise in circulating antibodies occurs during the first two weeks. In general,

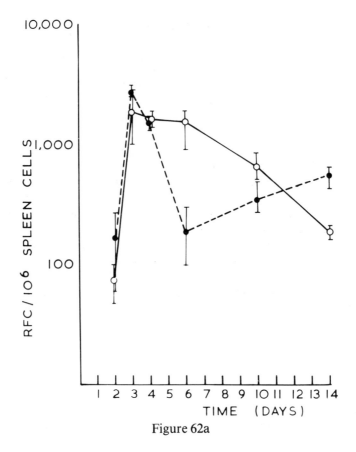

Figure 62a

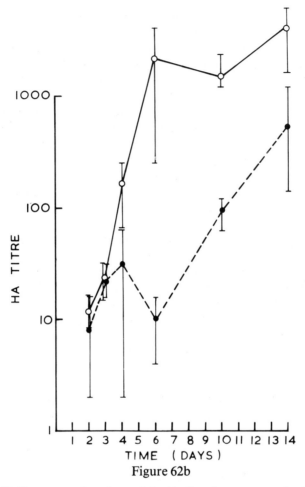

Figure 62b

Fig. 62. Temperature dependence of the kinetics of appearance of antibody-forming cells and circulating antibodies in the toad (*B. marinus*). Toads were immunized with horse erythrocytes. (A) Antibody-forming cells, identified by immunocytoadherence (rosette formation). o——o, cells from animals maintained at 37°C; •--•, cells from animals maintained at 25°C. (B) Antibodies, estimated by hemagglutination titer to horse erythrocytes. o—o, animals maintained at 37°C; •--•, animals kept at 25°C.

it was found that the magnitude of the early peak of RFC was independent of temperature but proportional to the dosage of antigen injected. The appearance of circulating antibodies was dependent upon both the dosage of antigen and the temperature. For example, animals maintained at 25°C and given six thousand million red cells produced serum antibody titers comparable to those exhibited by animals kept at 37°C and given the same number of red cells. However the same levels of antibody were achieved one week later at the lower temperature.

The timing of appearance of the two peaks of RFC is interesting in the light of Ruben's 1975 data, which was discussed in Chapter 11, on the appearance of putative helper RFC and secretory RFC in the carrier response of the newt. The first peak of toad RFC, which is temperature independent, corresponds in time to Ruben's helper RFC peak, while the temperature dependent RFC correspond to his secretory cells.

These data are consistent with the previous hypothesis of Bisset (1948), and provide some hints regarding possible mechanisms governing the temperature effect. Presumably, cells might divide more slowly at lower temperatures and this would suppress the rate of emergence of antibody-forming cells and circulating antibodies. In some situations this sort of general physiological effect may play the predominant role. However, in the results discussed here, the number of immunocytoadherent lymphocytes was independent of temperature, while the appearance of circulating antibodies was suppressed at lower temperatures. The nature of the antigen and the type of response given by mice may provide clues in the situation under consideration. Erythrocytes are thymus-dependent antigens in mice (Miller and Mitchell 1969) where the antibodies are synthesized by cells of the B-lymphocyte line but thymus-derived helper cells are necessary to initiate the response. If the immunocytoadherent cells of the toad are T lymphocytes and the cells that secrete antibodies are the progeny of B cells, the present results would be explained if collaboration between these two cell types was temperature dependent. Recent data obtained from *in vitro* studies of collaboration between helper T cells and B cells establishes that these T cells synthesize and secrete a specific collaborative factor which is IgM-type immunoglobulin (Feldmann, Cone, and Marchalonis 1973; Marchalonis et al. 1972; Marchalonis and Cone 1973a). Such a process of synthesis and secretion would most probably be retarded by lower temperatures. A similar conclusion would obtain if the role of the T cell was to synthesize and secrete molecules which activated B cells but showed no specificity for antigen (Katz and Benacerraf 1972). At present the identification of toad rosette-forming cells as homologous to mammalian T lymphocytes cannot be made conclusively, although the data described above suggest that the thymus in amphibians carries out a function comparable to that of the corresponding organ in mammals. Binding of red cells by toad lymphocytes is completely abolished by treatment with rabbit antiserum to toad immunoglobulin M, which is specific for light chains and μ chains. This observation does not distinguish between T cells and B cells, because rosette-forming cells of both types can be inhibited by antisera directed against murine light chains (Greaves and Hogg 1971; Marchalonis, Cone, and Rolley 1973) or $\mu$ chains (Greaves and Hogg 1971; Hogg and Greaves 1972). The exact interpretation of this phenomenon is an open question; but the fact that a simple treatment such as variation of temperature serves to dissociate discrete steps in the antigen-directed maturation of immunocompetent cells in ectothermic species may prove to be useful as a means of elucidating fundamental biochemical events which control immune processes.

## A tentative model for the temperature effect

All of the systems described give results consistent with the conclusion that the temperature-sensitive event in cellular and humoral immunity occurs after antigen processing and recognition. Avtalion et al. (1973) have shown that phagocytosis and clearance of both soluble and particulate antigens in carp occur at low temperatures (12-14°C). These workers have proposed a scheme for the chronological determination of the temperature-sensitive event. Their analysis is in accordance with that of Cone and Marchalonis (1972) and was independently derived for a different system. A modified version of these schemes is illustrated in Figure 63. Two temperature-sensitive stages are indicated. The first corresponds to either cellular interaction (T-B collaboration) or maturation of activated cells at 3-4 days following injection, and the second to antibody synthesis and release, or effector function of activated T cells in allograft rejection, at later times (about 8 days). This is only a provisional model, but it serves to illustrate the possible use of the temperature effect in ectotherms as a tool with which to analyze molecular events controlling discrete processes in the immune response.

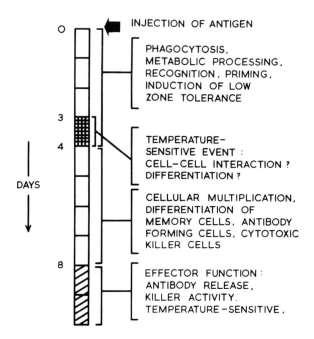

Fig. 63. Scheme illustrating possible sites at which temperature-sensitive events occur during the immune response. (Based upon Avtalion et al. 1973).

# 15 Antigenic Studies of Immunoglobulins of Eutherian Mammals

## *Immunological cross-reactions among mammalian immunoglobulins*

In this section I will consider antigenic comparison of immunoglobulins of other eutherian species with those of the human system. Although amino acid sequence data are available only for a small number of species, the identification of certain immunoglobulins with human or murine classes has been relatively simple because of antigenic cross-reactions. Nash and Mach (1971), for example, using rabbit and sheep antisera directed against human $\mu$, $\alpha$ and $\gamma$ chains, were able to identify immunoglobulins homologous to human IgM, IgA and IgG molecules in dolphins and sea lions. Such serological approaches, coupled with limited chemical analysis, establish that all eutherian mammals possess multiple classes of immunoglobulins.

Table 28 illustrates the multiplicity of immunoglobulin classes of nonprimate mammals. The asterisks represent the fact that the immunoglobulin indicated cross-reacts with the defining class of man. On this basis, it is readily apparent that IgA (Vaerman, Heremans and van Kerckhaven 1969; Vaerman and Heremans 1971; Mach and Pahud 1971; Nash et al. 1969) and IgE (Liakopoulou and Perelmutter 1971; Dobson, Rockey, and Soulsby 1971; Kanyerezi, Jaton, and Bloch 1971; Halliwell, Schwartzman, and Rockey 1972) immunoglobulins of diverse mammalian species share antigenic determinants with the $\alpha$ and $\varepsilon$ chains of human immunoglobulins respectively. Moreover, cross-reactions occur between human $\mu$ chain and $\mu$ chains of canine (Vaerman

TABLE 28: Immunoglobulin classes of mammalian orders.

| Order | Species | IgG | IgM | IgA | Other |
|-------|---------|-----|-----|-----|-------|
| Lagomorph | Rabbit[1] | + | + | +[a] | IgE |
| Rodent | Mouse[1] | γ2a, γ2b, γ1 | + | +[a] | |
| | Rat[1] | γ1, 2γ2 | + | + | IgE[a] |
| | Guinea pig[1] | γ1, γ2 | + | | IgE |
| | Hamster[1] | γ1, γ2 | + | | |
| Carnivore | Dog[1] | γ1, γ2a, γ2b, γ2c | +[a] | +[a] | IgE |
| | Mink[2] | γ1, γ2a, γ2b, γ2c | + | + | |
| | Cat[5, b] | + | + | + | |
| Artiodactyl | Pig[1] | γ1, γ2( )[6, a] | + | + | 3.1S IgG |
| | Cow[1] | γ1, γ2( )[6, a] | + | +[a] | 19S IgG |
| | Sheep[1] | γ1, γ2( )[6, a] | + | +[a] | |
| Perissodactyl | Horse[1] | γ2a, γ2b, γ2c( )[6, a] | + | | T component 10S γ1 |
| Cetacea | Dolphin[3] | +(2)[a] | +[a] | +[a] | |
| Pinnipeda | Sea lion[3] | +[a] | +[a] | +[a] | |
| Insectivora | Hedgehog[4] | +(2) | + | +[a] | |

*Sources:* (1) Greg 1969; (2) Tabel and Ingram 1972; (3) Nash and Mach 1971; (4) Vaerman and Heremans 1971; (5) Kehoe, Hurwitz, and Capra 1972; (6) Binaghi and Estevez 1972.

a. Shows antigenic relationship to the corresponding human immunoglobulin. See text for references to cross-reactions.

b. γ chain and λ chains of feline paraproteins cross-react with corresponding chains of human immunoglobulins.

and Heremans 1968), dolphin, and sea lion origin (Nash and Mach 1971). In addition, heavy chains of the IgG-type molecules of the latter two species as well as cats (Kehoe, Hurwitz, and Capra 1972) are antigenically related to the human γ chain. The basic antigenic similarity among γ chains of mammalian species was clearly demonstrated by Coe (1970), who immunized bullfrogs with rabbit IgG immunoglobulin. These amphibians produced antibodies directed against determinants located on the Fc piece of the rabbit γ chain. Such antiserum detected similar antigens on slow γ globulin of rodents (7Sγ2 of guinea pigs, hamster and *Peromyscus*) and man. It did not react with fast γ immunoglobulins (7Sγ1) or IgM-like antibodies of these species. The frog antiserum was also capable of distinguishing among human IgG subclasses. The best reaction was obtained with human IgG1 and the poorest with IgG4. The fact that IgG molecules of mammals possess a common Fc formation was shown in an indirect fashion by Kronvall and his colleagues (1970), who studied the capacity of the protein A of staphylococci to bind to serum immunoglobulins. This protein binds to the Fc portion of human and rabbit IgG immunoglobulin, but not to the other classes. Sera from the majority of mammals tested, including the echidna, possessed immunoglobulins which bound the A protein.

Table 28 also shows that many mammals possess subclasses of IgG immunoglobulin which are antigenically and electrophoretically distinct. Although the numerical designation corresponds in some cases, it cannot be inferred that the $\gamma$2a class of mouse, for example, is directly homologous to the $\gamma$2a class of the dog or the mink. Further studies to ascertain structural, functional, and immunological relatedness are required to establish such identifications. Although subclasses are generally present if specifically sought, the rabbit is an interesting exception, because present evidence based upon both amino acid sequence and antigenic analysis indicates that this species does not have $\gamma$ chain subclasses. The fact that a single sequence for the Fc piece could be obtained implies that one major subclass, if more than one is present, occurs in such a high proportion that the others constitute negligible fractions of the IgG immunoglobulin.

## Common immunoglobulin antigenic determinants in primates

Since definite antigenic similarities exist among immunoglobulins of man and those of lower mammalian species, a high degree of cross-reaction would be expected in comparisons among the primates. A variety of reagents capable of recognizing human immunoglobulin classes, subclasses, and allotypes have been applied to immunoglobulins of Hominoidea, Cercopithecoidea, and Prosimii (Alepa and Terry 1965; Shuster, Warner, and Fudenberg 1969). The application of reagents used to define the various human markers to nonhuman primates showed that the hominoids resemble man closely in classes, subclasses, and certain allotypes. The results, summarized in Table 29, indicate that the more advanced members of the hominoids are very similar to man in class and subclass antigenic determinants. The monkeys bear less relationship, particularly in the presence of particular IgG subclasses, and the prosimians, represented by the lemur and the galago, are definitely related to man only in $\kappa$ and $\lambda$ chain antigenic determinants. This table shows only the presence of specific immunological reactions; it does not illustrate quantitative differences which would be expected with evolutionary variation within the primate order. Shuster, Warner, and

TABLE 29: Distribution of human immunoglobulin class and subclass antigenic determinants among lower primates.

| Primate | $\varkappa$ | $\lambda$ | $\mu$ | $a$ | $\gamma 2$ | $\gamma 1$ | $\gamma 3$ |
|---|---|---|---|---|---|---|---|
| Chimpanzees (H)[a] | + | + | + | + | + | + | + |
| Orangutans (H) | + | + | + | + | 0 | + | + |
| Gibbons (H) | + | + | 0 | + | 0 | 0 | 0 |
| Baboons (C)[b] | + | + | + | + | 0 | 0 | 0 |
| Rhesus (C) | + | + | + | + | | | 0 |
| Lemur (P)[c] | + | + | 0 | + | | | ± |
| Galago (P) | + | 0 | ± | + | | | 0 |

*Sources:* Alepa and Terry 1965; Alepa 1969.
a. H = Hominids.
b. C = Cercopithecidae (Monkeys).
c. P = Prosimians.

Fudenberg (1969) have carried out quantitative analyses based upon inhibition of the precipitation of radioiodinated antigen by specific antisera. In order to illustrate the quantitative nature of phylogenetic variation in antigenic structure among primates, I have included their results obtained for $\alpha$ and $\lambda$ chains (Figures 64 and 65). All primates, including the prosimians, possess immunoglobulins which are capable of giving a high degree of inhibition of the specific reaction between rabbit antiserum to human $\alpha$ chain and the test $\alpha$ chain. In contrast, only the hominoids possess highly cross-reactive $\lambda$ chains. The $\lambda$ chains of cercopithecoides show a weak to moderate reaction. The lemur alone, of the prosimians tested, gave evidence of possessing a $\lambda$ chain which was antigenically similar to that of man (Figure 64). Light chains of feline IgG paraproteins also cross-react antigenically with human $\lambda$ chain (Kehoe, Hurwitz, and Capra 1972). These results suggest that some immunoglobulin chains have evolved more during primate evolution than others have. $\gamma$ chain antigenic determinants appear to be conserved within the primates and within eutherian mammals of all existing orders (Table 29). The genes encoding the $\lambda$ light chain and the $\mu$ and $\gamma 3$ heavy chains show major divergences even within the hominoids. The $\gamma 3$ antigenic determinant occurs in man, orangutan, chimpanzee and gorilla but is absent in the gibbon and in lower primates. The evolution of the primate $\mu$ chain is more complex, with human and chimpanzee chains being virtually identical. The other hominoids form a group that cross-reacts weakly with human and chimp $\mu$ chain. Interestingly, cercopithecoidean $\mu$ chains show antigenic

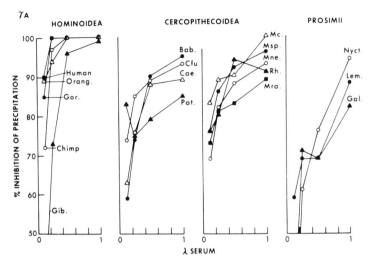

Fig. 64. Cross reactivity of nonhuman primate sera with alpha chain of human IgA immunoglobulin. The degree of cross reaction was assessed by using sera of the various species to inhibit the specific precipitation of human IgA by antiserum directed against the alpha chain. Gor. *Gorilla gorilla*; Gib., gibbon (*Hylobates lar*); Bab., baboon (*Papio doguera*); C fu., *Cercocebus fulginosus*; Cae., *Cercopithecus aethiops*; Pat., *Erythrocebus patas*; Mc., *Macaca cynamologous*; Msp., *Macaca speciosa*; Mne., *Macaca nemestrina*; Rh., *Macaca mulatta* (rhesus) Mra., *Macaca radiata*; Nyct., *Nycticebus coucang;* Lem., *Lemur fulvus*; Gal., *Galago senagalensis braccatus*; Chimp., chimpanzee. (From Shuster, Warner, and Fudenberg 1969).

similarity to the human $\mu$ chain, whereas the prosimians give virtually no reaction. Such pronounced antigenic variability may be correlated with, the emergence of subclasses of $\mu$ chains and quantitative differences in the expression of these in the various species.

The fact that hominoids also share certain human allotypes has been established in comparisons of human populations with chimpanzees (Table 30). The chimpanzees possess Gma which is expressed on the $\gamma$ chain and Gm 13 and 14. Moreover, one-quarter of the chimpanzee population possesses the Inv1 factor. For the purposes of this comparison, the exact structural definition of the $\gamma$ chain allotypic markers (Gm factors) is less important than is their observed distribution among human races and the chimpanzee. These data indicate that the genes encoding immunoglobulin chains continued to evolve during primate evolution, including the development of the races of man.

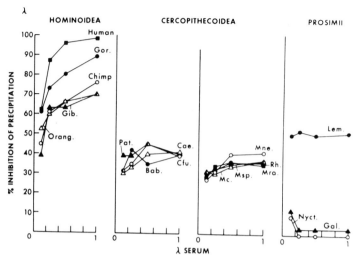

Fig. 65. Cross reactivity of nonhuman primate sera with human lambda light chain. The degree of cross reactivity was assessed by the capacity of the primate sera to inhibit specific combination of human lambda chain with antiserum directed against it. Abbreviations of species are the same as used in Figure 64. (From Shuster, Warner, and Fudenberg 1969).

## *Problems inherent in the construction of phylogenies from immunological data*

The preceding evidence of extensive antigenic cross-reactions among immunoglobulins of mammalian species is consistent with presently available amino acid sequence data indicating substantial homology among immunoglobulins of man, mouse, and rabbit. Although serological techniques provide a relatively simple approach to estimating the degree of relatedness and hopefully sequence homology among proteins of diverse species such as lysozymes (Prager and Wilson 1971) or lens proteins (Manski et al. 1967), the correlation between primary sequence and the formation of discrete antigenic determinants is not necessarily a simple one. Consequently, surprising results may occasionally be observed if limited numbers of antisera are used. Antigenic determinants can have a complex structural basis, as illustrated by conformational determinants, which depend upon the quaternary organization of light and heavy chains (Grey, Mannik, and Kunkel 1965). In addition, a small difference in amino acid sequence can bring about a major alteration in antigenicity, which may lead to paradoxical results. Nisonoff, Reichlin, and Margoliash (1970) observed that certain antisera

TABLE 30: Gm and InV factors in various races of man and in the chimpanzee.

| Population | Gm factors | | | | | | | InV factors | |
|---|---|---|---|---|---|---|---|---|---|
| New Designation | 1 | 2 | 3 | 5 | 6 | 13 | 14 | 1 | 3 |
| Old Designation | a | x | b² | b¹ | c | b³ | b⁴ | 1 | b |
| Caucasian | 57 | 20 | 90 | 90 | 0 | 90 | 90 | 20 | 99 |
| Mongoloid (Japanese) | 100 | 30 | 20 | 20 | 0 | 60 | 20 | 50 | 88 |
| Negroid (African) | 100 | 0 | 0 | 100 | 30-100 | 90 | 95 | 50 | 90 |
| Chimpanzee | 100 | 0 | 0 | 0 | 35 | 100 | 100 | 25 | |

*Source:* Data of Alepa 1969.

directed against human cytochrome c indicated that the human protein is antigenically closer to the corresponding kangaroo protein than it is to that of the rhesus monkey. Man and kangaroo both possess an isoleucine in residue position 58, which contributes predominantly to the antigenic determinant. Rhesus cytochrome c differs from that of man in only one residue position, but this occurs at the critical antigenic site. This protein contains a threonyl residue at position 58. Moreover, present knowledge of molecular biology makes amino acid sequence information a direct means of inferring the arrangement of nucleotide bases within the structural genes encoding the protein. Ideally, the elucidation of the amino acid sequences of the variable and constant regions of immunoglobulin chains of diverse vertebrate species is required to clarify the mechanisms responsible for the origin of V-region diversity and the evolutionary emergence of distinct C-region classes and subclasses.

## Synopsis

Extensive cross-reactions among immunoglobulin chains of eutherian mammals suggest that these proteins comprise a closely related group. Antigenic similarities among γ chains of eutherian mammals have been demonstrated by investigations employing frog antiserum to rabbit IgG immunoglobulin (Coe 1970). Moreover, fragmentary data suggest that immunoglobulin chains of prototherians and marsupials cross-react with those of eutherians. In the former case, γ chain of IgG immunoglobulin of the echidna showed a slight cross-reaction with that of man (Marchalonis and Atwell, unpublished observations). In the latter case, Bell (Bell, R. G., personal communication) observed that light chains of the quokka, a type of wallaby, share antigenic determinants with those of the guinea pig. Additional sharing of antigenic determinants has been reported below the level of mammals; Mehta, Reichlin, and Tomasi (1971) obtained evidence that the μ chain of chicken IgM immunoglobulin cross-reacts with that of man. The antigenic similarities presumably reflect regions of common amino acid sequence, while the fact that immunoglobulins of different species are not antigenically identical indicates that the genetic process of immunoglobulin evolution remains operative among existing mammals.

# 16 Molecules Possibly Related to Immunoglobulins

All vertebrates possess serum antibodies that exhibit structural and functional similarities to the immunoglobulins of mammals. In addition to these molecules, which are classed as immunoglobulins, many vertebrates tested contain proteins that are not obviously related to immunoglobulins but might, nevertheless, share a common ancestry with immunoglobulins. Further complexity has been introduced in studies of comparative immunity by the occasional practice of designating molecules as antibodies on the basis of functional, rather than structural criteria. In this chapter molecules that bind certain antigens but do not resemble immunoglobulins will be discussed, as well as evidence that a number of molecules which lack known antibody activity might belong to an extended immunoglobulin family, i.e., constitute a group of polypeptides homologous to immunoglobulins.

## "Antibodies" which are not immunoglobulins

In addition to antibodies possessing an essentially classical immunoglobulin polypeptide chain structure, many vertebrates, such as the lamprey, nurse shark, eel, and man, possess serum molecules exhibiting strong binding affinities for commonly studied antigens. These molecules have been studied in some detail, and their properties are listed in Table 31. The so-called antibodies of the eel (*Anguilla rostrata*) are naturally-occurring serum proteins that bind to human H(O) erythrocyte antigens (Bezekorovainy, Springer, and Desai 1971). They clearly do not resemble the IgM immunoglobulins of teleost fish in their physical

TABLE 31: Antibody-like proteins which are structurally dissimilar to immunoglobulins.

| Source | Binding Specificity | Molecular Weight (Intact) | Molecular Weight (Subunit) | Affinity Constant |
|---|---|---|---|---|
| See lamprey[1] (*Petromyzan marinus*) | Erythrocyte H (O) antigen | 320,000 (9S) | 75,000 N.C.[a] | N.M.[b] |
| Nurse shark[2] | Fructosans | 280,500 (10.6S) | 55,000–77,000 N.C. | $3.7 \times 10^4$ L/M |
| Eel (*Anguilla rostrata*)[3] | Erythrocyte H (O) antigen | 123,000 (7.2S) | 40,000 N.C. (4 × 10,000) | N.M. |
| Man (*C-reactive protein*)[4] | C-polysaccharide of pneumococcus (N-acetyl-galactosamine phosphate) | 129,000 (7.5S) | 21,500 | $3.0 \times 10^4$ L/M (uridine 5-mono-phosphate at 5°C) |

*Sources:* (1) Litman et al. 1970; (2) Harisdangkul et al. 1972a; (3) Bezkorovainy, Springer, and Desai 1971; (4) Gotschlich and Edelman 1965, 1967.

a. N.C. = noncovalently associated subunits.
b. N.M. = not measured.

properties. The former molecules have a molecular weight of 123,000 and consist of noncovalently bound subunits of mass 40,000. Teleost immunoglobulins, in contrast, exist in disulfide-bonded units of mass 700,000 daltons comprised of light chains (M. W. 22,500) and heavy chains (M. W. 70,000). Both the lamprey (Litman et al. 1970) and the nurse shark (Harisdangkul et al. 1972b) contain serum proteins characterized by a mass of approximately 300,000 daltons that are comprised of smaller units not covalently linked in the intact structure. The nurse-shark protein occurs in the absence of antigenic stimulation and binds to fructosans with an affinity of $3.7 \times 10^4$ liters per mole. The titer of lamprey antibody to human erythrocyte H(O) antigen is increased by immunization, but the protein might be more similar to nonimmunoglobulins, such as the fructosan-binding protein of the shark, rather than to lamprey IgM-like antibodies. In addition to the properties cited, these proteins differ markedly from immunoglobulins in their spectra when analyzed by circular dichroism. The C-reactive proteins of man and other mammals (Gotschlich and Edelman 1965, 1967) bind the C polysaccharide of pneumococcus and were once considered a type of antibody. These proteins bind glucosyl phosphates such as uridine 5-monophosphate with a binding constant of $3 \times 10^4$ 1/M. The intact protein possesses a molecular weight of 129,000, a value similar to that of the eel "antibody," and is composed of subunits of mass 21,500. Baldo and Fletcher (1973) recently reported the presence of a protein in plaice serum that exhibits binding properties similar to those of C-reactive protein. Such proteins have served as valuable tools for immunochemists engaged in unravelling the correlations between carbohydrate structure and blood group antigen specificity, but their place in the evolution of immunity is problematic and their function within the organism largely unknown.

It is necessary to keep an open mind regarding the possibility that such proteins, which are distinct from classical immunoglobulins, might be members of the immunoglobulin family. Recent work of Glenner's group (Glenner et al. 1971), for example, has shown conclusively that certain amyloid proteins consist of fragments of light chains. Furthermore, Peterson et al. (1972) have elucidated the complete primary sequence of an enigmatic serum protein $\beta_2$ microglobulin, finding it to be homologous to the internal duplications which occur within the $\gamma$ heavy chain. Although the evidence for C-reactive protein is not yet so firm, the possibility has been raised that this acute-phase protein is likewise related to immunoglobulin (Marchalonis and Weltman 1971). The carbohydrate-specific binding proteins considered here may prove to be remnants of a primitive recognition system concerned with carbohydrate antigens which are generally present on the surfaces of a variety of cells.

Such proteins probably serve some important function distinct from those mediated by antibodies, because they are present in species such as the shark and man, which possess classical immunoglobulin antibodies.

## Members of the extended immunoglobulin family

A system of proteins such as immunoglobulins, which are encoded by multiple genes undergoing translocation and recombination (Chapter 17), might be expected to generate a variety of proteins in addition to immediately recognizable immunoglobulins. This situation in fact exists, as recent studies of amyloid protein and $\beta_2$ microglobulin have clearly established. Amyloidosis is a pathological condition in which a unique protein, amyloid, is deposited in tissues. The accumulations of such material can reach the point at which the function of the organ is impaired and death may result. Glenner and his co-workers (1971) have shown by isolation of the deposited material and sequence analysis that the amyloid protein consists of fragments of immunoglobulin light chains. $\beta_2$ microglobulin ($\beta_2$M) is a protein of molecular weight 12,000 which appears in low amounts in serum, urine and cerebrospinal fluid of normal individuals. It is present in increased amounts in urine of patients suffering from renal tubular malfunction. In addition, it is found on the surfaces of a variety of cells, including lymphocytes (Evrin and Pertoft 1973) which can synthesize the molecules (Bernier and Fanger 1972). Peterson et al. (1972) have recently shown that this molecule is homologous to the $C\gamma3$ region of human $\gamma1$ heavy chains (27% of the residue positions were identical between these two proteins). It is interesting that this molecule binds to the surface of lymphocytes, because it suggests a possible effector role for the normal $C\gamma3$ domain within the intact $\gamma$ chain. $C\gamma2$ fixes complement (Kehoe and Fougereau 1969), but it has been difficult to ascribe a definite role to the third C-region domain. Similar results have been described for canine $\beta_2$ microglobulin (Smithies and Poulik 1972).

These two proteins have definitely been shown to belong to the extended immunoglobulin families. Other proteins that might share a common ancestral gene with V or C regions are haptoglobins (Black and Dixon 1968), $\alpha_1$-acid glycoproteins (Emura et al. 1971), and C-reactive protein. Table 32 presents a comparison of the amino-terminal sequences of $\alpha_1$-acid glycoprotein (Schmid et al. 1973) with those of the human $\kappa$ light chain subgroups and the $V_H$ sequence of IgG1 protein, Eu. Direct inspection shows that eight of the twenty-four residues of $\alpha_1$-acid glycoprotein ($\alpha_1$AG) tabulated are identical in position to corresponding amino acid residues of the immunoglobulin sequences. The optimal means for estimation of homology among proteins is a comparison of the nucleic acid codons that specify the amino acids; the data listed here provide a minimum estimate. Schmid and his colleagues (1973) found 27% identity in amino acid sequence when comparing the first 43 residues of $\alpha_1$AG with those of an unspecified human $\kappa$-type light chain. However, they report that the extent of homology between the two chains calculated when single point mutations were allowed was 80%.

In Chapter 3 C-reactive protein (CRP) was described as an acute phase protein which binds the C polysaccharide of Pneumococcus. Although the protein has solubility properties which have precluded sequence analysis at this time, comparison of its amino acid composition to that of immunoglobulin chains showed a striking similarity (Marchalonis and Weltman 1971). This molecule may prove to be of considerable evolutionary interest because it bears a number of similarities to invertebrate proteins, such as the hemagglutinin of the horseshoe crab, *Limulus polyphemus*. Both molecules consist of polymers comprised of subunits having a molecular weight of approximately 25,000, bind carbohydrates, and require $Ca^{++}$ for efficient binding (see Chapter 4). It is possible that both represent offshoots of the immunoglobulin line that diverged prior to the time at which V-region diversity was established. A listing of members of the extended immunoglobulin family is given in Table 33. As described above, the evidence for amyloid, $\beta_2$M and $\alpha_1$AG is firmly grounded in amino acid sequence, whereas the evidence for CRP is based upon comparisons of amino acid composition data. A case built largely upon circumstantial evidence of various sorts merely suggests a relationship between histocompatibility antigens and immunoglobulins, but this possibility deserves further attention because many proteins that are associated with cell membranes might share a common ancestry.

## Evolutionary relationships among the immunoglobulin extended family

The preceding data establish that $\beta_2$M is homologous, albeit distantly, to immunoglobulin. Nakamuro, Tanigaki, and Pressman (1973) have recently provided evidence that human $\beta_2$M might be structurally related to a common portion of human histocompatibility (HL-A) antigen. These workers isolated an HL-A fragment of molecular weight 11,000 which resembled $\beta_2$M in amino acid composition as analyzed by the different index method of Metzger et al. (1968), size, electrophoretic mobility, and distribution among the various tissues of the body. Although amino acid sequence data for the HL-A molecule are

TABLE 32: Comparison of N-terminal sequences of immunoglobulins and $a_1$ -acid glycoprotein.

|  | 1 | 2 | 3 | 4 | 5 | 6 | 7 | 8 | 9 | 10 | 11 |
|---|---|---|---|---|---|---|---|---|---|---|---|
| $V_x$ I | Asp | Ile Val | Gln | Met Leu | Thr | Gln | Ser | Pro | Ser | Ser Thr Phe | Leu |
| $V_x$ II | Glu Lys | Ile Met | Val | Leu Met | Thr | Gln | Ser | Pro | Gly Ala | Thr | Leu |
| $V_x$ III | Glu- Asp | Ile | Val | Leu Met | Thr | Gln | Ser Thr | Pro | Leu | Ser | Leu |
| $a_1$ AG | Glu | Ile | Pro | Leu | Cys | Ala | Asn | Leu | Val | Pro | Val |
| $V_H$ | PCA | Val | Gln | Leu | Val | Gln | Ser | Gly | ( ) | Ala | Glu |

required to establish the existence of homology between this protein and $\beta_2$M, the study of Nakamuro and his colleagues raises the exciting possibility that HL-A might be related in an evolutionary sense to immunoglobulin. Burnet (1970a) has stressed a symmetry between the variation of the strong histocompatibility antigens of man and mouse and the diversity of antibody molecules or receptors which recognize various antigenic determinants. Furthermore, the capacity to form antibodies to certain antigens, such as polymers of L-lysine in guinea pigs (Benacerraf, Green, and Paul 1967) and copolymers of tyrosine, glutamic acid, alanine, and lysine [(T,G)-A--L] in mice (Benacerraf and Mc-Devitt 1972), is affected by genes which are linked to genes encoding the major histocompatibility antigens in those species. In Chapter 17 I shall return to the concept of a functional relationship between histocompatibility antigens and immunoglobulins.

If the hypothesis of Nakamuro, Tanigaki, and Pressman (1973) that $\beta_2$M and HL-A are related is true, it follows logically that HL-A and immunoglobulin share common ancestral genes. The problem is to obtain, in the absence of primary sequence data, an estimate of the degree of homology among these three proteins and to assess the relatedness among members of the extended immunoglobulin family considered here. As I discussed in Chapter 5, various statistical means exist for comparing proteins in terms of amino acid composition. Harris and Teller (1973) pointed out that these approaches are independent of the molecular weight of the protein and the order of comparison. Moreover, within families of proteins where homologies have been determined on the basis of analysis of sequence data, the comparisons of amino acid composition data provide a reasonable first approximation to sequence

| 12 | 13 | 14 | 15 | 16 | 17 | 18 | 19 | 20 | 21 | 22 | 23 | 24 |
|----|----|----|----|----|----|----|----|----|----|----|----|----|
| Ser | Ala<br>Val | Ser | Val<br>Leu | Gly | Asp | Arg | Val<br>Ile | Thr | <u>Ile</u> | Thr<br>Ala | Cys | |
| Ser | Leu<br>Met | Ser | Pro | Gly | Glu<br>Asp | Arg | Ala | Thr | Leu | Ser | Cys | |
| <u>Pro</u> | Val | <u>Thr</u> | Pro | Gly | Glu | Pro | Ala | Ser | <u>Ile</u><br>Thr | Thr | Cys | |
| <u>Pro</u> | Ile | <u>Thr</u> | Asn | Ala | Thr | Leu | Asp | Arg<br>Gln | <u>Ile</u> | Thr | Gly | <u>Lys</u> |
| Val | Lys | Lys | Pro | Gly | Ser | Ser | Val | Thr | <u>Ile</u> | Thr | Cys | <u>Lys</u> |

homology. Since amino acid composition data are available for the $\beta_2$M, $\alpha_1$AG, HL-A, and immunoglobulin chains of man, it is feasible to compare them and construct a tentative phylogenetic tree depicting their evolutionary relationships. The comparison used here was performed using the S$\Delta$Q method devised by Marchalonis and Weltman (1971). Application of the other parameters gives parallel results, although the magnitudes of the estimates are different.

Table 34 presents S$\Delta$Q data for comparisons among human HL-A (Nakamuro, Tanigaki, and Pressman 1973), $\beta_2$M (Berggard and Bearn 1968), $\alpha_1$AG (Schmid et al. 1973) and immunoglobulin chains (Crumpton and Wilkinson 1963; Chaplin, Cohen, and Press 1965). A cell-surface proteoglycan of the sponge *Microciona partherna*, termed sponge aggregation factor (Henkart, Humphreys, and Humphreys 1973), is included to provide comparisons with an unrelated surface protein. This molecule serves in species-specific recognition among sponges (Chapter 4). In accordance with the calculation made by Nakamuro and his colleagues, HL-A and $\beta_2$M are quite similar in composition; these two proteins are as closely related by this approach as the immunoglobulin chains are to one another. On the other hand, $\alpha_1$AG and $\beta_2$M are markedly different in amino acid compostion, although the fact that the S$\Delta$Q value was less than 100 suggests that they probably share a common ancestry, since the observed frequency of two unrelated proteins showing an S$\Delta$Q comparison value of less than 100 was less than 5 percent. Immunoglobulins, particularly the light chain and the $\mu$ chain, differ from $\beta_2$M and HL-A by approximately similar values. A dendrogram constructed from the data in Table 34 provides a tentative scheme for the evolutionary relatedness of the proteins considered here

TABLE 33: Members of the extended family of immunoglobulins.

| Protein | Evidence for relatedness |
|---------|--------------------------|
| amyloid substance[1] | Amino acid sequence homology to light chain |
| $\beta_2$-microglobulin[2] | Amino acid sequence homology to $\gamma$ chain |
| haptoglobin[3] | Amino acid sequence homology to light chains |
| $a_1$-acid glycoprotein (an acute phase protein)[4] | Amino acid sequence homology to light and heavy chains |
| C-reactive protein (an acute phase protein)[5] | Similarity in amino acid composition |
| histocompatibility antigens[6] | speculation: possible distant relationship; similarity in amino acid composition |

*Sources:* (1) Glenner et al. 1971; (2) Peterson et al. 1972; Smithies and Poulik 1972; (3) Black and Dixon 1968; (4) Emura et al. 1971; (5) Marchalonis and Weltman 1971; (6) Nakamuro, Tanigaki, and Pressman 1973; Table 34.

TABLE 34: Comparisons among human immunoglobulins and putative members of the extended immunoglobulin family by amino acid composition analysis.[a]

| | HL-A | $a_1$-AG | $\mu$ | | L | $\beta$M | SAF |
|---|------|----------|-------|---|---|----------|-----|
| HL-A[1] | 0 | | | | | | |
| $a_1$-AG[2] | 84 | 0 | | | | | |
| $\mu$[3] | 63 | 54 | 0 | | | | |
| $\gamma$[4] | 79 | 106 | 29 | 0 | | | |
| L[5] | 65 | 77 | 26 | 24 | 0 | | |
| $\beta_2$-M[5] | 21 | 95 | 53 | 97 | 52 | 0 | |
| SAF[b, 6] | 129 | 111 | 121 | 184 | 177 | 170 | 0 |

*Sources:* (1) Data from Nakamuro, Tanigaki, and Pressman 1973; (2) Data from Schmid et al. 1973; (3) Data of Chaplin, Cohen, and Press 1965; (4) Data of Crumpton and Wilkinson 1963; (5) Data of Berggård and Bearn 1969; (6) Henkart, Humphries, and Humphries 1973.

a. Table lists $S_{\Delta}Q$ values for the comparisons indicated.

b. SAF = sponge aggregation factor.

(Figure 66). This interpretation is tentative and requires direct confirmation by primary amino acid sequence analysis. The model predicts that a gene once existed which was a common ancestor of the genes specifying HL-A, immunoglobulin, $\beta_2$M and $\alpha_1$AG. There was a duplication which created one branch leading to HL-A and $\beta_2$M and two other branches leading respectively to immunoglobulin and $\alpha_1$AG. Because of the relatively large S$\Delta$Q values in comparisons of HL-A with $\alpha_1$AG or immunoglobulins with both proteins, the genes encoding these three proteins must have diverged many millions of years ago. Since $\alpha_1$AG and immunoglobulins display about 25% homology in amino acid sequence, we can apply the same sorts of calculations used for hemoglobins (Zuckerkandl 1965) and estimate that the genes encoding them diverged over 300 million years ago.

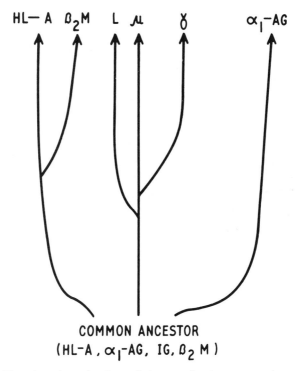

Fig. 66. Tentative scheme for the evolutionary relatedness among human immunoglobulins and other proteins probably homologous to immunoglobulins. L, light chain; $\mu$, mu chain; $\gamma$, gamma chain; HL-A, human histocompatibility antigen (common portion of Nakamuro, Tanigaki, and Pressman 1973); $\beta_2$M, beta$_2$ microglobulin; $\alpha_1$AG, alpha$_1$ acid glycoprotein.

## Evolution of histocompatibility antigens

The histocompatibility antigens of man and mouse are a genetically diverse group of proteins encoded by multiple alleles at two loci (Hildemann 1970; Shreffler et al. 1971; Snell, Cherry, and Demant 1971). It is worthwhile considering the evolution of these genetic loci in some detail because: (1) the preceding data suggest that these molecules might be homologous to immunoglobulins; (2) variation in histocompatibility antigens may be functionally correlated (Burnet, 1970a) or casual (Jerne 1971) in the generation of antibody diversity; (3) certain genes involved in immune responsiveness are linked to genes encoding the major histocompatibility antigens in mice and guinea pigs (Benacerraf and McDevitt 1972). The latter two points will be discussed in Chapter 17 in the contexts of the origin of antibody diversity and cell-surface receptors for antigen. The H-2 system of mice is considered to be homologous to the HL-A histocompatibility antigen system of man, and studies with inbred strains of mice have enabled Shreffler and his coworkers (1971) to propose a model for the origin of the genes encoding H-2 antigens. The scheme for the evolution of the H-2 system of mice described by Shreffler et al. (1971) is given in Figure 67. Initially, one H-2 gene was situated close to an unrelated marker Ss. It is interesting that Ss and another gene closely linked to it (Slp) control the levels of complement in mice (Demant et al. 1973). The H-2 proto-type gene underwent tandem duplication giving two copies, each of which could then mutate independently of the other. A process of inverted duplication and translocation then occurred such that the Ss marker mapped in between two sets of H-2 gene pairs. Individual mutation and recombination occurred within each cistron. The scheme somewhat resembles immunoglobulin evolution, inasmuch as tandem duplication and translocation were involved and multiple alleles exist at the different H-2 loci. However, somatic diversification probably would not occur normally within this system unless the gene products were nonantigenic or mimicked existing self-specificities. Otherwise they would be recognized as nonself and eliminated. As I shall discuss below, a competition between variation of surface antigens and the immune surveillance system within the internal milieu of an individual organism may provide a dynamic driving force for the generation of antibody variation. Jerne (1971) argued that the selective pressure forcing antibody to diversify was the requirement for immunoglobulin to mutate away from the possession of reactivity towards self-histocompatibility antigens. In any case, the histocompatibility antigens and the immunoglobulins may form a functionally linked system which operates to generate diversity in immune recognition. The possibility cannot be discounted that histocompatibility antigens and immuno-

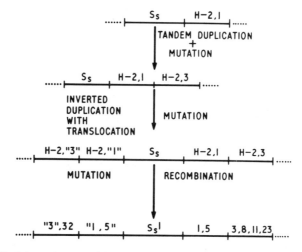

Fig. 67. Scheme for the origin of the genetic loci encoding mouse histocompatibility antigens (H-2) and linked markers (Ss). This system, like that proposed for immunoglobulins, involves complex events such as duplication and translocation. (From Shreffler et al. 1971).

globulins, both cell-surface-associated glycoproteins, share common ancestral genes. This, however, is not a necessary condition for their functional interaction. The formal similarity between the evolution of immunoglobulins and histocompatibility antigens may only reflect the fact that gene duplication is a major mechanism of protein evolution in eukaryotic species (Ingram 1963; Watts and Watts 1968).

## Synopsis

Evidence has been presented in this chapter suggesting that serum proteins or cell-surface-associated proteins that are not obviously similar to immunoglobulins might nevertheless be homologous to the immunoglobulins. Amino acid sequence data show that $\beta_2$ microglobulin, $\alpha_1$-acid glycoprotein and amyloid substance show considerable homology to immunoglobulins. Possible homologies to immunoglobulin were indicated for C-reactive protein and histocompatibility antigens on the basis of similarities in amino acid composition and suggested functional interactions (for histocompatibility antigens). In addition, many vertebrate species possess serum proteins which bind to foreign erythrocytes or carbohydrates but do not exhibit the classical immunoglobulin-type structure. The relationship between these proteins and immunoglobulins remains to be determined.

# 17 Evolutionary Origins of Immunity

The capacity of lymphocytes to recognize foreign antigen and to react in an immune fashion is an integral part of vertebrate development. Although the evolutionary data are presently incomplete, it is possible to draw certain general speculative conclusions regarding the nature of the molecules involved in the recognition of antigen and, furthermore, to outline the events responsible for the emergence and evolution of these molecules. Since these receptors are synthesized and carried by cells, it is also necessary to consider the appearance of the cells involved and the evolutionary variation in the system which they comprise.

## Immunoglobulins of lower vertebrates

Even the most primitive of living vertebrates possess lymphocytes and can form immunoglobulins to a wide variety of antigens. The latter property of immunoglobulins suggested that immunoglobulins of primitive vertebrates might show V-region heterogeneity comparable to that exhibited by mammalian antibodies. Electrophoretic analyses and limited amino acid sequence data of light chains support this conclusion. This observation implies that the mechanism which generates antibody diversity is operative in the most primitive of true vertebrates. It is possible, nevertheless, that certain protochordates from which the immediate ancestors of vertebrates descended might represent a stage at which limited variability exists within the recognition molecules of the immune system. I use the term recognition molecules because species such as tunicates possess lymphocyte-like blood cells (Overton 1966) and exhibit a limited capacity to reject allografts (Oka and Watanabe 1960) but may not have antibody-like molecules in the blood (Tanaka 1973). In any case, we must consider that the immunoglobulins of lower vertebrates, in particular those of elasmobranchs and higher classes, show a strong resemblance to those of mammals in polypeptide chain structure and diversity of these chains. This obvious similarity is reflected in chemical properties such as amino acid compositions of the light and heavy chains (Marchalonis 1972) and in physical properties manifested by circular dichroic spectra (Litman et al. 1971a) and electron microscopy (Feinstein and Munn 1969).

Although the more primitive vertebrates demonstrate a broad range of immunological competence, their immune systems are less complex than those of mammals in organization of lymphoid tissue and in the number of immunoglobulin classes present. Cyclostomes, elasmobranchs, and most Osteichthyes possess a single class of immunoglobulins which most closely resembles the IgM class of mammals. In contrast, man, as a representative mammal, possesses five main classes and certain of these classes can be factored into two or more subclasses. The ancestors of teleost fishes diverged early from the ancestors of higher vertebrates and substantial evolutionary variation occurred within both groups. Such intraclass evolution is reflected in the observations of Trump (1970) and Gitlin, Perricelli, and Gitlin (1973) which imply that IgM subclasses may exist in the teleosts and elasmobranchs.

Immunoglobulin classes distinct from the IgM type exist within the Dipnoi and in all higher vertebrate classes. A low molecular weight heavy chain characterized by mass of approximately 38,000 daltons first appears at the phylogenetic level of Dipnoi, and possible homologues are found in reptiles and birds. Although the exact homology among these chains of the three separate vertebrate classes remains open, I shall tentatively refer to them as IgN type. Amphibians possess at least two types of immunoglobulins; IgM-like molecules are present in all amphibians, while anurans contain an additional class which bears certain similarities to the IgG class in mammals. The precise relationship between the anuran IgG-like class and the true IgG molecules remains questionable because the primitive urodele amphibians possess only IgM immunoglobulins. Either the urodeles lost the capacity to form IgG-like molecules or the genetic event responsible for the emergence of the amphibian $\gamma$-like heavy chain occurred subsequent to the divergence of the anurans and urodeles. If the latter alternative is correct, these molecules cannot be directly homologous to the mammalian $\gamma$ chain. This position is strengthened by repeated observations showing that amphibian $\gamma$-like heavy chains migrate significantly more slowly in polyacrylamide gel electrophoresis than do $\gamma$ chains of mammals. Reptiles and their relatively recent descendants, birds, possess IgM and IgN-type immunoglobulins. The latter molecular species constitutes the predominant immunoglobulin class in turtles and ducks. Chickens, in contrast, have a class of molecules generally referred to as IgG as their major serum component. The identification of this immunoglobulin as IgG has been questioned by Leslie and Clem (1969), who reported that the molecular weight of its heavy chain was 65,000 rather than the mass of 50,000 daltons characteristic of mammalian $\gamma$ chains. In accordance with common usage, I shall refer to this class as IgG(Y). Lebacq-Verheyden, Vaerman, and Heremans (1972) have recently reported the existence of an immunoglobulin resembling IgA in the bile of chickens. The appearance of immunoglobulins possessing clear-cut similarity to the predominant IgG class of eutherian mammals occurs in the prototherian mammals as represented by the echidna. Although the situation has not yet been clarified for metatherians, all eutherian mammals possess a variety of immunoglobulin classes and subclasses, and evidence for the presence of IgM, IgG, IgA, and IgE is often based upon functional, physical and antigenic criteria.

The above immunoglobulin data have been given a molecular interpretation in Figure 68. This schematic diagram is based upon sequence studies which establish that light chains, $\gamma$ chains, and $\mu$ chains consist of internal domains which are homologous to each other because they arose by tandem duplication from a common ancestral gene. This domain arrangement is critical to the function of the particular immunoglobulin class, because each domain possesses a definite function, and emergence of distinct classes appears to be correlated with the addition or deletion of units corresponding to domains (about 110

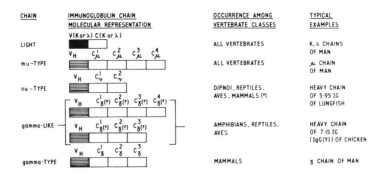

Fig. 68. Molecular interpretation of immunoglobulin chains of vertebrate species. Each unit represents a domain consisting of approximately 110 amino acid residues.

amino acids). Light chains consist of two domains which are $V_\kappa C_\kappa$ for $\kappa$ chains and $V_\lambda$ $C_\lambda$ for $\lambda$ chains. $\gamma$ chains consist of one $V_H$ region and three constant domains $C_\lambda 1$, $C_\lambda 2$, and $C_\lambda 3$ (Edelman and Gall 1969). $V_H$ contains amino acid residues contributing to the antigen-combining site in combination with $V_\lambda$ or $V_\kappa$ in quaternary folding (see Fig. 17, Chapter 2). $C\gamma 1$ interacts with $C_\lambda$ or $C_\kappa$ probably to form a steric structure, which stabilizes the antigen-combining site. The residues which fix complement are located in $C\gamma 2$ (Kehoe and Fougereau 1969) and the function of $C\gamma 3$ is probably to allow the molecules to attach to cell surfaces (Peterson et al. 1972; Smithies and Poulik 1972). The $\mu$ chain possesses an analogous structure with five domains, one variable and four constant (Putnam et al. 1972). The $\nu$ chain is drawn by analogy to the others represented. The molecular scheme illustrated here provides a basis for the discussion of the origin of V-region diversity and its interrelationship to the evolution of C regions, which correlates with the emergence of immunoglobulin classes. Moreover, the domain arrangement provides an explanation for the conservation of quaternary structures (Figure 17) and possibly for similar sequences among immunoglobulins of diverse species. Corresponding domains carry out similar functions, and selective pressures have operated to retain structures consistent with these functions (Marchalonis 1972).

## The origin of antibody diversity

Two approaches provide evidence that the mechanism responsible for the generation of antibody diversity was an ancient phylogenetic development. One line of evidence was discussed above. It consists of the capacity of lower vertebrates to respond to a variety of antigens and the electrophoretic heterogeneity of their immunoglobulins. The second piece of evidence is the paucity of identical residues between variable and constant regions of immunoglobulin chains. Although this homology might be obscured by the presence of V-region variability itself, $V_\kappa$ and $C_\kappa$ regions of the IgG1 myeloma protein Eu differ in over 90% of their residue positions (Edelman and Gall 1969). If such domains are the products of genes that had a common ancestor, their divergence occurred in the distant past. Applying the observed mutation rate for mammalian immunoglobulins (approximately 20 residues per 100 residues per $10^8$ years), the genes encoding $V_\kappa$ and $C_\kappa$ diverged over 400 million years ago. This time corresponds generally to the separation of vertebrates and invertebrates, and therefore V-region diversity would constitute an integral aspect of vertebrate immunity.

Antibody diversity is a basic facet of immunity, and it throws into sharp relief other key attributes of the immune system. This problem has recently been reviewed at great length (Hood and Talmage 1970; Gally and Edelman 1972; Smith 1973). Basically, the immune system is faced with the problem of recognizing and reacting to an enormous collection of foreign structures. This recognition is carried out by molecules which are antibodies associated with the lymphocyte plasma membrane. Two classes of theories were proposed to account for this capacity. The first type, known as instructive theories, states that the antigen provides special information to the cells which direct the synthesis of molecules specifically complementary to it. The clearest presentation of an instructive hypothesis was that of Pauling (1940), who argued that the antigen interferes with folding of the newly-synthesized antibody so that the protein molecule would be sterically complementary to it. This hypothesis predicts that: (a) all antibodies would possess the same amino acid sequence, (b) antigen must be present in all antibody-forming cells, and (c) a given cell can make antibodies of any specificity. Prediction a clearly was not confirmed. Predictions b and c also were inconsistent with recent data, particularly that of Nossal and his colleagues, who provided evidence that antibody-forming plasma cells contain fewer than 30 molecules of antigen (Nossal and Ada 1971) and that one antibody-forming cell makes antibodies of one specificity (Nossal et al. 1964). One cell can produce antibodies of different classes, however, as discussed in regard to the IgM to IgG switch in Chapter 3.

The second class of explanations is termed selective. It argues that the information for synthesis of a given antibody is present in the genome of a cell, and that contact of the antigen with the cell induces production of that antibody. In principle, the selection by antigen can operate at two levels, namely at a clonal level or at an internal (subcellular) level. The second method of selection assumes that each cell possesses the genetic information to synthesize the entire range of antibodies, and the one it synthesizes is merely a reflection of the particular antigen which happens to impinge upon it at a given time when the particular receptor is exposed. Although this theory has not been conclusively disproven, it has two serious drawbacks which render it unlikely. In the first place, the method of recognition of antigen and regulation of antibody synthesis must be complicated. Secondly, this approach does not readily explain the fact that one cell makes only one antibody.

During the course of this text, I have described various aspects of the sequences of antibodies and will not reiterate them here. When such information is required in reference to the present discussion, the relevant data are available in preceding chapters which deal with the nature of the variable regions of immunoglobulins. A complete theory must account for (a) the selective nature of antibody production and (b) the existence and properties of variable regions. The amino acid substitutions which occur in V regions derive chiefly from single base changes in the nucleotide triplet codons. This event, which is the standard mechanism of evolutionary variation in hemoglobins and cytochromes, is responsible for over 90% of intrasubgroup interchanges in light chain V regions. Moreover, not every position varies. Cysteines, for example, tend to be invariant within subgroups from myeloma protein to myeloma protein. In addition, some positions may vary more than others. The latter considerations imply that selection plays an important role in this mechanism. Both mutation and selection are operative in the generation of V-region diversity, as they are in the evolution of proteins.

The amino acid sequence data establish that many V-region sequences can be associated with one constant region sequence. In the case of human $\kappa$ chains, there exist three $V_\kappa$ subgroups or basic sequences which cannot be converted readily into one another, but only one $C_\kappa$ region. In practice, one hundred distinct V-region sequences have been found for $\kappa$ chains, although the C regions differed only at position 191, depending on the Inv allotype. Analogous situations obtain for $\lambda$ chains and the heavy chain genetic system. If separate genes exist for V and C regions, it is legitimate to ask at what level of protein synthesis they are joined. According to the central dogma of molecular biology, the information encoded in nuclear genetic material, DNA, is transcribed into messenger RNA, which acts as the template upon which amino acids are incorporated into protein. The process of protein synthesis according to the mRNA template is termed translation. In principle, the joining of V and C genes, or their products, could occur at any level of this scheme, but a wealth of present evidence suggests these two genes are joined at the DNA level. These data can be summarized as follows: (a) immunoglobulin is synthesized as a single unit, growing from the amino terminal to the carboxyl terminal (Knopf, Parkhouse and Lennox 1967); (b) the size of the polyribosomes which synthesize light chains and heavy chains is consistent with the presence of mRNAs of the proper size to code for intact light or heavy chains (Scharff 1967); (c) immunoglobulins made by myeloma tumors are often stable for many cell generations (Harris, A. W. personal communication), which suggests that there is a firm union of V and C region genetic information at the germ-line level; and (d) certain deletions occur among myeloma heavy chains, which begin in the V region and overlap into the C region. A myeloma $\gamma$ chain described by Connell et al. (1970) and Smithies et al. (1970), proceeds normally from the N terminus to residue 19, and then has a deletion encompassing the rest of the V region and extending to residue 217. If the genes encoding V and C regions were not linked at the DNA level, this deletion would have had to occur independently in separate V and C genes. Points c and d strongly suggest that V and C genes form a contiguous stretch of DNA. The first two points imply only that the union of V and C gene products does not occur at the translation level. However, no advantage is to be gained if the union occurs at the mRNA stage, so the logical conclusion is to consider that light chains and heavy chains synthesized by a given cell must be read from complete genes comprised of V and C genes joined at the DNA level.

The preceding points of clonal selection and the fact that multiple V genes must be joined to one of a lesser number of C genes implies that a special mechanism such as translocation (Lennox and Cohn 1967; Gally and Edelman 1970, 1972) must occur. Consider for example, a tandem array of V genes $V_1$ ——$V_n$ and C region $C_a$. In order for a polypeptide of $V_j C_a$ to be synthesized, the genes for $V_j$ and $C_a$ must be contiguously juxtaposed so that messenger RNA could be read from the complete gene. Since $V_j$, most probably, would not be next to $C_a$, it would have to be translocated to the proper position. Moreover, if a cell could make only those immunoglobulin polypeptides for which it had complete genes, the random translocation of a particular V gene to the constant gene would give it the capacity to make only one antibody. Taken on a population basis, this translocation mechanism would result in clonal selection, a concept introduced in Chapter 3. Furthermore, the switch from IgM to IgG antibodies of the same specificity would be allowed because $V_H$ is shared by all heavy chain types and there is no a priori reason to suppose that the translocation mechanism would cease to function after a cell was stimulated. An alternative process, resulting in the proper alignment of a particular V gene with the C gene, was proposed by Smithies (1970). He proposed that a branched network of V genes, each attached to the C gene, exists. During RNA synthesis, the RNA polymerase can make a choice at a given branch point which V gene will be transcribed. After transcription of the V gene, the RNA polymerase goes to read the C gene. The implications for clonal selection and differentiation are similar to those imparted by translocation.

How then can variation occur in V regions? One mechanism which has been proposed is a translational model which presupposes the misreading of messenger RNA by transfer RNAs (Mach, Koblet, and Gross 1967). However, present evidence is strongly against such a proposal (Lennox and Cohn 1967). This is summarized as follows: (a) light chain made from mouse mRNA by a rabbit reticulocyte system is still the same light chain (Stavnezer and Huang 1971); (b) a translational mechanism could not give predictable interchanges; i.e., it could not account for the fact that the majority of amino acid interchanges are the result of single nucleotide base changes (Lennox and Cohn 1967); (c) the regulation of such a system would be extremely difficult.

Two main types of theories are presently considered. One is based on the premise of the existence of a large number of germ-line genes coding for V region. The other proposes that a small number of V-region genes are carried in the germ line, but that this group is expanded by somatic events. Both explanations incorporate translocation to explain the clonal restriction of antibody snythesis. Figure 69 presents schematic diagrams of these concepts. The germ line multiple-gene hypothesis contends that large numbers of V genes are inherited. The evidence in favor of multiple genes comes from the existence of subgroups which represent a total of over 10 genes, and the fact that comparisons of V-region sequences show that the interchanges involved resemble those produced by classical mutation and selection. Although a cell might carry $10^4$ light chain V genes and $10^4$ heavy chain V regions, giving the possibility of $(10^4)(10^4) = 10^8$ distinct antibody combining sites, this would constitute only 0.2% of human genetic material (Hood and Talmage 1970). Furthermore, translocation of $V_j$ to $C_a$ would guarantee clonal restriction and allelic exclusion of antibodies expressed by a given lymphocyte. Such a multiple gene mechanism is feasible but suffers certain weaknesses because it cannot account for divergence and selection of V genes. Since V genes would be greater than 90% identical and antigen selection cannot act at the germ-line level, the $10^4$ V genes should drift until only one remains. Another drawback is that inelegant explanations must be offered to account for the existence of residues in amino acid sequence which are characteristic of a particular species.

Somatic hypotheses maintain that a few (less than 100) germ-line V genes are expanded during ontogeny by somatic mutation (Smithies 1967; Lennox and Cohn 1967; Cohn et al. 1974) and/or recombination (Gally and Edelman 1970). Such explanations circumvent the problem of genetic drift and readily explain the observed patterns of sequence variability. A weakness of a strictly mutational model is that an abnormally high mutation rate might be required to account for the degree of variability encountered. Another somatic theory proposes that somatic recombination occurs among a set of tandemly arranged inherited V genes which differ from each other in a small number of point mutations (Gally and Edelman 1970). Recombination would be favored by the degree of homology among the members of the array. Translocation, which most probably is required to ensure formation of an active V-C genetic unit, would also function in the generation of new sequences. The existence of tandem arrays of V genes is likely, since heavy chain constant region genes (Natvig, Kunkel, and Litwin 1967;

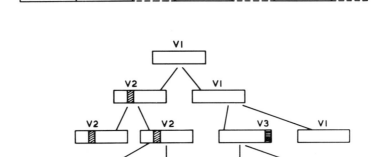

Fig. 69. Two possible genetic mechanisms for the generation of antibody diversity. (A) Multiple variable region genes in the germ line. The particular one which is expressed by a given lymphocyte is the result of differentiation. (B) Somatic mutation. Diversity of variable region genes occurs because the small number of inherited germ line genes can undergo mutation during the course of division of lymphocyte precursors.

Natvig and Kunkel 1973) and hemoglobin nonalpha chains of man (Ingram 1963) form such genetic complexes. As stated in Chapter 3, the translocon arrangement proposed by Gally and Edelman (1972) is an array of tandemly linked V genes and the constant region genes with which they are associated. Cell somatic mechanisms are wasteful because large numbers of cells must be generated under conditions in which only an extremely small fraction might be used. The body can, however, tolerate such expenditures. For example, Metcalf and Moore (1971) have estimated that more than 95% of the lymphocytes generated by division within the thymus die within that organ.

None of the present models are perfect, but some are more attractive than others. Somatic models are favored by recent studies aimed at estimating the number of related sequences which were present in the family of DNA molecules which hybridized with messenger RNA for immunoglobulin (Delovich and Baglioni 1973; Tonegawa et al. 1974). These workers concluded that 20-30 or fewer copies of the DNA were present, rather than the $10^4$ predicted for germ line theories. In contrast, Premkumar, Shoyab, and Williamson (1974) arrived at an estimate of approximately 5,000 copies of V-region genes using hybridization approaches. Wasserman, Kehoe, and Capra (1974) argue that the presence

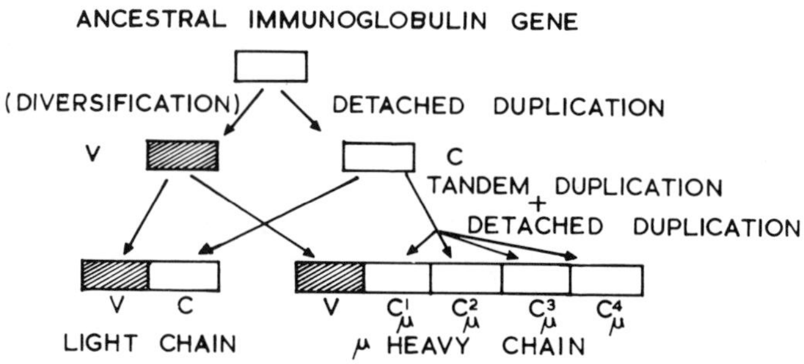

Fig. 70. Genetic scheme for the origin of vertebrate immunoglobulins. The events depicted here occurred below the phylogenetic level of cyclostomes.

is required to cause cell lysis. The detached duplication which resulted in formation of C-region genes may have been a means of removing recognition units from the cell surface and allowing them to function in solution, where they would first combine with antigen and then mediate in the inactivation and/or destruction of the antigen. As effector functions of the molecule distinct from antigen binding (e.g., complement fixation) arose, additional domains were added to carry out these activities. The creation of light chains, possibly of both $\kappa$ and $\lambda$ types, and $\mu$ heavy chains by fusion of V and C genes and tandem duplication of constant homology regions occurred over 300 million years ago, preceding the divergence of the ancestors of elasmobranchs and mammals. The common evolutionary origins of $V_H$, $V_\kappa$ and $V_\lambda$ is shown by the presence of a conserved pentapeptide clearly present in $V_\kappa$ and $V_\lambda$ and showing pronounced similarities to certain $V_H$ peptides (Stanton et al. 1974). Other immunoglobulin classes probably arose by detached duplication, coupled with deletion of homology units, of the gene encoding the constant segment of the $\mu$ chain.

A composite diagram mapping the emergence of distinct V and C genes upon the geologic time is given in Figure 71. This scheme incorporates the following observations: (a) divergence of V and C genes occurred at a time comparable to or preceding the origin of vertebrates; (b) $\kappa$ and $\lambda$ light chains and $\mu$-type heavy chains existed in the lower fishes; (c) the first heavy polypeptide chain to diverge from the $\mu$ chain was the $v$ chain, which was present in higher bony fishes; (d) the $\gamma$-like chain of anuran amphibians may represent the results of a gene duplication independent of that which produced the $\gamma$ chain of mammals; (e) the $\alpha$ chain of avian species is directly homologous to the $\alpha$ chains of mammals (although recent evidence on the partial sequence of human

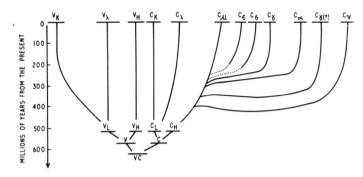

Fig. 71. Hypothetical scheme for the evolution of cistrons encoding immunoglobulin polypeptides. This model assumes a common ancestry for V
genes and C genes and the early development of a means of joining
these genes at the DNA level. Heavy chains of distinct classes are
represented as emerging by detached duplication from the gene specifying the $\mu$ chain constant region. Subclasses, which represent recent
duplications within a particular class, are not shown in this
diagram.

$\alpha$ chain suggests that $\alpha$ chain might be of mammalian origin (Chuang,
Capra, and Kehoe 1973); and (f) the gene encoding the true $\gamma$ chain of
higher mammals arose in the therapsid ancestors of monotremes, marsupials, and eutherian mammals.

A variety of more recent events which continue the mechanisms of
tandem duplication resulted in the appearance of mammalian immunoglobulin subclasses. The IgG1 subclass of human immunoglobulins
shows pronounced similarity to the $\gamma$ chain of rabbit, thereby indicating
that the genes involved diverged subsequent to the emergence of mammals. The appearance of human subclasses, however, probably occurred
late within the evolution of the primates. The arrangement illustrated
here is a tentative model, which is subject to alterations and expansions
as amino acid sequence data for immunoglobulins of lower species
become available.

An important general observation of the evolution of the immune
system is that primitive developments were not discarded with increasing
phylogenetic complexity, although they are sometimes difficult to
recognize in higher species. The IgM immunoglobulins are present in all
vertebrates studied, and their heavy chains show a high degree of
similarity. Substantial conservation of light chain structure is also observed. Even the IgN molecule of the lungfish may be present in reptiles,
birds, and mammals. The emergence of distinct immunoglobulin classes
by duplication of heavy chain C-region genes probably correlates with
the presence of selective pressures which could better be answered by

immunoglobulins distinct from IgM. Low molecular weight immuno-globulins could, for example, distribute between vascular and extra-vascular spaces of the body, whereas high molecular weight IgM would be restricted to vascular regions. The evolutionary pressures which necessitated the proliferation of immunoglobulin classes and subclasses remain problematic. A major selective force in maintaining the primitive IgM immunoglobulin may be deduced from recent observations that this immunoglobulin probably functions as the lymphocyte surface receptor for antigen (Greaves and Hogg 1971; Marchalonis and Cone 1973a; Marchalonis 1974b; Warner 1974) on both B and T lymphocytes.

## *The lymphocyte receptor problem in evolutionary perspective*

According to (John) Northrop, there is a common pattern in all con-troversies. . . . "There is a complicated hypothesis, which usually entails an element of mystery and several unnecessary assumptions. This is opposed by a more simple explanation, which contains no unnecessary assumptions. The complicated one is always the popular one at first, but the simpler one, as a rule, eventually is found to be correct. This process frequently requires 10 to 20 years. The reason for this long time lag was explained by Max Planck. He remarked that 'scientists never change their minds, but eventually die.'" (Florkin and Stotz 1972).

Some sort of capacity to distinguish self from nonself is a characteris-tic of all animal species. In primitive multicellular animals, such as sponges, recognition appears to be dependent on cell surface glyco-proteins of unique structure. It would be fanciful to ascribe any sort of immunological basis to this recognition. Cell-cell recognition in many cases might be explained by enzyme-substrate interactions (Roth, McGuire, and Roseman 1971) at the cell surface or by interaction of surface glycoproteins totally distinct from immunoglobulins or their precursors. Nevertheless, vertebrate immunological mechanisms may well have evolved out of such nonimmunologic surface recognition phenomena. Furthermore, the distinct possibility exists that vertebrate lymphocytes retain the capacity to demonstrate recognition responses more primitive than immune recognition, differentiation, and effector function (Hildemann 1974). Lymphocytes carrying out functions at-tributed to T lymphocytes and B lymphocytes of mammals occur in all vertebrates investigated. The fact that these animals form antibodies per se demonstrates the presence of B-type cells.

Recognition and reaction phenomena indicative of T-cell functions occur in all vertebrates, and also protochordates and echinoderms

(Hildemann 1974; Hildemann and Reddy 1973; Hildemann, Dix, and Collins 1974). Antigen-specific immune collaboration among lymphocytes, resembling T-B collaboration in mice, has been reported in studies which suggest a carrier effect in the secondary responses of cyclostomes, teleosts, and amphibians (Cohen forthcoming; Ruben 1975). Investigations using newts (Ruben 1975) provide the best evidence for collaboration between carrier-specific helper cells and hapten-specific antibody-secreting cells. The discreteness of the functional lymphocyte populations and the thymic origin of the helper cells have not been formally established. The importance of the thymus in cellular immunity and antibody formation to some antigens has been clearly shown in amphibians (Cohen forthcoming) by the classical means of removing this organ early in the life of the animal and then observing a decreased level of response.

The existence of other responses of T cells has been established in lower species. Peripheral blood lymphocytes of teleosts and elasmobranchs respond to certain phytomitogens (Lopez, Sigel, and Lee 1974). In addition, peripheral blood cells of the amphibians *Bufo marinus* (Cohen 1971) and *X. laevis* give *in vitro* mixed lymphocyte reactions. The former blastogenic response is clearly not immunological but serves as a preliminary index of T-cell function. The MLR responses also might not represent a true immunological reaction, but a more primitive recognition system analogous to the hyperplastic contact reactions seen in coelenterates (Hildemann and Reddy 1973; Hildemann 1974). Such reactions are evoked upon admixture of two populations of viable, antigenically disparate cells of the same species, but presensitization is unnecessary and the system lacks a memory component. Furthermore, such a response can be given by a hematopoietic stem cell. It is noteworthy that the genes conditioning MLR are linked to genes encoding histocompatibility antigens in mice. These observations lead to an important generalization regarding T-cell function. T cells can carry out at least two types of recognition roles: (a) an antigen specific function, where the recognition factor is immunoglobulin, e.g., helper T cells (Marchalonis and Cone 1973a) and (b) blastogenic responses to certain alloantigens, where the receptor molecule is not known (Crone, Koch, and Simonsen 1972) and the process of proliferation does not necessarily result in the generation of immune effector cells, e.g., MLR and graft versus host (GVHR). Figure 72 presents a scheme illustrating the evolution of lymphocytes carrying out general nonimmune recognition (e.g., lectin binding), quasi-immune T cell mediated recognition (e.g., MLR), specific T cell mediated recognition (e.g., delayed type hypersensitivity or rejection of strong histocompatibility antigens), and B-cell antibody production. As depicted here, the general recognition

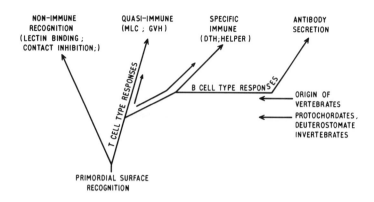

Fig. 72. Schematic model illustrating the evolutionary emergence of cells performing immune and nonimmune recognition and effector function. The model is based upon the presumption that all lymphocytes had a common evolutionary precursor.

and quasi-immune cell mediated immunity might have preceded the emergence of vertebrates and the appearance of B cells.

One of the main concepts permeating this text is that antibodies, i.e., immunoglobulins, function as the lymphocyte receptor for antigen. This hypothesis was first proposed by Ehrlich in 1900 and is now widely accepted for cells of the B-lymphocyte series (Marchalonis 1974b, 1975; Warner 1974). Furthermore, it is well documented in mammalian (Marchalonis 1974b, 1974c; Warner 1974) and avian (Theis and Thorbecke 1973; Rouse and Warner 1972; Szenberg, Cone, and Marchalonis 1974) systems that this generalization also applies to at least some T cells. The particular type of immunoglobulin considered to constitute the B-cell surface receptor resembles the 7S subunit of IgM (Marchalonis and Cone 1973a; Vitetta, and Uhr 1973). The antigen receptor story on T cells is not so clear because it is difficult to demonstrate surface Ig on mammalian T cells by some techniques, and not all T-cell functions, notably MLR and GVHR, were reliably inhibited by treatment with antisera to Ig (Crone, Koch, and Simonsen 1972). Surface IgM-like immunoglobulins of T cells of chickens (Szenberg, Cone, and Marchalonis 1974) and various mammals (Marchalonis, Cone, and Atwell 1972; Moroz and Hahn 1973; Moroz and Lahat 1974; Ladoulis et al. 1974; Boylston and Mowbray 1973, 1974; Rieber and Riethmuller 1974; Smith et al. 1975) have now been isolated, and antigen binding (Cone et al. 1972; Cone and Marchalonis 1973; Feldmann, Cone, and Marchalonis 1973; Rollinghoff et al. 1973) and collaborative function (Feldmann and Nossal 1973; Feldmann, Cone, and Marchalonis 1973) have been demonstrated for 7S IgM of specifically activated T-cell populations. Phylogenetic evidence is beginning to appear which supports the presence and putative receptor role of IgM immunoglobulin on lymphocytes of vertebrate species ranging from fish to amphibians. Surface immunoglobulin (IgM) has been detected by binding of fluorescein-labeled antisera on thymus cells of larval *Xenopus* (Du Pasquier et al. 1972), thymus cells of trout (Chiller 1974) and carp (Emmrich, Richter, and Ambrosius 1975) and all peripheral blood lymphocytes (presumably a mixture of T- and B-type cells) of sharks (Clem, L. W., personal communication). Moreover, thymus lymphocytes as well as spleen lymphocytes of *Xenopus* and the two teleost fish were reported to synthesize their surface IgM immunoglobulin. It should be emphasized that recent studies (Marchalonis 1975) show that the heavy chain of T-cell immunoglobulins of various species is not identical to the $\mu$ chain of serum IgM or B cells.

Antigen-binding lymphocytes have been observed in forms as primitive as teleosts (Diener 1970) and amphibians (Diener and Marchalonis 1970; Cone and Marchalonis 1972; Ruben, van der Hoven, and Dutton 1973). In experiments using antiglobulins to determine whether immunoglobulins might be involved in the specific combination of cells with antigen, it was found that antiserum to *B. marinus* IgM completely inhibited the binding to horse erythrocytes of lymphocytes from toads immunized with this antigen. Moreover, putative helper rosettes of *R. pipiens*, which appear well before detectable circulating antibody, were completely inhibited by rabbit antiserum to IgM of this frog (Edwards et al. 1975). These observations concur with reports that specific antigen binding by both mammalian B and T cells is inhibited by antisera to certain immunoglobulins (Ashman and Raff 1973; Bach and Dardenne 1972; Dwyer, Warner, and McKay 1972; Hogg and Greaves 1972; Marchalonis et al. 1973; Marchalonis 1974b; Warner 1974). Results with amphibians further emphasize the dichotomy of T-cell response outlined in Figure 72. Although thymus lymphocytes express IgM and specific rosette formation, including helper rosettes, is completely suppressed by antisera to this protein, the MLR reaction of *Xenopus* is not inhibited by these antisera (Du Pasquier 1973). The preceding data are consistent with the proposal that surface immunoglobulin, especially the primitive IgM immunoglobulin, serves as the surface receptor for antigen on B cells and T cells of all vertebrate species. I would emphasize, however, that inhibition of antigen binding by antisera specific for immunoglobulins cannot conclusively establish the receptor hypothesis, because the blocking might result from simple steric hindrance if immunoglobulin were located close to some yet unknown receptor molecule. Alternative explanations of inhibition by such reagents also can be proposed; however, they are not germane here.

A salient conclusion linking receptor studies and evolutionary data is that the primitive immunoglobulin in evolution, IgM, serves as the receptor for antigen on the surfaces of mammalian lymphocytes. This observation implies that recognition of antigen by cell-bound immunoglobulin was one of the original functions of the IgM molecules when it emerged early in vertebrate evolution, and that this function was retained despite the appearance of a variety of immunoglobulin classes distinct from IgM antibodies.

The nature of the receptor in the allogeneic reactions I referred to as quasi-immune (Hildemann and Reddy 1973; Lafferty 1973) in Figure 72 remains to be determined. The successes of Warner and his colleagues (Mason and Warner 1970; Rouse and Warner 1972) and Riethmuller, Rieber, and Seeger (1971) in blocking GVHR with antisera to light chains establish the presence of immunoglobulin on the reacting cells. For reasons described in the preceding paragraph, it does not demonstrate that immunoglobulin is the actual receptor. It has been vehemently argued that the recognition system in the mixed lymphocyte reaction cannot be immunoglobulin (Crone, Koch, and Simonsen 1972). This type of cellular reaction is especially important from a comparative point of view because cellular phenomena resembling allograft rejection have been reported in species such as annelid worms (Cooper 1970) and coelenterates (Hildemann and Reddy 1973), which lack immunoglobulins. Arguments based upon studies of inheritance patterns of immune responsiveness in mice and guinea pigs have raised the possibility that a set of recognition molecules, alternative to immunoglobulin, exists on T lymphocytes of mammals. McDevitt (McDevitt and Landy 1972) has hypothesized that this recognition system, which functions in the graft versus host reaction and the mixed lymphocyte reaction, represents a set of receptors more primitive than immunoglobulin. It is puzzling from a phylogenetic perspective that vertebrates more primitive than amphibians do not give an *in vitro* mixed lymphocyte reaction (Cohen 1975; and Chapter 11), but unequivocally reject allografts and synthesize IgM-like antibodies. Negative findings of course must be considered with some skepticism because improved experimental procedures or removal of inhibitory factors might disclose MLR in primitive species.

Two types of molecules, namely histocompatibility antigens and products of immune response genes linked to H-antigen genes, have been proposed as T-lymphocyte nonimmunoglobulin receptor molecules. It must also be remembered that T lymphocytes possess a number of surface molecules which initiate diverse physiological responses. Examples of such molecules are receptors for insulin (Gavin, Buell, and

Roth 1972) and glycoproteins that bind lectins. It now appears that certain immune response gene products (Ia antigens) are expressed primarily on B lymphocytes (Hämmerling et al. 1974; McDevitt et al. 1974; Goding et al. 1975) where they may serve, among other things, as stimulator molecules in MLR (Boehmer and Sprent 1974; Cheers and Sprent 1972). Moreover, Katz and Benacerraf (1974) have proposed that such molecules play a vital role in T-B collaboration by serving as B-cell receptors for a distinctive molecule on specifically activated T cells. If they are present on T cells, such immune response gene products might be sterically close to the immunoglobulin receptor and allow conjoint recognition of complex macromolecular determinants (Warner 1974) or serve in the process in which T cells, having recognized antigen, become effective helper cells (Marchalonis, Morris, and Harris 1974). Propinquity between T-cell antigen receptors and H antigens is suggested by experiments in which specific combination of T cells with antigen was blocked by antisera to these antigens (Hämmerling and McDevitt 1974; McKenzie 1974; Basten et al. 1971; Basten, Miller, and Abraham 1975) as well as antisera to immunoglobulins. This interpretation is not conclusive because of the inadequacies of inhibition approaches discussed above. Moreover, antisera to the Θ (Thy 1) antigen will inhibit binding T cells in some cases (Hämmerling and McDevitt 1974; McKenzie 1975), and antisera to H-2 antigens have been reported to inhibit the binding of antigen by B cells. H antigens and Ia molecules represent two physically distinct (H antigen M. W. 45,000; Ia M. W. 35,000) membrane glycoproteins (Cullen et al. 1974) that are specified by closely linked genes. They provide definite markers of biological individuality and may be involved in the mediation of primitive recognition and differentiation responses of certain sets of T lymphocytes. Detailed investigations of the allogenic reactions in protochordates and certain invertebrates should provide information on the surface markers which serve as receptors and as stimulators in quasi-immune responses. This information should elucidate further the evolutionary relationships among immunoglobulins, H antigens, and other members of the extended immunoglobulin family.

## The origin of adaptive immunity

The preceding information leads us to the crucial problem of determining why adaptive immunity emerged. We know that this phenomenon arose within the vertebrates or their immediate ancestors and is correlated with the presence of lymphocytes, a thymus, immunoglobulins, and the capacity to reject allografts. A variety of environmental factors can be proposed as candidates for the selective forces which prompted the genesis of vertebrate immunity. Foremost among these is resistance to infectious diseases. However, invertebrates are subject to a variety of viral and bacterial pathogens and metazoan parasites, but have not developed clear-cut immunity. Generally, they handle invading organisms by phagocytosis and encapsulation (Salt 1963). Another factor that could act as a stimulus in the generation of the type of diversity characteristic of vertebrate immunity is a trend toward malignancy correlated with heritable genetic changes in somatic cells. If the possibility existed that cells would mutate to a neoplastic state during somatic division, it was an evolutionary necessity that some mechanism should exist for eliminating or inactivating such potentially dangerous mutant cells. In the case that the transformation to malignancy correlated with changes in surface antigens on these cells, the mechanism of elimination might be one based upon immunological surveillance (Thomas 1959; Burnet 1970b). Basically, allograft rejection is not a natural situation, because skin or organ transplants require the intervention of human agencies. The closest natural situation to an allograft is that in the development of a mammalian fetus bearing alloantigens of paternal origin that are distinct from those of the mother (Beer and Billingham 1971). Mutation of genes encoding cell surface antigens might well occur somatically, and the immune system would have to be able to generate sufficient recognition potential to keep abreast of the mutating potentially malignant cells. Although the histocompatibility antigen system can be exceedingly complex, consistent with the definition of antigenic individuality, somatic variation in this system would not usually be tolerated. Surface molecules that have undergone such mutation would be recognized as antigenic. Moreover, the new surface configurations could be correlated with malignant transformation because the aberrant clone of cells might not be subject to normal recognition and regulatory processes such as contact inhibition. In contrast, somatic variation of antibodies would be relatively unhindered because V regions are generally not antigenic. Two competing systems would thus be generated. The first is a group of cell surface proteins in which variants arising by somatic mutation must be eliminated; the second comprises a surveillance system that must possess the capacity to recognize nonself variants of the former. The pressures of the immune recognition system would force the mutants in the histocompatibility system to resemble the normal self-configurations because major deviations from self would be readily recognized as foreign. If the minor variants of the histocompatibility antigens were still possessed by malignant cells, the immune system would be driven to recognize slight differences in molecular structure.

This concept can be generalized to incorporate resistance to infectious diseases and to parasites. Such an expansion is justifiable because bacteria, parasites, and mammalian tissues share common surface antigenic determinants (Zabriskie 1967; Rapaport 1972). Moreover, the form of antigenic variability (Beale and Wilkinson 1961) and host immune response to certain parasitic infections follows that predicted from the above. In essence, large antigenic differences would elicit a strong immune response that would virtually eliminate parasites bearing such antigens. A residual population bearing less obvious antigens could then flourish until an immune reaction was evoked to it. This process would then continue in terms of oscillations between superiority of parasite or immune system until an acceptable balance was attained. Again, as in the malignant transformation situation, the parasite would be driven to mimic the host in antigenic structure either by variation in genetic expression (Damian 1964; Sprent 1962) or by protective incorporation of host antigens (e.g., serum proteins) into its surface (Damian 1967). Brown (1971) has provided evidence that malaria parasites, *Plasmodia*, survive within the hostile immunological environment of the host by frequently changing the expression of certain antigens. *Plasmodia* are antigenically complex, and as many as 30 distinct antigens have been observed (Brown, Brown, and Hills 1968; McGregor 1972). The frequency of antigenic variation within populations of such protozoa in the presence of variant specific antibodies is striking. Antigenic variation has been found in populations consisting of as few as $10^3$ *Plasmodia* (Brown, Brown, and Hills 1968; Brown 1971). Although the mechanism of this antigenic variability is not known, this variation would certainly provide a continually evolving set of new antigens to test the immune capacity of the infected animal.

Recent studies indicate that immunity to allografts (World Health Organization 1969a) and to protozoan and metazoan (Dineen and Adams 1971; Ogilvie and Jones 1969) parasites is carried out chiefly by T lymphocytes. Investigations of lower vertebrates establish that this sort of immunity is of ancient phylogenetic development. All vertebrates, in addition, can form circulating antibodies, although the levels in certain cyclostomes are quite low. This observation raises a provocative issue for speculations on the origin of circulating antibodies; both cellular and humoral immunity must have arisen together, since immunoglobulins serve as recognition units on both T cells and B cells. It is conceivable that vigorous antibody production might lag phylogenetically behind cellular responses because immune T cells alone are capable of destroying foreign cells. The function of antibodies would be to amplify the effects of cellular immunity by promoting phagocytosis and complement-dependent lysis of altered cells or pathogens. In order to accomplish this objective, duplication of constant region domains was required to allow the stabilized receptor to leave the cell surface and to develop domains capable of carrying out effector functions subsequent to binding antigen by the Fab fragment of the antibody molecule.

The interplay between a varying aggressor system and an adaptive protector system within the ecosystem of the vertebrate body can be divided into three possible situations. Aggressor advantage might exist early in the course of an infection because the number of antigens expressed exceeds the number of antibodies initially present. Such a situation would facilitate proliferation of the infectious agent. However, two results favorable to the host can subsequently occur: one of balance, in which there is a one to one correspondence between the nonself antigens and the antibodies capable of recognizing them; the second in which the immune system of the host can recognize more antigenic specificities related to those of the aggressor system than are presently expressed. The latter situation would represent immunological memory for two reasons; (a) the number of cells capable of reacting to previously encountered aggressor antigens is amplified and (b) the antibodies present have higher binding affinities than those produced early in the response. For this reason, the antibodies would have a greater range of cross-reactivities. The immune system, thus, can be considered a learning system that is constantly under challenge by a variety of antigenic forms which it must recognize and eliminate. We can question why such a system did not arise within the protostomate invertebrates. Certain of these organisms are susceptible to infections, parasitemias, and to the development of neoplasms (Dawe and Harshberger 1969), but the protective systems have remained generally simple. The genetic system generating antibody diversity may have developed only within the immediate ancestors of vertebrates and been maintained afterward because of its adaptive value as a generalized surveillance mechanism. The above "accidental" hypothesis is not very satisfying because some invertebrate species, e.g., oligochaete annelids, appear to be capable of manifesting a type of allograft immunity. The phylogenetic distribution of such responses and the lack of cells of the lymphoid series in these species, however, render it unlikely that such responses are directly homologous to those of vertebrate immunity. We may be forced to reject teleology and accept the conclusion that the genetic mechanism underlying immune diversification arose by chance within the protochordate ancestors of vertebrates. Once the improbable event had occurred, however, it was of such evolutionary advantage that the system was retained in recognizable form in all living vertebrates.

# 18 Continuing Problems in the Evolution of Immunity

Immunity is a dynamic process. Evidence obtained from chemical studies of antibodies establishes that these molecules evolved during the emergence of vertebrate classes and species. Furthermore, studies of changes of binding affinities of antibodies as a function of time within individuals shows that variation and selection occur within the individual immunologically competent animal. Just as the immune system is a process in biological evolution, our concepts of immune mechanisms and the experimental approaches we use to test these concepts vary as a process in intellectual evolution. In this section I will briefly reiterate a selection of issues which have been discussed because these issues presently constitute the expanding frontier of knowledge in evolutionary studies of immunity and bear directly upon major immunological problems that are usually approached only in mammalian systems.

The basic patterns of immune recognition and antibody formation were established early in vertebrate evolution. All vertebrates possess cells which can be termed lymphocytes and exhibit the capacity to mount both cell mediated and humoral immune responses, although the antibodies found in the most primitive vertebrates consist only of one class which resembles the IgM molecules of mammals. Antibodies of lower vertebrates are clearly homologous to those of mammals, and speculations regarding their similarity to the immunoglobulin classes of man have been made on the basis of a variety of physicochemical properties. Sufficient similarities exist among the IgM-like immunoglobulins of all vertebrates to indicate that the heavy chains present in these immunoglobulins share a common ancestor in the primordial $\mu$ chains. Detailed primary sequence data are required to confirm this hypothesis and to allow theoretical reconstruction of the sequence of the primordial gene from the observed sequences available. Such analyses must also be carried out on the non-IgM immunoglobulins which occur among Dipnoi, amphibians, reptiles, and birds. Another approach to the problems of homologies among immunoglobulins of vertebrate classes, orders, and species is the use of antisera to establish schemes of antigenic relationships. Highly sensitive and quantitative techniques including radioimmunoassay and microcomplement fixation are now available for this purpose. In essence, the basic pattern of phylogenetic emergence of immunoglobulin light and heavy polypeptide chains has been sketched in broad outline, but further investigation is still required to establish the precise relationships among the immunoglobulin classes and subclasses of nonmammalian vertebrates.

Both cell mediated and humoral immunity occur in cyclostomes which possess lymphocytes and IgM-like antibodies. Although no data exist at this time regarding the molecular nature of the lymphocyte receptor for antigen in these species, it is tempting to conclude that the IgM antibodies, probably existing in the form of a monomer composed of two light chains and two heavy chains, serve as this receptor. The protochordates possess cells similar to lymphocytes and are capable of limited cellular recognition but are apparently incapable of synthesizing circulating antibodies. The nature of the putative receptor for antigen on the surface of protolymphocytes of tunicates is a problem which bears directly upon the origins of vertebrate immunity. It is conceivable that this recognition structure is directly homologous to the primordial immunoglobulin variable region. Constant regions of light chains and $\mu$-like chains are clearly present in cyclostomes and elasmobranchs, but may not occur below the phylogenetic level of true vertebrates. Sensitive techniques including immunocytoadherence and binding of radioiodinated antigens to cells can be applied to the question of enumerating antigen-binding cells in protochordates. Furthermore, plasma membrane proteins of living cells of protochordates can be labelled with $^{125}$I-iodide using the enzyme lactoperoxidase to catalyze the reaction (Marchalonis unpublished), and such proteins can be investigated to determine physicochemical properties and binding affinities for antigens.

The division of lymphocytes into thymus-derived lymphocytes and bone-marrow-derived lymphocytes has been extensively studied in mice, where distinct markers are available for the cell types and the existence of isogenic strains allows detailed genetic analyses of T- and B-cell function. I reviewed evidence obtained for cyclostomes, fish, and amphibians that suggests that functions attributable to T cells as well as B cells occur in these vertebrates. However, insufficient data are presently available to prompt the firm conclusion that distinct T- and B-cell populations exist in such species. Solution to this problem is hindered by the lack of inbred strains, which would facilitate experiments involving transfer of cell populations in most lower vertebrates. Inbred strains of leopard frogs are maintained by G. Nace of the University of Michigan, but have not yet been used in investigations of putative T-B collaborative responses in amphibian species. The chicken, which possesses a primary B-cell-producing lymphoid organ, the bursa of Fabricius, provides detailed information on T and B lymphocytes in avian species. The pronounced structural and functional separation between T and B cells of chickens is usually taken as a clear-cut model for the properties of such cells in mammals. However, caution must be observed in this direct identification because the chicken is phylogenetically quite distinct from mammals and, moreover, lacks immunoglobulins strictly homologous to IgG antibodies. The question of the phylogenetic generality of T cells and B cells and their functional interactions is a crucial issue and, no doubt, will be pursued with zeal using amphibian and avian model systems.

The issues discussed above arise from evolutionary considerations and are relevant to an understanding of the basic events of immune function in all vertebrates. Two other areas which are unique to lower vertebrates should also prove useful in the elucidation of the mechanisms of immune activation and differentiation. These are (1) the effect of temperature upon the immune responses of ectothermic vertebrates, and (2) the prolonged free-living larval stages of amphibians. The first property allows the investigator to control the rate of antibody production or allograft rejection and emphasizes that antibody formation can be dissected into a series of discrete cellular events. A large body of literature which was initiated over 100 years ago on the development of amphibia provides information on the morphological changes that these readily-available embryos undergo and, furthermore, illustrates the variety of technical approaches for the experimental manipulation of such development. A number of investigators such as Volpe, Ruben, Cohen, and Du Pasquier are now exploiting knowledge of the developmental properties of larval amphibians to obtain basic information on the emergence and functional properties of immunologically competent lymphocytes.

## Concluding remarks

This book has been directed towards the description and analysis of data which identify and clarify the major trends in the evolution of immunity. This process arose within the vertebrates or their immediate ancestors, and the basic patterns of antibody structure and cellular immunity were established within the lower fishes. The genetic mechanisms implicated in the initial appearance of immunity continued to operate throughout vertebrate phylogeny. Although many gaps remain in our knowledge of the phylogeny of immunity, the field has progressed to the stage where certain lower vertebrate systems now provide theoretical and experimental models for the elucidation of central issues of mammalian immunity. Moreover, many puzzling developments in human and mammalian immunity are clarified when analyzed in an evolutionary context. In particular, certain lower vertebrate systems exist which make possible new approaches to problems of immune recognition, tolerance, and collaboration between thymus-derived and bone-marrow-derived lymphocytes. The study of the phylogeny of immunity still retains the excitement of new discoveries and contributes directly toward the applied questions of immunology.

# References

Abdou, N. I., and Abdou, N. L. 1972. Bone marrow: the bursa equivalent in man? *Science 175:* 446-8.

Abel, C. A., and Grey, H. M. 1968. Studies on the structure of mouse γA myeloma proteins. *Biochemistry 7:* 2682-88.

Acton, R. T., Bennett, J. C., Evans, E. E., and Schrohenloher, R. E. 1969. Physical and chemical characterization of an oyster hemagglutinin. *J. Biol. Chem. 4:* 4128-35.

Acton, R. T., Weinheimer, P. F., Wolcott, M., Evans, E. E., and Bennett, J. C. 1970. N-terminal sequences of immunoglobulin heavy and light chains from three species of lower vertebrates. *Nature 228:* 991-92.

Acton, R. T., Weinheimer, P. F., Dupree, H. K., Evans, E. E., and Bennett, J. C. 1971a. Phylogeny of immunoglobulins. Characterization of a 14S immunoglobulin from the gar, *Lepisosteus osseus*. *Biochemistry 10:* 2028-36.

Acton, R. T., Weinheimer, P. F., Dupree, H. K., Russell, T. R., Wolcott, M., Evans, E. E., Schrohenloher, R. E., and Bennett, J. C. 1971b. Isolation and characterization of the immune macroglobulin from the paddlefish, *Polyodon spathula*. *J. Biol. Chem. 246:* 6760-69.

Acton, R. T., Weinheimer, P. F., Hall, S. J., Niedermeier, W., Shelton, E., and Bennett, J. C. 1971c. Tetrameric immune macroglobulins in three orders of bony fishes. *Proc. Natl. Acad. Sci. (U.S.A). 68:* 107-11.

Acton, R. T., Evans, E. E., Weinheimer, P. F., Niedermeier, W., and Bennett, J. C. 1972a. Purification and characterization of two classes of immunoglobulins from the marine toad, *Bufo marinus.* *Biochemistry 11:* 2751-59.

Acton, R. T., Niedermeier, W., Weinheimer, P. F., Clem, L. W., Leslie, G. A., and Bennett, J. C. 1972b. The carbohydrate composition of immunoglobulins from diverse species of vertebrates. *J. Immunol. 109:* 371-81.

Acton, R. T., Weinheimer, P. F., Shelton, E., Niedermeier, W. and Bennett, J. C. 1972c. Phylogeny of immunoglobulins—purification and physicochemical characterization of the immune macroglobulin from the turtle, *Pseudemus seripta. Immunochemistry 9:* 421-33.

Ada, G. L., Nossal, G. J. V., Pye, J., and Abbot, A. 1964. Antigens in immunity. I. Preparation and properties of flagellar antigens from *Salmonella adelaide. Aust. J. Exp. Biol. Med. Sci. 42:* 267-82.

Alcock, D. M. 1965. Antibody production in the common frog. *Rana temporaria. J. Pathol. Bacteriol. 90:* 31-43.

Alepa, F. P. 1969. Antigenic factors characteristic of human immuno-globulin G detected in the sera of nonhuman primates. *Ann. N.Y. Acad. Sci. 162:* 170-76.

Alepa, F. P., and Terry, W. D. 1965. Genetic factors and polypeptide chain subclasses of human immunoglobulin G detected in chimpan-zee serums. *Science 150:* 1293-94.

Alkan, S. S., Williams, E. B., Nitecki, D. E., and Goodman, J. W. 1972. Antigen recognition and the immune response. Humoral and cellular responses to small molecules and bifunctional antigen molecules. *J. Exp. Med. 135:* 1228.

Allen, C. P. and Gilmour, D. G. 1962. The B blood group systems of chickens. III. The effects of two heterozygous genotypes on the survival and egg production of multiple crosses. *Genetics 47:* 1711-18.

Aloj, S., Salvatore, G., and Roche, J. 1967. Isolation and properties of a native subunit of lampreys' thyroglobulin. *J. Biol. Chem. 242:* 3810-14.

Ambrosius, H. and Drössler, K. 1972. Spezifische zellvermittelte Immunitet bei Froschlurchen: I. Quantitative Nachweistechnik mit dem Makrophagen-Migrations-Hemmtest für Peritonealeysudat-zellen und Milzstücken. *Acta Biol. Med. Ger. 29:* 437-40.

Ambrosius, H. and Fiebig, H. 1972. Evolution of antibody affinity. In *L'étude Phylogenique et Ontogenique de la Réponse Immunitaire et son Apport a la Théorie Immunologique,* ed. P. Liacopoulos and J. Panijel, pp. 135-46. Boulogne: Inserm.

Ambrosius, H., Hemmerling, J., Richter, R., and Schimke, R. 1970. Immunoglobulins and the dynamics of antibody formation in poikilothermic vertebrates (*Pisces, Urodela, Reptilia*). In *Devel-*

*opmental Aspects of Antibody Formation and Structure*, ed. J. Sterzl and I. Riha, pp. 727-44. Prague: Czechoslovak Academy of Sciences.

Andersson, B. and Blomgren, H. 1971. Evidence for thymus-independent humoral antibody production in mice against polyvinylpyrollidone and *E. coli* lipopolysaccharide. *Cell. Immunol. 2:* 411-24.

Archer, O. K., Sutherland, D. E. R., and Good, R. A. 1964. The developmental biology of lymphoid tissue in the rabbit. Consideration of the role of thymus and appendix. *Lab. Invest. 13:* 259-71.

Armstrong, W. D., Diener, E. and Shellam, G. 1969. The antigen-reactive cell in normal, immune and tolerant mice. *J. Exp. Med. 129:* 393-410.

Ashman, R. F. and Raff, M. C. 1973. Direct demonstration of theta-positive antigen-binding cells, with antigen-induced movement of thymus-dependent cell receptors. *J. Exp. Med. 137:* 69-84.

Atwell, J. L. and Marchalonis, J. J. 1975. Phylogenetic emergence of immunoglobulin classes distinct from IgM. *J. Immunogenet. 1:* 367-91.

Atwell, J. L., Marchalonis, J. J., and Ealey, E. H. M. 1973. Major immunoglobulin classes of the echidna, *Tachyglossus aculeatus. Immunology 25:* 835-46.

Avtalion, R. R. 1969. Temperature effect on antibody production and immunological memory in carp (*Cyprinus carpio*) immunized against bovine serum albumin (BSA). *Immunology 17:* 927-31.

Avtalion, R. R., Wojdani, A., Malik, Z., Shahrabani, R. and Duczyminer, M. 1973. Influence of environmental temperature on the immune response in fish. *Curr. Top. Microbiol. and Immunol. 61:* 1-35.

Bach, J.-F. and Dardenne, M. 1972. Antigen recognition by T lymphocytes. I. Thymus and marrow dependence of spontaneous rosette forming cells in the mouse. *Cell. Immunol. 3:* 1-10.

Bach, F. H., Widmer, M. B., Bach, M. L., and Klein, J. 1972. Serologically defined and lymphocyte defined components of the major histocompatibility complex in the mouse. *J. Exp. Med., 136:* 1430-44.

Baculi, B. S., Cooper, E. L., and Brown, B. A. 1970. Lymphomyeloid organs of amphibia. V. Comparative histology in diverse anuran species. *J. Morphol. 131:* 315-28.

Bailey, S., Miller, B. J., and Cooper, E. L. 1971. Transplantation immunity in annelids. II. Adoptive transfer of the xenograft reaction. *Immunology 21:* 81-86.

Baldo, B. A. and Fletcher, T. C. 1973. C-reactive protein-like precipitins in plaice. *Nature 246:* 145-46.

Baldwin, W. M. III, and Cohen, N. 1970. Liver-induced immunosuppression of allograft immunity in urodele amphibians. *Transplantation 10:* 530-37.

Baldwin, W. M. III, and Cohen, N. 1971. Effects of diverse tissue implants on the survival of subsequent skin allografts transplanted across weak histocompatibility barriers in newts and mice. *Transplant. Proc. 3:* 217-19.

Bang, F. B. 1967. Serological responses among invertebrates other than insects. *Fed. Proc. 26:* 1680-84.

Bankhurst, A. D., Warner, N. L., and Sprent, J. 1971. Surface immunoglobulins on thymus and thymus-derived lymphoid cells. *J. Exp. Med. 134:* 1005-15.

Barrington, E. J. W. 1965. *The Biology of Hemichordata and Protochordata.* Edinburgh: Oliver and Boyd.

Basten, A., Miller, J. F. A. P. and Abraham, R. 1975. Relationship between Fc receptors, antigen-binding sites on T and B cells, and H-2 complex-associated determinants. *J. Exp. Med. 141:* 547-60.

Basten, A., Miller, J. F. A. P., Warner, N. L., and Pye, J. 1971. Specific inactivation of thymus-derived (T) and non-thymus-derived (B) lymphocytes by $^{125}$I-labelled antigen. *Nature [New Biol.] 231:* 104-06.

Beale, G. H. and Wilkinson, J. F. 1961. Antigenic variation in unicellular organisms. *Annu. Rev. Microbiol. 15:* 263-96.

Beard, J. 1894. The development and probable function of the thymus. *Anat. Anz. 9:* 476-86.

Beard, J. 1900. The source of leucocytes and the true function of the thymus. *Anat. Anz. 18:* 550-73.

Beer, A. E. and Billingham, R. E. 1971. Immunobiology of mammalian reproduction. *Adv. Immunol. 14:* 1-84.

Bell, R. G., Stephens, C. J. and Turner, K. J. 1974. Marsupial immunoglobulins: an immunoglobulin molecule resembling eutherian IgA in serum and secretions of *Setonix brachyurus* (Quokka). *J. Immunol. 113:* 371-78.

Benacerraf, B. and McDevitt, H. O. 1972. Histocompatibility-linked immune response genes. *Science 175:* 273-79.

Benacerraf, B., Green, I. and Paul, W. E. 1967. The immune response of guinea pigs to hapten-poly-L-lysine conjugates as an example of the genetic control of the recognition of antigenicity. *Cold Spring Harbor Symp. Quant. Biol. 32:* 569-75.

Benedict, A. A. and Pollard, L. W. 1972. Three classes of immunoglobulins found in the sea turtle, *Chelonia mydas. Folia Microbiol.* (Praha) *17:* 75-78.

Bennich, H. and Johansson, S. G. O. 1971. Structure and function of human immunoglobulin E. *Adv. Immunol. 13:* 1-55.

Berggård, I. and Bearn, A. G. 1968. Isolation and properties of a low molecular weight beta-2-globulin occurring in human biological fluids. *J. Biol. Chem. 243:* 4095-5103.

Bernier, G. M. and Fanger, M. W. 1972. Synthesis of microblogulin by stimulated lymphocytes. *J. Immunol. 109:* 407-09.

Berrill, N. J. 1955. *The Origin of Vertebrates.* New York: Oxford University Press.

Bezkorovainy, A., Springer, G. F., and Desai, P. R. 1971. Physicochemical properties of the eel anti-human blood group H(O) antibody. *Biochemistry 10:* 3761-64.

Billingham, R. E., Brent, L., and Medawar, P. B. 1956. Quantitative studies on tissue transplantation immunity. III. Actively acquired tolerance. *Philos. Trans. R. Soc. Lond.* (Biol. Sci.) *239:* 357-414.

Binaghi, R. et Estevez, M. B. 1972. Homologies entre les immunoglobulines de differentes especes animales. *L'étude Phylogenique et Ontogenique de la Réponse Immunitaire et son Apport a la Théorie Immunologique,* ed. P. Liacopoulos and J. Panijel, pp. 147-51. Boulogne: Inserm.

Bisset, K. A. 1948. The effect of temperature upon antibody production in cold-blooded vertebrates. *J. Pathol. Bacteriol. 60:* 87-92.

Bitter-Suermann, D., Dierich, M., Konig, W., and Hadding, U. 1972. Bypass activation of the complement system starting with C3. I. Generation and function of an enzyme from a factor of guinea pig serum and cobra venom. *Immunology 23:* 267-81.

Black, J. A. and Dixon, G. H. 1968. Amino-acid sequence of alpha chains of human haptoglobins. *Nature 218:* 736-41.

Block, M. 1964. The blood forming tissues and blood of the newborn opossum (*Didelphys virginiana*). I. Normal development through about the one hundredth day of life. *Ergeb. Anat. Entwicklungsgesch. 37:* 237-366.

Block, M. 1967. The 'fetal' opossum as an experimental tool in ontogeny of immunologic competence. In *Ontogeny of Immunity,* ed. R. T. Smith, R. A. Good, and P. A. Miescher, pp. 150-63. Gainesville: University of Florida Press.

Boehmer, H. von and Sprent, J. 1974. Expression of M. locus differences by B cells but not by T cells. *Nature 249:* 363-65.

Boffa, G. A., Fine, J. M., Drilhon, A., and Amouch, P. 1967. Immunoglobulins and transferrin in marine lamprey sera. *Nature 214:* 700-702.

Bordet, J. and Gargou, O. 1901. Sur l'existence de substances sensibilsatrices dans la plupart des sérum antimicrobiens. *Ann. Inst. Pasteur Lille 15:* 289-303.

Borysenko, M. 1969. Skin allograft and xenograft rejection in the snapping turtle, *Chelydra serpentina. J. Exp. Zool. 170:* 341-58.

Borysenko, M. and Tulipan, P. 1973. The graft-versus-host reaction in the snapping turtle *Chelydra serpentina. Transplantation 16:* 496-504.

Boylston, A. W. and Mowbray, J. F. 1973. Cited in Graves, M. F., Owen, J. J. T., and Raff, M. C. *T and B Lymphocytes*, p. 116. Amsterdam: Excerpta Medica Foundation.

Boylston, A. W. and Mowbray, J. F. 1974. Surface immunoglobulin of a mouse T-cell lymphoma. *Immunology 27:* 855-61.

Bradshaw, C. M., Clem, L. W., and Sigel, M. M. 1969. Immunologic and immunochemical studies on the gar, *Lepisosteus platyrhincus. J. Immunol. 103:* 496-504.

Bradshaw, C. M., Clem, L. W., and Sigel, M. M. 1971. Immunologic and immunochemical studies on the gar, *Lepisosteus platyrhincus.* II. Purification and characterization of immunoglobulin. *J. Immunol. 106:* 1480-87.

Bradshaw, C. M., Richards, A. S., and Sigel, M. M. 1971. IgM antibodies in fish mucus. *Proc. Soc. Exp. Biol. Med. 136:* 1122-24.

Braunitzer, G. 1966. Phylogenetic variation in the primary structure of hemoglobins. *J. Cell. Physiol. 67* (suppl. 1): 1-19.

Briggs, R. and King, T. J. 1952. Transplantation of living nuclei from blastula cells into enucleated frogs' eggs. *Proc. Nat. Acad. Sci. (U.S.A.) 38:* 455-63.

Brown, I. N., Brown, K. N., and Hills, L. A. 1968. Immunity to malaria: the antibody response to antigenic variation by *Plasmodium knowlesi. Immunology* 14: 127-38.

Brown, K. N. 1971. Protective immunity to malaria provides a model for the survival of cells in an immunologically hostile environment. *Nature 230:* 163-67.

Bruton, O. C. 1952. Agammaglobulinemia. *Pediatrics, 9:* 722-28.

Buchner, H. 1889. Uber die bakterientödlende Wirkung des Zellenfreien Blutserums. *Centralbl. Bakt. Parasitol. 5:* 817-23.

Burnet, F. M. 1959. *The Clonal Selection Theory of Acquired Immunity.* Nashville, Tennessee: Vanderbilt University Press.

Burnet, F. M. 1970a. A certain symmetry: histocompatibility antigens compared with immunocyte receptors. *Nature 226:* 123-26.

Burnet, F. M. 1970b. The concept of immunological surveillance. *Prog. Exp. Tumor Res. 13:* 1-27.

Burnet, F. M. 1971. Self-recognition in colonial marine forms and flowering plants in relation to the evolution of immunity. *Nature 232:* 230-35.

Campbell, R. D. and Bibb, C. 1970. Transplantation in coelenterates. *Transplant. Proc. 2:* 202-11.

Cantacuzene, J. 1923. Le problème de l'immunité chez les Invertebres. *Célébration du 75me anniversaire de la foundation de la société de biologie. C. R. Soc. Biol. (Paris)* pp. 48-119.

Cauldwell, C. B., Henkart, P., and Humphreys, T. 1973. Physical properties of sponge aggregation factor. A unique proteoglycan complex. *Biochemistry 12:* 3051-55.

Cebra, J. J., Ray, A., Benjamin, D., and Birshtein, B. 1971. Localization of affinity label within the primary structure of γ2 chain from guinea pig IgG (2). In *Progress in Immunology*, ed. B. Amos, pp. 269-84. New York: Academic Press.

Cerny, J. and Ivanyi, J. 1966. The dose of antigen required for the suppression of the IgM and IgG antibody response in chickens. II. Studies at the cellular level. *Folia Biol. Praha 12:* 343-53.

Cerny, J., Ivanyi, J., Madar, J., and Hraba, T. 1965. The nature of the delay in the immune response after administration of large doses of protein antigen in chicks. *Folia Biol. Praha 11:* 402-05.

Chaplin, H., Cohen, S. and Press, E. M. 1965. Preparation and properties of the peptide chains of normal human 19S gamma-globulin (IgM). *Biochem. J. 95:* 256-61.

Chapuis, R. M. and Koshland, M. E. 1974. Mechanism of IgM polymerization. *Proc. Natl. Acad. Sci. (U.S.A.) 71:* 657-61.

Charlemagne, J. 1972. Les réactions immunitairies chez les amphibians urodeles. I. Résultats acquis et possibilities expérimentales. In *L'étude Phylogenique et Ontogenique de la Réponse Immunitaire et son Apport a la Théorie Immunologique*, ed. P. Liacopoulos et J. Panijel, pp. 89-96. Boulogne: Inserm.

Chartrand, S. L., Litman, G. W., Lapointe, N., Good, R. A., and Frommel, D. 1971. The evolution of the immune response. XII. The immunoglobulins of the turtle. Molecular requirements for biologic activity of 5.7S immunoglobulin. *J. Immunol. 107:* 1-11.

Chase, M. W. 1959. Models for hypersensitivity studies. In *Cellular and Humoral Aspects of the Hypersensitive States*, ed. H. S. Lawrence, pp. 251-78. New York: Hoeber-Harper.

Chateaureynaud-Duprat, Pierrette P. 1970. Specificity of allograft reaction in *Eisenia foetida. Transplant Proc. 2:* 222-25.

Cheers, C. and Sprent, J. 1972. B-lymphocytes as stimulators of a mixed lymphocyte reaction. *Transplantation 15:* 336-37.

Cheng, T. C. 1970. Immunity in Mollusca with special reference to reaction to transplants. *Transplant. Proc. 2:* 226-30.

Chiller, J. M. 1974. Reported at Workshop on Phylogeny of Immunity at the Second International Congress of Immunologists, Brighton, England.

Chiller, J. M., Hodgins, H. O., and Weiser, R. S. 1969. Antibody response in rainbow trout *(Salmo gairdneri).* II. Studies on the kinetics of development of antibody-producing cells and on complement and natural hemolysin. *J. Immunol. 102:* 1202-07.

Chiller, J. M., Hodgins, H. O., Chambers, V. C., and Weiser, R. S. 1969. Antibody response in rainbow trout (*Salmo gairdneri*). I. Immunocompetent cells in the spleen and anterior kidney. *J. Immunol. 102:* 1193-1201.

Ching, Y.-C., and Wedgwood, R. J. 1967. Immunologic responses in the axolotl, *Siredon mexicanum. J. Immunol. 99:* 191-200.

Choi, Y. S. and Good, R. A. 1971. New immunoglobulin-like molecules in the serum of bursectomized irradiated chickens. *Proc. Natl. Acad. Sci. (U.S.A.) 68:* 2083-86.

Chuang, C.-Y., Capra, J., and Kehoe, J. M. 1973. Evolutionary relationship between carboxyterminal region of a human alpha chain and other immunoglobulin chain constant regions. *Nature 244:* 158-68.

Claman, H. N. and Chaperon, E. A. 1969. Immunologic complementation between thymus and marrow cells—a model for the two-cell theory of immunocompetence. *Transplant. Rev. 1:* 92-113.

Clawson, C. C., Finstad, J. and Good, R. A. 1966. Evolution of the immune response. V. Electron microscopy of plasma cells and lymphoid tissue of the paddlefish. *Lab. Invest. 153:* 1830-47.

Clem, L. W. 1971. Phylogeny of immunoglobulin structure and function. IV. Immunoglobulins of the giant grouper, *Epinephelus itaira. J. Biol. Chem. 246:* 9-15.

Clem, L. W. and Leslie, G. A. 1969. Phylogeny of immunoglobulin structure and function. In *Immunology and Development*, ed. M. Adinolfi, pp. 62-88. London: Spastics International Medical Publications.

Clem, L. W. and Leslie, G. A. 1971. Production of 19S IgM antibodies with restricted heterogeneity from sharks. *Proc. Natl. Acad. Sci. (U.S.A.) 68:* 139-41.

Clem, L. W. and Sigel, M. M. 1966. Immunological and immunochemical studies on holostean and marine teleost fishes immunized with bovine serum albumin. In *Phylogeny of Immunity*, ed. R. T. Smith, P. A. Miescher, and R. A. Good, pp. 209-17. Gainesville: University of Florida Press.

Clem, L. W. and Small, P. A., Jr. 1967. Phylogeny of immunoglobulin structure and function. I. Immunoglobulins of the lemon shark. *J. Exp. Med. 125:* 893-920.

Clem, L. W. and Small, P. A., Jr. 1970. Phylogeny of immunoglobulin structure and function. V. Valences and association constants of teleost antibodies to a haptenic determinant. *J. Exp. Med. 132:* 385-400.

Coe, J. E. 1970. Specificity of antibody produced in the bullfrog (*Rana catesbeiana*). *J. Immunol. 104:* 1166-74.

Cohen, E. 1967-68. Immunologic observations of the agglutinins of the hemolymph of *Limulus polyphemus* and *Birgus latro. Trans. N. Y. Acad. Sci. 30:* 427-43.

Cohen, E., Rowe, A. W., and Wissler, F. C. 1965. Heteroagglutinins of the horseshoe crab *Limulus polyphemus. Life Sci. 4:* 2009-16.

Cohen, N. 1969. Immunogenetic and developmental aspects of tissue transplantation immunity in urodele amphibians. In *Biology of Amphibian Tumors*, ed. M. Mizell, pp. 153-68. New York: Springer-Verlag.

Cohen, N. 1971a. Amphibian transplantation reactions: A review. *Am. Zool. 11:* 193-205.

Cohen, N. 1971b. Reptiles as models for the study of immunity and its phylogenesis. *J. Am. Vet. Med. Assoc. 159:* 1662-71.

Cohen, N. 1975. Phylogenetic emergence of lymphoid tissues and cells. In *The Lymphocyte: Structure and Function*, ed. J. J. Marchalonis. New York: Marcel Dekker, Inc.

Cohen, N. and Borysenko, M. 1970. Acute and chronic graft rejection possible phylogeny of transplantation antigens. *Transplant. Proc. 2:* 333-36.

Cohen, N. and Hildemann, W. H. 1968. Population studies of allograft rejection in the newt, *Diemictylus viridescens. Transplantation 6:* 208-17.

Cohen, P. P. 1970. Biochemical differentiation during amphibian metamorphosis. *Science 168:* 533-43.

Cohen, S. and Porter, R. R. 1964. Structure and biological activity of immunoglobulins. *Adv. Immunol. 4:* 287-349.

Cohn, M. 1972a. Antibody diversification: The somatic mutation model revisited. In *The Biochemistry of Gene Expression in Higher Organisms*, ed. J. W. Lee and J. K. Pollack, pp. 574-92. Sydney: Australia and New Zealand Book Co.

Cohn, M. 1972b. Evaluation and summary. In *Genetic Control of Immune Responsiveness*, ed. H. O. McDevitt and M. Landy, pp. 370-448. New York: Academic Press.

Cohn, M., Blomberg, B., Geckeler, W., Raschke, W., Riblet, R., and Weigert, M. 1974. First order considerations in analyzing the generator of diversity. In *The Immune System: Genes, Receptors, Signals*, ed. E. E. Sercarz, A. R. Williamson and C. F. Fox, pp. 89-117. New York: Academic Press.

Cole, G. J. and Morris B. 1971. Homograft rejection and hypersensitivity reactions in lambs thymectomized in utero. *Aust. J. Exp. Biol. Med. Sci. 49:* 75-88.

Cone, R. E. and Marchalonis, J. J. 1972. Cellular and humoral aspects of the influence of environmental temperature of the immune response of poikilothermic vertebrates. *J. Immunol. 108:* 952-57.

Cone, R. E. and Marchalonis, J. J. 1973. Antigen binding specificity of cell surface immunoglobulin isolated from T (helper) cells. *Aust. J. Exp. Biol. Med. Sci. 51:* 689-700.

Connell, G. E., Dorrington, K. J., Lewis, A. F., and Parr, D. M. 1970. The conformation of an atypical IgG myeloma protein and its papain fragments. *Can. J. Biochem. 48:* 784-89.

Cooper, A. J. 1971. Ammocoete lymphoid cell populations *in vitro*. In *Proceedings of the Fourth Annual Leucocyte Culture Conference*, ed. O. R. McIntyre, pp. 137-47, New York: Appleton-Century-Crofts.

Cooper, E. L. 1970. Transplantation immunity in helminths and annelids. *Transplant. Proc. 2:* 216-21.

Cooper, E. L. and Hildemann, W. H. 1965. The immune response of larval bullfrogs (*Rana catesbeiana*) to diverse antigens. *Ann. N.Y. Acad. Sci. 126:* 647-61.

Cooper, E. L., Brown, B. A., and Baculi, B. S. 1971. New observations on lymph gland (LM1) and thymus activity in larval bullfrogs, *Rana catesbeiana*. In *Morphological and Functional aspects of immunity*, ed. K. Lindahl-Kiessling, G. Alm and M. G. Hanna, Jr. *Adv. Exp. Med. Biol. 12:* 1-10.

Cooper, M. D., Gabrielsen, A. E. and Good, R. A. 1967. Role of the thymus and other central lymphoid tissues in immunological disease. *Ann. Rev. Med. 18:* 113-38.

Cooper, M. D., Peterson, R. D. A., and Good, R. A. 1965. Delineation of the thymic and bursal lymphoid systems in the chicken. *Nature 205:* 143-46.

Cooper, M. D., Peterson, R. D. A., South, M. A., and Good, R. A. 1966. The functions of the thymus system and the bursa system in the chicken. *J. Exp. Med. 123:* 75-102.

Cowden, R. R. and Dyer, R. F. 1971. Lymphopoietic tissue and plasma cells in amphibians. *Am. Zool. 11:* 183-92.

Crewther, P. and Warner, N. 1972. Serum immunoglobulins and antibodies in congenitally athymic (nude) mice. *Aust. J. Exp. Biol. Med. Sci. 50:* 625-35.

Crone, M., Koch, C., and Simonsen, M. 1972. The elusive T cell receptor. *Transplant. Rev. 10:* 36-56.

Crumpton, M. J. and Wilkinson, J. M. 1963. Amino acid compositions of human and rabbit gamma-globulins and of the fragments produced by reduction. *Biochem. J. 88:* 228-34.

Cullen, S. E., David, C. S., Shreffler, D. C., and Nathenson, S. G. 1974. Membrane molecules determined by the H-2 associated immune response region: Isolation and some properties. *Proc. Natl. Acad. Sci. (U.S.A.) 71:* 648-52.

Curtis, S. K. and Volpe, E. P. 1971. Modification of responsiveness to allografts in larvae of the leopard frog by thymectomy. *Dev. Biol. 25:* 177-97.

Damian, R. T. 1964. Molecular mimicry. Antigen sharing by parasite and host and its consequences. *Am. Naturalist 98:* 129-49.

Damian, R. T. 1967. Common antigens between adult *Schistosoma mansoni* and the laboratory mouse. *J. Parasitol. 53:* 60-64.

Davis, B. D., Dulbecco, R., Eisen, H. N., Ginsberg, H. S., and Wood, W. B., Jr. 1969. *Microbiology* (Fifth printing with corrections). New York: Harper and Row (Hoeber Medical Division).

Dawe, C. J. and Harshbarger, J. C., eds. 1969. *Neoplasms and related disorders of invertebrate and lower vertebrate animals.* Bethesda, Maryland: *Natl. Cancer. Inst. Monogr.* 31.

Day, N. K. B., Geiger, H., Finstad, J., and Good, R. A. 1972. A starfish hemolymph factor which activates vertebrate complement in the presence of cobra venom factor. *J. Immunol. 109:* 164-67.

Day, N. K. B., Gewurz, H., Johannsen, R., Finstad, J., and Good, R. A. 1970. Complement and complement-like activity in lower vertebrates and invertebrates. *J. Exp. Med. 132:* 941-50.

Dayhoff, M. O. 1969. *Atlas of Protein Sequence and Structure.* Vol. 4. Silver Spring, Md.: National Biomedical Research Foundation.

Delovitch, T. L. and Baglioni, C. 1973. Estimation of light-chain gene reiteration of mouse immunoglobulin by DNA-RNA hybridization. *Proc. Natl. Acad. Sci. (U.S.A.) 70:* 173-78.

Demant, P., Capkova, J., Henzova, E., and Voracova, B. 1973. The role of the histocompatibility—2—linked Ss-SIp region in the control of mouse complement. *Proc. Natl. Acad. Sci. (U.S.A.) 70:* 863-64.

Deutsch, H. F. and Morton, J. I. 1957. Dissociation of human serum macroglobulins. *Science 125:* 600-01.

Diener, E. 1968. A new method for the enumeration of single antibody-producing cells. *J. Immunol. 100:* 1062-70.

Diener, E., 1970. Evolutionary aspects of immunity and lymphoid organs in vertebrates. *Transplant. Proc. 2:* 309-17.

Diener, E. and Marchalonis, J. J. 1970. Cellular and humoral aspects of the primary immune response of the toad, *Bufo marinus. Immunology 18:* 279-93.

Diener, E. and Nossal, G. J. V. 1966. Phylogenetic studies on the immune response. I. Localization of antigens and immune response in the toad, *Bufo marinus. Immunology 10:* 535-42.

Diener, E., Ealey, E. H. M., and Legge, J. S. 1967. Phylogenetic studies on the immune response. III. Autoradiographic studies on the lymphoid system of the Australian echidna, *Tachyglossus aculeatus. Immunology 13:* 339-47.

Diener, E., O'Callaghan, F., and Kraft, N. 1971. Immune response *in vitro* to Salmonella H-antigens not affected by anti-theta serum. *J. Immunol. 107:* 1775-77.

Diener, E., Shortman, K., and Russell, P. 1970. Induction of immunity and tolerance *in vitro* in the absence of phagocytic cells. *Nature 225:* 731-32.

Diener, E., Wistar, R. and Ealey, E. H. M. 1967. Phylogenetic studies on the immune response. II. The immune response of the Australian echidna. *Tachyglossus aculeatus. Immunology, 13:* 329-38.

Dineen, J. K. and Adams, D. B. 1971. The role of the recirculating thymus-dependent lymphocyte in resistance to *Trichostrongylus colubriformis* in the guinea pig. *Immunology 20:* 109-13.

Dobson, C., Rockey, J. H., and Soulsby, E. J. L. 1971. Immunoglobulin E antibodies in guinea pigs: characterization of monomeric and polymeric components. *J. Immunol. 107:* 1431-39.

Douglas, T. C. 1972. Occurrence of a theta-like antigen in rats. *J. Exp. Med. 136:* 1054-62.

Dreesman, G., Larson, C., Pinckard, R. N., Groyon, R. M., and Benedict A. A. 1965. Antibody activity in different chicken globulins. *Proc. Soc. Exp. Biol. Med., 118:* 292-96.

Du Pasquier, L. 1970. Ontogeny of the immune response in animals having less than one million lymphocytes: the larvae of the toad *Alytes obstetricans. Immunology 19:* 353-62.

Du Pasquier, L. 1973. Ontogeny of the immune response in cold-blooded vertebrates. *Curr. Top. Microbiol. Immunol. 61:* 37-88.

Du Pasquier, L., Weiss, N., and Loor, F. 1972. Direct evidence for immunoglobulins on the surface of thymus lymphocytes of amphibian larvae. *Eur. J. Immunol. 2:* 366-70.

Dwyer, J. M., Warner, N. L. and Mackay, I. R. 1972. Specificity and nature of the antigen-combining sites on fetal and mature thymus lymphocytes. *J. Immunol. 108:* 1439-46.

East, E. M. and Mangelsdorf, A. J. 1925. A new interpretation of the hereditary behavior of self-sterile plants. *Proc. Natl. Acad. Sci. (U.S.A.) 11:* 166-71.

Edelman, G. M. 1959. Dissociation of γ-globulin. *J. Am. Chem. Soc. 81:* 3155-56.

Edelman, G. M. 1970. The covalent structure of a human γG-immuno-globulin. XI. Functional implications. *Biochemistry 9:* 3197-205.

Edelman, G. M. and Gall, W. E. 1969. The antibody problem. *Annu. Rev. Biochem. 38:* 415-66.

Edelman, G. M. and Gally, J. A. 1962. The nature of Bence-Jones proteins. Chemical similarities to polypeptide chains of myeloma globulins and normal γ globulins. *J. Exp. Med. 116:* 207-27.

Edelman, G. M. and Gally, J. A. 1967. Somatic recombination of duplicate genes: an hypothesis on the origin of antibody diversity. *Proc. Natl. Acad. Sci. (U.S.A.) 57:* 353-58.

Edelman, G. M. and Poulik, M. D. 1961. Studies on structural units of the γ-globulins. *J. Exp. Med. 113:* 861-84.

Edwards, B. F., Ruben, L. N., Marchalonis, J. J. and Hylton, C. 1975. Surface characteristics of spleen cell erythrocyte rosette formation in the grass frog *Rana pipiens. Adv. Exp. Med. Biol.,* in press.

Ehrlich, P. 1900. On immunity with special references to cell life. *Proc. R. Soc. Lond. (Biol.) 66:* 424-48.

Eisen, H. N. and Siskind, G. W. 1964. Variations in affinities of antibodies during the immune response. *Biochemistry 3:* 996-1008.

Eisen, H. N., Simms, E. S., and Potter, M. 1968. Mouse myeloma proteins with antihapten antibody activity. The protein produced by plasma cell tumor MOPC-315. *Biochemistry 7:* 4126-34.

Elek, S. D., Rees, T. A., and Gowing, N. F. C. 1962. Studies on the immune response in a poikilothermic species (*Xenopus laevis* Daudin). *Comp. Biochem. Physiol. 7:* 255-67.

Emmrich, F., Richter, R. F. and Ambrosius, H. 1975. Immunoglobulin determinants on the surface of lymphoid cells of carp. *Eur. J. Immunol.,* in press.

Emura, J., Ikenaka, T., Collins, J. H., and Schmid, K. 1971. The constant and variable regions of the carboxyl-terminal CNBr fragment of $α_1$-acid glycoprotein. *J. Biol. Chem. 246:* 7821-23.

Endean, R. 1960. The blood cells of the ascidian, *Phallusia mammillata. Q. J. Microsc. Sci. 101:* 177-97.

Engle, R. L. Jr., Woods, K. R., Paulsen, E. C., and Pert, J. H. 1958. Plasma cells and serum proteins in marine fish. *Proc. Soc. Exp. Biol. Med. 98:* 905-09.

Evans, E. E., Kent, S. P., Attleberger, M. H., Seibert, C., Bryant, R. E., and Booth, B. 1965. Antibody synthesis in poikilothermic vertebrates. *Ann. N.Y. Acad. Sci. 126:* 629-46.

Evans, E. E., Kent, S. P., Bryant, R. E., and Moyer, M. 1966. Antibody formation and immunological memory in the marine toad. In *Phylogeny of Immunity*, ed. R. T. Smith, P. A. Miescher and R. A. Good, pp. 218-26. Gainesville: University of Florida Press.

Evans, E. E., Painter, B., Evans, M. L., Weinheimer, P., and Acton, R. T. 1968. An induced bactericidin in the spiny lobster, *Panulirus argus. Proc. Soc. Exp. Biol. Med. 128:* 394-98.

Everhart, D. L. and Shefner, A. M. 1966. Specificity of fish antibody. *J. Immunol. 97:* 231-34.

Evrin, P. E. and Pertoft, H. 1973. $\beta_2$-microglobulin in human blood cells. *J. Immunol. 111:* 1147-54.

Feinstein, A. and Munn, E. A. 1969. Confirmation of the free and antigen-bound IgM antibody molecules. *Nature 224:* 1307-09.

Feldmann, M. 1972. Cell interactions in the immune response *in vitro*. V. Specific collaboration via complexes of antigen and thymus-derived cell immunoglobulin. *J. Exp. Med. 136:* 737-60.

Feldmann, M. and Nossal, G. J. V. 1972. Tolerance, enhancement and the regulation of interactions between T cells, B cells and macrophages. *Transplant. Rev. 13:* 3-34.

Feldmann, M., Cone, R. E., and Marchalonis, J. J. 1973. Cell interactions in the immune response *in vitro*. VI. Mediation by T-cell surface IgM. *Cell. Immunol. 9:* 1-11.

Felton, L. D. 1949. The significance of antigen in animal tissue. *J. Immunol.* 61: 107-17.

Fernandez-Moran, H., Marchalonis, J. J., and Edelman, G. M. 1968. Electron microscopy of a hemagglutinin from *Limulus polyphemus*. *J. Mol. Biol. 32:* 467-69.

Fidler, J. E., Clem, L. W., and Small, P. A. Jr., 1969. Immunoglobulin synthesis in neonatal nurse sharks *(Ginglymostoma cirratum)*. *Comp. Biochem. Physiol. 31:* 365-71.

Finstad, J. and Good, R. A. 1966. Phylogenetic studies of adaptive immune responses in the lower vertebrates. In *Phylogeny of Immunity*, ed. R. T. Smith, P. A. Miescher, and R. A. Good, pp. 173-88. Gainesville: University of Florida Press.

Finstad, C. L., Good, R. A., and Litman, G. W. 1974. The erythrocyte agglutinin from *Limulus polyphemus* hemolymph: Molecular structure and biological function. *Ann. N.Y. Acad. Sci., 234:* 170-80.

Finstad, C. L., Litman, G. W., Finstad, J., and Good, R. A. 1972. The evolution of the immune response. XIII. The characterization of purified erythrocyte agglutinins from two invertebrate species. *J. Immunol. 108:* 1704-11.

Fischberg, M., Gurdon, J. B., and Elsdale, T. R. 1958. Nuclear transplantation in *Xenopus laevis. Nature 181:* 424.

Fitch, W. M. 1966. An improved method of testing for evolutionary homology. *J. Mol. Biol. 16:* 9-16.

Fitch, W. M. and Margoliash, E. 1967. Construction of phylogenetic trees. *Science 155:* 279-84.

Fleischman, J. B., Pain, R. H., and Porter, R. R. 1962. Reduction of γ globulins. *Arch. Biochem. Biophys.* (Suppl. 1): 174-80.

Fletcher, T. C. and Baldo, B. A. 1974. Immediate hypersensitivity responses in flatfish. *Science 185:* 360-61.

Fletcher, T. C. and Grant, P. T. 1969. Immunoglobulins in the serum and mucus of the plaice (*Pleuronectes platessa*). *Biochem. J. 115:* 65P.

Fletcher, T. C. and White, A. 1973. Antibody production in the plaice (*Pleuronectes platessa L.*) after oral and parenteral immunization with *Vibrio angeillarum* antigens. *Aquaculture 1:* 417-28.

Florkin, M. and Stotz, E. H. 1972. *A History of Biochemistry*, p. 180. New York: American Elsevier.

Fodor, J. von 1886. Die Fähigkeit des Blutes Bakterien zu vernichten. *Dtsch. Med. Wochenschr. 13:* 745-47.

Foster, A. B. and Webber, J. M. 1960. Chitin. *Adv. Carbohydr. Chem. 15:* 371-93.

Fougereau, M. et Houdayer, M. 1968. Immunoglobulins et réponse immunitaire chez l'axolotl (*Ambystoma mexicanum*). *Ann. Inst. Pasteur Lille, 115:* 968.

Fougereau, M., Houdayer, M., et Dorson, M. 1972. Réponse immunitaire et structure multicatenaire des immunoglobulines d'une teleostein: la truite arc-en-ceil (*Salmo gairdneri*) et d'un amphibien urodele l'axolotl (*Ambystoma mexicanum*). In *L'étude Phylogenique et Ontogenique de la Réponse Immunitaire et son Apport a la Théorie Immunologique*, ed. P. Liacopoulos and J. Panijel, pp. 121-33. Boulogne: Inserm.

Freeman, G. 1970. The reticuloendothelial system of tunicates. *J. Reticuloendothel. Soc. 7:* 183-194.

Frenzel, E.-M. und Ambrosius, H. 1971. Anti-Hapten-Antikörper bein neideren Wirbeltieren. *Acta Biol. Med. Ger. 26:* 165-71.

Frommel, D., Perey, D. Y. E., Massoyeff, R., and Good, R. A. 1970. Low molecular weight serum immunoglobulin M in experimental trypanosomiasis. *Nature 228:* 1208-10.

Frommel, D., Litman, G. W., Finstad, J., and Good, R. A. 1971. The evolution of the immune response. XI. The immunoglobulins of the horned shark, *Heterodontus francisci:* Purification, characterization and structural requirement for antibody activity. *J. Immunol.* 106: 1234-43.

Gally, J. A. and Edelman, G. M. 1970. Somatic translocation of antibody genes. *Nature 227:* 314-48.

Gally, J. A. and Edelman, G. M. 1972. The genetic control of immunoglobulin synthesis. *Ann. Rev. Gene.* 6, 1-46.

Garber, B. B. and Moscona, A. A. 1972. Reconstruction of brain tissue from cell suspensions. II. Specific enhancement of aggregation of embryonic cerebral cells by supernatant from homologous cell cultures. *Dev. Biol. 27:* 235-43.

Gavin, J. R. III, Buell, D. N., and Roth, J. 1972. Water-soluble insulin receptors from human lymphocytes. *Science 178:* 168-69.

Geczy, C. L., Green, P. C., and Steiner, L. A. 1973. Immunoglobulins in the developing amphibians, *Rana catesbeiana. J. Immunol. 111:* 1261-67.

Gershon, R. K. and Kondo, K. 1971. Infectious immunological tolerance. *Immunology* 21: 903-14.

Gershon, R. K., Gery, I. and Waksman, B. H. 1974. Suppressive effects of in vivo immunization on PHA responses *in vitro. J. Immunol. 112:* 215-21.

Gewurz, H., Finstad, J., Muschel, L. H., and Good, R. A. 1966. Phylogenetic inquiry into the origins of the complement system. In *Phylogeny of Immunity*, ed. R. T. Smith, P. A. Miescher, and R. A. Good, pp. 105-16. Gainesville: University of Florida Press.

Gigli, I. and Austen, K. F. 1971. Phylogeny and function of the complement system. *Annu. Rev. Microbiol. 25:* 309-32.

Gilden, R. V. and Rosenquist, G. L. 1963. Duration of response to a soluble antigen. *Nature 199:* 87.

Gingrich, R. E. 1964. Acquired humoral immune response of the large milkweed bug, *Oncopeltus fasciatus* Dallas, to injected materials. *J. Insect. Physiol. 10:* 179-94.

Gitlin, D., Perricelli, A., and Gitlin, J. D. 1973. Multiple immunoglobulin classes among sharks and their evolution. *Comp. Biochem. Physiol.* (B) *44:* 225-39.

Givol, D., Strausbauch, P. H., Hurwitz, E., Wilchek, M., Haimovich, J., and Eisen, H. N. 1971. Affinity labeling and cross linking of the heavy and light chains of a myeloma protein with anti-2, 4-dinitrophenyl activity. *Biochemistry 10:* 3461-66.

Glenner, G. G., Terry, W., Harada, M., Isersky, C., and Page, D. 1971. Amyloid fibril proteins: proof of homology with immunoglobulin light chains by sequence analyses. *Science 172:* 1150-51.

Glick, B., Chang, T. S., and Jaap, R. G. 1956. The bursa of Fabricius and antibody production. *Poult. Sci. 35:* 224-25.

Glossmann, H. and Nevill, D. M., Jr. 1971. Glycoproteins of cell surfaces. A comparative study of three different cell surfaces of the rat. *J. Biol. Chem. 246:* 6339-46.

Goding, J. W., Nossal, G. J. V., Shreffler, D. C. and Marchalonis, J. J. 1975. Ia antigens on murine lymphoid cells: Distribution, surface movement and partial characterization. *J. Immunogenet. 2:* 41-57.

Goetzl, E. J. and Metzger, H. 1970. Affinity labeling of a mouse myeloma protein which binds nitrophenyl ligands. Sequence and position of a labeled tryptic peptide. *Biochemistry, 9:* 3862-71.

Goldschneider, I. and Cogen, R. B. 1973. Immunoglobulin molecules on the surface of activated T lymphocytes in the rat. *J. Exp. Med. 138:* 163-75.

Goldshein, S. J. and Cohen, N. 1972. Phylogeny of immunocompetent cells. I. In vitro blastogenesis and mitosis of toad (*Bufo marinus*) splenic lymphocytes in response to phytohemagglutinin and in mixed lymphocyte cultures. *J. Immunol. 108:* 1025-33.

Good, R. A. and Papermaster, B. W. 1964. Ontogeny and phylogeny of adaptive immunity. *Adv. Immunol. 4:* 1-115.

Good, R. A., Finstad, J., Pollara, B., and Gabrielsen, A. E. 1966. Morphologic studies on the evolution of the lymphoid tissues among the lower vertebrates. In *Phylogeny of Immunity*, ed. R. T. Smith, P. A. Miescher, and R. A. Good, pp. 149-68. Gainesville: University of Florida Press.

Goodner, K. 1926. Studies in anaphylaxis. IV. Allergic manifestations in frogs. *J. Immunol. 11:* 335-41.

Gorczynski, R. M., Miller, R. G., and Phillips, R. A. 1973. Reconstitution of T cell-depleted spleen cell populations by factors from T cells. I. Conditions for the production of active T cell supernatants. *J. Immunol. 110:* 968-83.

Gotschlich, E. C. and Edelman, G. M. 1965. C-reactive protein: a molecule composed of subunits. *Proc. Natl. Acad. Sci. (U.S.A.) 54:* 558-66.

Gotschlich, E. C. and Edelman, G. M. 1967. Binding properties and specificity of C-reactive protein. *Proc. Natl. Acad. Sci. (U.S.A.) 57:* 706-12.

Graff, R. J., Silvers, W. K., Billingham, R. E., Hildemann, W. H., and Snell, G. D. 1966. The cumulative effect of histocompatibility antigens. *Transplantation 4:* 605-17.

Greaves, M. F. and Hogg, N. M. 1971. Immunoglobulin determinants on the surface of antigen binding T and B lymphocytes in mice. In *Prog. Immunol.,* ed. B. Amos, pp. 111-26. New York: Academic Press.

Greaves, M. F., Möller, E. and Möller, G. 1970. Studies on antigen-binding cells. II. Relationships to antigen-sensitive cells. *Cell. Immunol. 1:* 386-403.

Grey, H. M. 1966. Structure and kinetics of formation of antibody in the turtle. In *Phylogeny of Immunity*, ed. R. T. Smith, P. A. Miescher, and R. A. Good, pp. 227-33. Gainesville: University of Florida Press.

Grey, H. M. 1967. Duck immunoglobulins. I. Structural studies on a 5.7S and 7.8S γ-globulin. *J. Immunol. 98:* 811-19.

Grey, H. M. 1969. Phylogeny of immunoglobulins. *Adv. Immunol. 10:* 51-104.

Grey, H. M., Mannik, M. and Kunkel, H. G. 1965. Individual antigenic specificity of myeloma proteins. Characteristics and localization of subunits. *J. Exp. Med. 121:* 561-75.

# REFERENCES

Grey, H. M., Abel, C. A., Yount, W. J., and Kunkel, H. G. 1968. A subclass of human γA-globulins (γA2) which lacks the disulfide bonds linking heavy and light chains. *J. Exp. Med. 128:* 1223-36.

Gross, E. and Witkop, B. 1961. Selective cleavage of the methionyl peptide bonds in ribonuclease with cyanogen bromide. *J. Am. Chem. Soc., 83:* 1510-11.

Grubb, R. 1970. *The Genetic Markers of Human Immunoglobulins*, New York: Springer-Verlag.

Gurdon, J. B. and Brown, D. D. 1965. Cytoplasmic regulation of RNA synthesis and nucleolus formation in developing embryos of *Xenopus laevis. J. Mol. Biol. 12:* 27-35.

Hadji-Azimi, I. 1969. Electrophoretic study of the serum proteins of normal and "lymphoid tumour"-bearing *Xenopus. Nature 221:* 264-65.

Hadji-Azimi, I. 1971. Studies on *Xenopus laevis* immunoglobulins. *Immunology 21:* 463-74.

Haimovich, J., Eisen, H. N., Hurwitz, E., and Givol, D. 1972. Localization of affinity-labeled residues on the heavy and light chains of two myeloma proteins with anti-hapten activity. *Biochemistry 11:* 2389-98.

Hala, K., Hasek, M., Hlozanek, F., Hort, J., Knizetova, F., and Mervartovna, H. 1966. Synergetic lines of chickens. II. Inbreeding and selection within the M, W, and I lines and crosses between the C, M. and W lines. *Folia Biol., Praha, 12:* 407-21.

Halliwell, R. E. W., Schwartzman, R. M., and Rockey, J. H. 1972. Antigenic relationship between human IgE and canine IgE. *Clin. Exp. Immunol. 10:* 399-407.

Halpern, M. S. and Koshland, M. E. 1970. Novel subunit in secretory IgA. *Nature 228:* 1276-78.

Hämmerling, G. J. and McDevitt, H. O. 1974. Antigen binding T and B lymphocytes. II. Studies on the inhibition of antigen binding to T cells and B cells by anti-immunoglobulin and anti-H-2 sera. *J. Immunol. 112:* 1734-40.

Hämmerling, G. J., Deak, B. D., Mauve, G., Hämmerling, U., and McDevitt, H. O. 1974. B lymphocyte alloantigens controlled by the I region of the major histocompatibility complex in mice. *Immunogenetics 1:* 68-81.

Hanna, N., Bhan, I., and Leskowitz, S. 1973. Hapten-specific helper affects in antibody formation. *Nature 244:* 569-70.

Harisdangkul, V., Kabat, E. A., McDonough, R. J., and Sigel, M. M. 1972a. A protein in normal nurse shark serum which reacts specifically with fructosans. I. Purification and immunochemical characterization. *J. Immunol. 108:* 1244-58.

Harisdangkul, V., Kabat, E. A., McDonough, R. J., and Sigel, M. M. 1972b. A protein in normal nurse shark serum which reacts specifically with fructosans. II. Physicochemical studies. *J. Immunol. 108:* 1259-70.

Harris, C. E. and Teller, D. D. 1973. Estimation of primary sequence homology from amino acid composition of evolutionary related proteins. *J. Theor. Biol. 38:* 347-62.

Hartline, H. K. 1968. Visual receptors and retinal interaction. In *Les Prix Nobel en 1967*, pp. 242-59. Stockholm: P. A. Norstedt & Soner.

Henkart, P., Humphreys, S., and Humphreys, T. 1973. Characterization of sponge aggregation factor. A unique proteglycan complex. *Biochemistry 12:* 345-50.

Herd, Z. L. and Ada, G. L. 1969. Distribution of [125]I-immunoglobulins IgG subunits and antigen-antibody complexes in rat lymph nodes. *Aust. J. Exp. Biol. Med. Sci. 47:* 73-80.

Hildemann, W. H. 1957. Scale homotransplantation in goldfish *(Carassius auratus). Ann. N.Y. Acad. Sci. 64:* 775.

Hildemann, W. H. 1970a. *Immunogenetics.* San Francisco: Holden-Day.

Hildemann, W. H. 1970b. Transplantation immunity in fishes: Agnatha, Chondrichthyes and Osteichthyes. *Transplant. Proc. 2:* 253-59.

Hildemann, W. H. 1972. Phylogeny of transplantation reactivity. In *Transplantation Antigens—Markers of Biological Individuality,* ed. B. D. Kahan, and R. A. Reisfeld, pp. 3-73. New York: Academic Press.

Hildemann, W. H. 1974. Some new concepts in immunological phylogeny. *Nature 250:* 116-20.

Hildemann, W. H. and Dix, T. G. 1972. Transplantation reactions of tropical Australia echinoderms. *Transplantation 4:* 624-33.

Hildemann, W. H. and Haas, R. 1959. Homotransplantation immunity and tolerance in the bullfrog. *J. Immunol. 83:* 478-85.

Hildemann, W. H. and Reddy, A. L. 1973. Phylogeny of immune responsiveness: marine invertebrates. *Fed. Proc. 32:* 2188-94.

Hildemann, W. H. and Thoenes, G. H. 1969. Immunological responses of pacific hagfish. I. Skin transplantation immunity. *Transplantation 7:* 506-21.

Hildemann, W. H., Dix, T. G., and Collins, J. D. 1974. Tissue transplantation in diverse marine invertebrates. *Curr. Top. Immunobiol. 4:* 141-50.

Hill R. L., Delaney, R., Fellows, R. E., Jr., and Lebovitz, H. E. 1966. The evolutionary origins of the immunoglobulins. *Proc. Natl. Acad. Sci. (U.S.A.) 56:* 1762-69.

Hilschmann, N. and Craig, L. C. 1965. Amino acid sequence studies with Bence-Jones proteins. *Proc. Natl. Acad. Sci. (U.S.A)* 53: 1403-09.

Hink, W. F. 1970. Immunity in insects. *Transplant. Proc. 2:* 233-35.

Hirschhorn, K. 1967. Derepression and differentiation of human peripheral lymphocytes *in vitro.* In *Ontogeny of Immunity,* ed. R. T. Smith, R. A. Good, and P. A. Miescher, pp. 49-55. Gainesville: University of Florida Press.

Hodgins, H. O., Weiser, R. S., and Ridgway, G. J. 1967. The nature of antibodies and the immune response in rainbow trout (*Salmo gairdneri*). *J. Immunol.* 99: 534-44.

Hofstad, M. S. 1953. Immunization of chickens against Newcastle Disease by formalin-inactivated vaccine. *Am. J. Vet. Res. 14:* 586-89.

Hogg, D. McC. and Jago, G. R. 1970. The antibacterial action of lactoperoxidase. The nature of the bacterial inhibitor. *Biochem. J. 117:* 779-90.

Hogg, N. M. and Greaves, M. F. 1972. Antigen-binding thymus-derived lymphocytes. II. Nature of the immunoglobulin determinants. *Immunology 22:* 967-80.

Hood, L. and Talmage, D. W. 1970. Mechanism of antibody diversity: germ line basis for variability. *Science 168:* 325-34.

Hood, L., Gray, W. R., Sanders, B. G., and Dreyer, W. J. 1967. Light chain evolution. *Cold Spring Harbor Symp. Quant. Biol. 32:* 133-46.

Hood, L., Eichmann, K., Lackland, H., Krause, R. M., and Ohms, J. J. 1970. Rabbit antibody light chains and gene evolution. *Nature 228:* 1040-44.

Horton, J. D. 1971. Ontogeny of the immune system in amphibians. *Am. Zool. 11:* 219-28.

Horton, J. D. and Manning, M. J. 1972. Response to skin allografts in *Xenopus laevis* following thymectomy at early stages of lymphoid organ maturation. *Transplantation 14:* 141-54.

Howard, J. G., Christie, G. H., Courtenay, B. M., Leuchars, E., and Davies, A. J. S. 1971. Studies on immunological paralysis. VI. Thymic independence of tolerance and immunity to type III pneumococcal polysaccharide. *Cell. Immunol. 2:* 614-26.

Huff, C. G. 1940. Immunity in invertebrates. *Physiol. Rev. 20:* 68-88.

Hume, D. *Enquiries Concerning the Human understanding and Concerning the Principles of Morals.* Reprinted from the posthumous edition of 1777 and edited with introduction, comparative tables, and analytical index by L. A. Selby-Bigge, 1963. Oxford: Clarendon Press.

Humphreys, T. D. 1970. Specificity of aggregation in porifera. *Transplant. Proc. 2:* 194-98.

Hunter, P., Munro, A. and McConnell, I. 1972. Properties of educated T cells for rosette formation and cooperation with B cells. *Nature* [*New Biol.*] *236:* 52-53.

Hurvitz, A. L., Kehoe, J. M., and Capra, J. D. 1971. Characterization of three homogeneous canine immunoglobulins. *J. Immunol. 107:* 648-54.

Ingram, V. M. 1963. *The Hemoglobulins in Genetics and Evolution.* New York: Columbia University Press.

Ivanyi, J. and Cerny, J. 1965. The effect of protein antigen dosage on its elimination from the blood and organs. *Folia Biol. Praha 11:* 335-48.

Jamieson, G. A. and Greenwalt, T. J., eds. 1971. *Glycoproteins of Blood Cells and Plasma.* Philadelphia: J. B. Lippincott.

Jensen, J. A. 1969. A specific inactivator of mammalian C′4 isolated from nurse shark (*Geriglymostoma cirratum*) serum. *J. Exp. Med. 130:* 217-41.

Jerne, N. K. 1971. The somatic generation of immune recognition. *Europ. J. Immunol. 1:* 1-9.

Jerry, L. M., Kunkel, H. G., and Grey, H. M. 1970. Absence of disulfide bonds linking the heavy and light chains: a property of a genetic variant of γA2 globulins. *Proc. Natl. Acad. Sci. (U.S.A.) 65:* 557-63.

Johnston, W. H., Jr., Acton, R. T., Weinheimer, P. F., Niedermeier, W., Evans, E. E., Shelton, E., and Bennett, J. C. 1971. Isolation and physicochemical characterization of the 'IgM-like' immunoglobulin from the stingray, *Dasyatis americana. J. Immunol 107:* 782-93.

Jordan, H. E. 1938. Comparative hematology. In *Handbook of Hematology.* Vol. II, ed. H. Downey, pp. 703-862. New York: Hoeber.

Kabat, D. 1972. Gene selection in hemoglobin and in antibody-synthesizing cells. *Science 175:* 134-40.

Kallman, K. D. 1970. Genetics of tissue transplantation in teleosts. *Transplant. Proc. 2:* 263-71.

Kalmutz, S. E. 1962. Antibody production in the opossum embryo. *Nature 193:* 851-53.

Kanyerezi, B., Jaton, J.-C., and Bloch, K. J. 1971. Human and rat γE: serological evidence of homology. *J. Immunol. 106:* 1411-12.

Karush, F. 1962. Immunologic specificity and molecular structure. *Adv. Immunol. 2:* 1-40.

Kassin, L. F. and Pevnitskii, L. A. 1969. Detection of antibody-forming cells in the turtle spleen by a modified method of local hemolysis in gel. *Bull. Exp. Biol. Med. 67:* 287-90.

Katz, D. H. and Benacerraf, B. 1972. The regulatory influence of activated T cells on B-cell responses to antigen. *Adv. Immunol. 15:* 1-94.

Katz, D. H. and Benacerraf, B. 1974. The role of histocompatibility gene products in cooperative cell interactions between T and B lymphocytes. In *The Immune System: Genes, Receptors, Signals,* ed. E. E. Sercarz, A. R. Williamson, and C. F. Fox, pp. 569-96. New York: Academic Press.

Kehoe, J. M. and Capra, J. D. 1972. Sequence relationships among the variable regions of immunoglobulin heavy chains from various mammalian species. *Proc. Natl. Acad. Sci. (U.S.A.) 69:* 2052-55.

Kehoe, J. M. and Fougereau, M. 1969. Immunoglobulin peptide with complement fixing activity. *Nature 224:* 1212-13.

Kehoe, J. M., Hurvitz, A. I., and Capra, J. D. 1972. Characterization of three feline paraproteins. *J. Immunol. 109:* 511-16.

Kent, S. P., Evans, E. E., and Attleberger, M. H. 1964. Comparative Immunology. Lymph nodes in the amphibian, *Bufo Marinus. Proc. Soc. Exp. Biol. Med. 116:* 456-59.

Kettman, J. R. and Dutton, R. W. 1971. Radioresistance of the enhancing effect of cells from carrier-immunized mice in an *in vitro* primary immune response. *Proc. Natl. Acad. Sci. (U.S.A.). 68:* 699-703.

Kimura, M. and Ohta, T. 1971. Protein polymorphism as a phase of molecular evolution. *Nature 229:* 467-69.

Kincade, P. W. and Cooper, M. D. 1971. Development and distribution of immunoglobulin-containing cells in the chicken. An immunofluorescent analysis using purified antibodies to $\mu$, $\gamma$ and light chains. *J. Immunol. 106:* 371-82.

Kincade, P. W., Lawton, A. R., Bockman, D. E., and Cooper, M. D. 1970. Suppression of immunoglobulin G synthesis as a result of antibody-mediated suppression of immunoglobulin M synthesis in chickens. *Proc. Natl. Acad. Sci. (U.S.A.) 67:* 1918-25.

Kindmark, C.-O. 1971. Stimulating effect of C-reactive protein on phagocytosis of various species of pathogenic bacteria. *Clin. Exp. Immunol. 8:* 941-48.

Klaus, G. G. B., Nitecki, D. E. and Goodman, J. W. 1971. Amino acid sequences of free and blocked N-termini of leopard shark immunoglobulins. *J. Immunol. 107:* 1250-58.

Klaus, G. G. B., Halpern, M. S., Koshland, M. E. and Goodman, J. W. 1971. A polypeptide chain from leopard shark 19S immunoglobulin analogous to mammalian J chain. *J. Immunol. 107:* 1785-87.

Klein, E. M. and Eskeland, T. 1971. Surface IgM on lymphoid cells. In *Cell Interactions and Receptor Antibodies in Immune Responses,* ed. O. Mäkelä, A. Cross and T. U. Kosunen, pp. 91-97. New York: Academic Press.

Klein, J. and Park, J. M. 1973. Graft-versus-host reaction across different regions of the H-2 complex of the mouse. *J. Exp. Med. 137:* 1213-55.

Klebanoff, S. J. 1967. Iodination of bacteria: a bactericidal mechanism. *J. Exp. Med. 126:* 1063-78.

Knopf, P. M., Parkhouse, R. M. E., and Lennox, E. S. 1967. Biosynthetic units of an immunoglobulin heavy chain. *Proc. Natl. Acad. Sci. (U.S.A.) 58:* 2288-95.

Koene, R., McKenzie, I. F. C., Painter, E., Sachs, D. H., Winn, H. J., and Russell, P. S. 1971. Soluble mouse histocompatibility antigens. *Transplant. Proc. III:* 231-33.

Kohler, H., Shimizu, A., Paul, C., Moore, V., and Putnam, F. W. 1970. Three variable-gene pools common to IgM, IgG and IgA immunoglobulins. *Nature 227:* 1318-20.

Kownatzki, E. 1973. Reassociation of IgM subunits in the presence and absence of J chain. *Immunol. Commun. 2:* 105-13.

Kraft, N., Shortman, K., and Marchalonis, J. 1971. Density distribution analysis of a primary antibody-forming cell response. *Immunology 20:* 919-30.

Krause, R. M. 1970. The search for antibodies with molecular uniformity. *Adv. Immunol. 12:* 1-56.

Kronvall, G., Seal, U. S., Finstad, J., and Williams, R. C., Jr. 1970. Phylogenetic insight into evolution of mammalian Fc fragment of $\gamma$G globulin using staphylococcal protein A. *J. Immunol. 104:* 140-47.

Krueger, R. G. and Twedt, R. M. 1963. Cellular demonstration of antibody production in *Rana pipiens. J. Immunol. 90:* 952-55.

Kubo, R. T., Rosenblum, I. Y., and Benedict, A. A. 1970. The unblocked N-terminal sequence of chicken IgG $\lambda$-like light chains. *J. Immunol. 105:* 534-36.

Kubo, R. T., Rosenblum, I. Y., and Benedict, A. A. 1971. Amino terminal sequences of heavy and light chains of chicken anti-dinitrophenyl antibody. *J. Immunol. 107:* 1781-84.

Kunkel, H. G. 1954. Zone electrophoresis. *Methods Biochem. Anal. 1:* 141-70.

Lachmann, P. J. and Thompson, R. A. 1970. Reactive lysis: the complement-mediated lysis of unsensitized cells. II. The characterization of activated factor as C56 and the participation of C8 and C9. *J. Exp. Med. 131:* 643-57.

Ladoulis, C. T., Gill, T. J. III, Chen, S.-H., and Misra, D. N. 1974. The structure and metabolism of lymphocyte membranes. *Prog. Allergy 18:* 205-88.

Lafferty, K. J. 1973. An alternate mechanism for immune recognition. In *The Biochemistry of Gene Expression in Higher Organisms,* eds.

J. K. Pollak and J. W. Lee, pp. 593-605. Sydney: Australia and New Zealand Book Co.

Landsteiner, K. 1962. *The Specificity of Serological Reactions.* New York: Dover Publications Inc.

Landsteiner, K. and van der Scheer, J. 1936. On cross reactions of immune sera to azoproteins. *J. Exp. Med. 63:* 325-39.

Langmuir, I. 1916. The constitution and fundamental properties of solids and liquids. *J. Am. Chem. Soc. 38:* 2221-95.

Lawrence, D. A., Spiegelberg, H. L., and Weigle, W. O. 1973. 2,4-Dinitrophenyl receptors on mouse thymus and spleen cells. *J. Exp. Med. 137:* 470-82.

Lebacq-Verheyden, A-M., Vaerman, J. P., and Heremans. J. F. 1972. A possible homologue of mammalian IgA in chicken serum and secretions. *Immunology 22:* 165-75.

Lefevre, M. E., Reincke, U., Arbas, R., and Gennaro, J. F. 1973. Lymphoid cells in the turtle bladder. *Anat. Rec. 176:* 111-20.

Legler, D. W. and Evans, E. E. 1966. Hemolytic complement in amphibia. *Proc. Soc. Exp. Biol. Med. 121:* 1158-62.

Legler, D. W., Evans, E. E., and Dupree, H. K. 1967. Comparative immunology: Serum complement of freshwater fishes. *Trans. Am. Fisheries Soc. 96:* 237-42.

Lennox, E. S. and Cohn, M. 1967. Immunoglobulins. *Annu. Rev. Biochem. 36:* 365-406.

Lerch, E. G., Huggins, S. E., and Bartel, A. H. 1967. Comparative immunology. Active immunization of young alligators with hemocyanin. *Proc. Soc. Exp. Biol. Med. 124:* 448-51.

Leslie, G. A. and Clem, L. W. 1969. Phylogeny of immunoglobulin structure and function. III. Immunoglobulins of the chicken. *J. Exp. Med. 130:* 1377-52.

Leslie, G. A. and Clem, L. W. 1972. Phylogeny of immunoglobulin structure and function. VI. 17S, 7.5S and 5.7S anti-DNP of the turtle, *Pseudamys scripta. J. Immunol. 108:* 1656-64.

Leslie, G. A., Clem, L. W., and Rowe, D. 1971. The molecular weight of human IgD heavy chains. *Immunochemistry 8:* 565-68.

Levin, A. S., Fudenberg, H. H., Hopper, J. E., Wilson, S. K., and Nisonoff, A. 1971. Immunofluorescent evidence for cellular control of synthesis of variable regions of light and heavy chains of immunoglobulins G and M by the same gene. *Proc. Natl. Acad. Sci. (U.S.A.) 68:* 169-71.

Liakopoulou, A. and Perelmutter, L. 1971. Antigenic relationship between human IgE immunoglobulin and rat homocytotropic antibody. *J. Immunol. 107:* 131-37.

Lin, H. H., Caywood, B. E., and Rowlands, D. T., Jr. 1971. Primary and secondary immune responses of the marine toad (*Bufo marinus*) to bacteriophage f2. *Immunology 20:* 373-80.

Litman, G. W., Frommel, D., Finstad, J., Howell, J., Pollara, B. W., and Good, R. A. 1970. The evolution of the immune response. VIII Structural studies of the lamprey immunoglobulin. *J. Immunol. 105:* 1278-85.

Litman, G. W., Frommel, D., Finstad, J., and Good, R. A. 1971a. The evolution of the immune response IX. Immunoglobulins of the bowfin: Purification and characterization. *J. Immunol. 106:* 747-54.

Litman, G. W., Frommel, D., Finstad, J., and Good, R. A. 1971b. Evolution of the immune response. X. Immunoglobulins of the bowfin: subunit nature. *J. Immunol. 107:* 881-86.

Litman, G. W., Frommel, D., Rosenberg, A., and Good, R. A. 1971c. N-terminal amino-acid sequence of African lungfish immunoglobulin light chains. *Proc. Natl. Acad. Sci. (U.S.A.) 68:* 2321-24.

Litman, G. W., Rosenberg, A., Frommel, D., Pollara, B., Finstad, J. and Good, R. A. 1971d. Biophysical studies of the immunoglobulins. The circular dichroic spectra of the immunoglobulins—a phylogenetic comparison. *Int. Arch. Allergy Appl. Immunol. 40:* 551-75.

Litwin, S. D. and Kunkel, H. G. 1967. The genetic control of $\gamma$-globulin heavy chains. Studies of the major heavy chain subgroup utilizing multiple genetic markers, *J. Exp. Med. 125:* 847-62.

Lopez, D. M., Sigel, M. M., and Lee, J. C. 1974. Phylogenetic studies on T cells. I. Lymphocytes of the shark with differential response to phytohemagglutinin and Concanavalin A. *Cell. Immunol. 10:* 287-93.

Lykakis, J. J. 1969. The production of two molecular classes of antibody in the toad, *Xenopus laevis*, homologous with mammalian $\gamma$M (19S) and $\gamma$G (7S) immunoglobulins. *Immunology 16:* 91-98.

Lynch, N. R. and Turner, K. J. 1974. Immediate hypersensitivity responses in the marsupial *Setonix brachyurus* (the quokka). Characterization of the homocytotropic antibody. *Aust. J. Exp. Biol. Med. Sci. 52:* 755-66.

Macela, A. and Romanovsky, A. 1969. The role of temperature in separate stages of the immune response in anurans. *Folia Biol. Praha 15:* 157-60.

Mach, B., Koblet, H., and Gros, D. 1967. Biosynthesis of immunoglobulin in a cell-free system. *Cold Spring Harbor Symp. Quant. Biol. 32:* 269-75.

Mach, J. P. and Pahud, J.-J. 1971. Secretory IgA, a major immunoglobulin in most bovine external secretions. *J. Immunol. 106:* 552-63.

MacLeod, C. M. and Avery, O. T. 1941. The occurrence during acute infections of a protein not normally present in the blood. III. Immunological properties of the C-reactive protein and its differentiation from normal blood proteins. *J. Exp. Med. 73:* 191-200.

Maniatis, G. M. and Ingram, V. M. 1971. Erythropoiesis during amphibian metamorphosis. II. Immunochemical study of larval and adult hemoglobins of *Rana catesbeiana. J. Cell. Biol. 49:* 380-89.

Maniatis, G. M., Steiner, L. A., and Ingram, V. M. 1969. Tadpole antibodies against frog hemoglobin and their effect on development. *Science 165:* 67-69.

Mann, D. L., Fahey, J. L., and Nathenson, S. G. 1970. Molecular comparisons of papain solubilized H-2 and HL-A alloantigens. In *Histocompatibility Testing,* ed. P. I. Terasaki, pp. 461-67. Copenhagen: Munksgaard.

Manning, M. J. and Horton, J. D. 1969. Histogenesis of lymphoid organs in larvae of the South African clawed toad, *Xenopus laevis* Daudin. *J. Embryol. Exp. Morph. 22:* 265-77.

Manski, W., Halbert, S. P., Javier, P., and Auerbach-Pascal, T. 1967. On the use of antigenic relationships among species for the study of molecular evolution. III. The lens proteins of the late Actinopterygii. *Int. Arch. Allergy Appl. Immunol. 31:* 529-45.

Marchalonis, J. J. 1969. Isolation and characterization of immunoglobulin-like proteins of the Australian lungfish *(Neoceratodus forsteri). Aust. J. Exp. Biol. Med. Sci. 47:* 405-19.

Marchalonis, J. J. 1970. Phylogenetic origins of antibody structure. *Transplant. Proc. 2:* 318-20.

Marchalonis, J. J. 1971a. Isolation and partial characterization of immunoglobulin of goldfish (*Carassius auratus*) and carp (*Cyprinus carpio*). *Immunology 20:* 161-73.

Marchalonis, J. J. 1971b. Immunoglobulins and antibody production in amphibians. *Am. Zool. 11:* 171:81.

Marchalonis, J. J. 1972. Conservation in the evolution of immunoglobulin $\mu$ chains. *Nature (New Biol.) 236:* 84-6.

Marchalonis, J. J. 1974a. Antibodies and surface immunoglobulins of immunized congenitally athymic (nu/nu) mice. *Aust. J. Exp. Biol. Med. Sci. 52:* 535-47.

Marchalonis, J. J. 1974b. Lymphocyte receptors for antigen. *J. Med. (Basel) 5:* 329-67.

Marchalonis, J. J. 1974c. Phylogenetic origins of antibodies and immune recognition. In Progress in Immunology II. ed. L. Brent and J. Holborow, vol. 2, pp. 249-59. Amsterdam: North-Holland.

Marchalonis, J. J. 1975. Lymphocyte surface immunoglobulins: Molecular properties and functions as receptors for antigen. *Science, 190:* 20-29.

Marchalonis, J. J. and Atwell, J. L. 1972. Phylogenetic emergence of distinct immunoglobulin classes. In *L'étude phylogenique et Ontogenique de la Réponse Immunitaire et son Apport a la Théorie Immunologique,* ed. P. Liacopoulos et J. Panijel, pp. 153-62. Boulogne: Inserm.

Marchalonis, J. J. and Cohen, N. 1973. Isolation and partial characterization of immunoglobulin from a urodele amphibian (*Necturus maculosis*). *Immunology 24:* 395-407.

Marchalonis, J. J. and Cone, R. E. 1973a. Biochemical and biological properties of lymphocyte surface immunoglobulin. *Transplant. Rev. 14:* 3-49.

Marchalonis, J. J. and Cone, R. E. 1973b. The phylogenetic emergence of vertebrate immunity. *Aust. J. Exp. Biol. Med. Sci. 51:* 461-88.

Marchalonis, J. J. and Edelman, G. M. 1965. Phylogenetic origins of antibody structure. I. Multichain structure of immunoglobulins in the smooth dogfish (*Mustelus canis*). *J. Exp. Med. 122:* 610-18.

Marchalonis, J. J. and Edelman, G. M. 1966a. Polypeptide chains of immunoglobulins from the smooth dogfish (*Mustelus canis*). *Science. 154:* 1567-68.

Marchalonis, J. J. and Edelman, G. M. 1966b. Phylogenetic origins of antibody structure. II. Immunoglobulins in the primary immune response of the bullfrog. *Rana catesbeiana. J. Exp. Med. 124:* 901-13.

Marchalonis, J. J. and Edelman, G. M. 1968a. Isolation and characterization of a hemagglutinin from *Limulus polyphemus. J. Mol. Biol. 32:* 453-65.

Marchalonis, J. J. and Edelman, G. M. 1968b. Phylogenetic origins of antibody structure. III. Antibodies in the primary immune response of the sea lamprey, *Petromyzon marinus. J. Exp. Med. 127:* 891-914.

Marchalonis, J. J. and Germain, R. N. 1971. Tolerance to a protein antigen in a poikilotherm, the marine toad, *Bufo marinus. Nature 231:* 321-22.

Marchalonis, J. J. and Gledhill, V. X. 1968. An elementary stochastic model for the induction of immunity and tolerance. *Nature 220:* 608-11.

Marchalonis, J. J. and Schonfeld, S. A. 1970. Polypeptide chain structure of stingray immunoglobulin. *Biochim. Biophys. Acta 221:* 604-11.

Marchalonis, J. J. and Weltman, J. K. 1971. Relatedness among proteins:

A new method of estimation and its application to immunoglobulins. *Comp. Biochem. Physiol. (B) 38:* 609-25.

Marchalonis, J. J., Allen, R. B., and Saarni, E. S. 1970. Immunoglobulin classes of the clawed toad, *Xenopus laevis. Comp. Biochem. Physiol. (L) 35:* 49-56.

Marchalonis, J. J., Atwell, J. L., and Haustein, D. 1974. Molecular properties of isolated surface immunoglobulins of human chronic lymphocytic leukemia cells. *Biochim. Biophys. Acta 351:* 99-112.

Marchalonis, J. J., Cone, R. E., and Atwell, J. L. 1972. Isolation and partial characterization of lymphocyte surface immunoglobulins. *J. Exp. Med. 135:* 956-71.

Marchalonis, J. J., Cone, R. E. and Rolley, R. T. 1973. Amplification of thymus-influenced lymphocytes by poly (A:U): Inhibition of antigen binding by antiserum to immunoglobulin light chain. *J. Immunol. 110:* 561-66.

Marchalonis, J. J., Ealey, E. H. M., and Diener, E. 1969. Immune response of the tuatara, *Sphenodon punctatum. Aust. J. Exp. Biol. Med. Sci. 47:* 367-80.

Marchalonis, J. J., Morris, P. J. and Harris, A. W. 1974. Speculations on the function of immune response gene products. *J. Immunogenet. 1:* 63-77.

Marchalonis, J. J., Cone, R. E., Atwell, J. L., and Rolley, R. T. 1973. Structure and function of lymphocyte surface immunoglobulin. In *The Biochemistry of Gene Activation in Higher Organisms*, ed. J. W. Lee and J. K. Pollack, pp. 629-47. Sydney: Australian and New Zealand Publishing Co.

Margoliash, E., Schenck, J. R., Hargie, M. P., Burokas, S., Richter, W. R., Barlow, G. H., and Moscona, A. A. 1965. Characterization of specific cell-aggregating materials from sponge cells. *Biochem. Biophys. Res. Commun. 20:* 383-88.

Martin, R. G. and Ames, B. N. 1961. A method for determining the sedimentation behavior of enzymes: Application to protein mixtures. *J. Biol. Chem. 236:* 1372-79.

Mason, S. L. and Warner, N. L. 1970. The immunoglobulin nature of the antigen recognition site on cells mediating transplantation immunity and delayed hypersensitivity. *J. Immunol. 104:* 762-65.

Maung, R. T. 1963. Immunity in the tortoise, *Testudo ibera. J. Pathol. Bact. 85:* 51-66.

McCullach, P. 1973. The transfer of immunological tolerance with tolerant lymphocytes. *Aust. J. Exp. Biol. Med. Sci. 51:* 445-59.

McDevitt, H. O. 1972. In *Genetic Control of Immune Responsiveness*, ed. H. O. McDevitt and M. Landy, pp. 92-101. New York: Academic Press.

McDevitt, H. O. and Landy, M. 1972. *Genetic Control of Immune Responsiveness.* New York: Academic Press.

McDevitt, H. O., Bechtol, K. B., Hammerling, G. J., Lonai, P., and Delovitch, T. L. 1974. Ir genes and antigen recognition. In *The Immune System: Genes, Receptors, Signals*, ed. E. E. Sercarz, A. R. Williamson, and C. F. Fox, pp. 597-632. New York: Academic Press.

McGregor, L. A. 1972. Immunology of malarial infection and its possible consequences. *Br. Med. Bull. 28:* 22-27.

McKay, D. and Jenkin, C. R. 1970. Immunity in the invertebrates. The role of serum factors in phagocytosis of erythrocytes by haemocytes of the freshwater crayfish (*Parachaeraps bicarinatus*). *Aust. J. Exp. Biol. Med. Sci. 48:* 139-50.

McKenzie, I. F. C. 1975. Ly 4.2: A cell membrane antigen of murine B lymphocytes. II. Functional studies. *J. Immunol., 114:* 856-62.

McKinney, E. C. and Sigel, M. M. 1974. Cell migration inhibition in the gar, *Lepisosteus platyrhincus. Abstr. Ann. Meeting Am. Soc. Microbiol. M 400.*

Mehta, P. D., Reichlin, M., and Tomasi, T. B., Jr. 1972. Comparative studies of vertebrate immunoglobulins. *J. Immunol. 109:* 1272-77.

Merler, E., Karlin, L., and Matsumoto, S. 1968. The valency of human immunoglobulin antibody. *J. Biol. Chem. 243:* 386-90.

Mestecky, J., Zikan, J., and Butler, W. T. 1971. Immunoglobulin M and secretory immunoglobulin A: presence of a common polypeptide chain different from light chains. *Science 171:* 1163-65.

Metcalf, D. and Moore, M. A. S. 1971. *Haemopoietic cells.* Amsterdam: North-Holland Publishing Company.

Metchnikoff, E. 1884. Uber eine Sprospilzkrank-hert der Daphmin Beitrag, zur Lehre uber den Kampf der Phagocyten gegen Krankheitserreger. *Virchow's Arch. (Zellpathol.) 96:* 177-95.

Metchnikoff, E. 1905. *Immunity in Infective Diseases.* Cambridge: Cambridge University Press.

Metzger, H. 1970. Structure and function of $\gamma$M macroglobulins. *Adv. Immunol. 12:* 57-116.

Metzger, H., Shapiro, M. B., Mosimann, J. E., and Vinton, J. E. 1968. Assessment of compositional relatedness between proteins. *Nature 219:* 1166-68.

Michelson, E. H. 1963. Development and specificity of miracidial immobilizing substances in extracts of the snail, *Australorbis glabratus*, exposed to various agents. *Ann. N.Y. Acad. Sci. 113:* 486-91.

Miller, F. and Metzger, H. 1965. Characterization of human macroglobulin. I. The molecular weight of its subunit. *J. Biol. Chem. 240:* 3325-33.

Miller, J. F. A. P. 1961. Immunological function of the thymus. *Lancet,* 2: 748-49.

Miller, J. F. A. P. 1972. Lymphocyte interactions in antibody responses. *Int. Rev. Cytol. 33:* 77-130.

Miller, J. F. A. P. and Mitchell, G. F. 1969. Thymus and antigen-reactive cells. *Transplant. Rev. 1:* 3-42.

Miller, J. F. A. P., Block, M., Rowlands, D. T. Jr., and Kind, P. 1965. Effect of thymectomy on hematopoietic organs of the opossum "embryo." *Proc. Soc. Exp. Biol. Med. 118:* 916-21.

Mitchell, G. F. 1974. T-cell mediated regulation of antibody production *in vivo.* In *Progress in Immunology II.* vol. 3, ed. L. Brent and J. Holborow, pp. 89-98, Amsterdam: North Holland.

Mitchison, N. A. 1964. Induction of immunological paralysis in two zones of dosage. *Proc. R. Soc. London (B.) 161.* 275-92.

Mitchison, N. A. 1971a. The carrier effect in the secondary response to hapten-carrier conjugates. I. Measurement of the effect with transferred cells and objections to the local environment hypothesis. *Eur. J. Immunol. 1:* 10-17.

Mitchison, N. A. 1971b. The relative ability of T and B lymphocytes to see protein antigens. In *Cell interactions and receptor antibodies in immune responses,* ed. O. Makela, A. Cross, and T. U. Kosunen, pp. 249-60. New York: Academic Press.

Mitchison, N. A., Rajewsky, K., and Taylor, R. B. 1970. Cooperation of antigenic determinants and of cells in the induction of antibodies. In *Developmental Aspects of Antibody Formation and Structure,* ed. J. Sterzl and I. Riha, pp. 547-61. Prague: Czechoslovak Academy of Sciences.

Möller, E., Bullock, W. W., and Mäkelä, O. 1973. Affinity of T and B lymphocyte receptors for hapten determinants. *Eur. J. Immunol. 3:* 172-79.

Möller, G., Andersson, J., Pohlit, H., and Sjöberg, O. 1973. Quantitation of the number of mitogen molecules activating DNA synthesis in T and B lymphocytes. *Clin. Exp. Immunol. 13:* 89-99.

Moroz, C. and Hahn, J. 1973. Cell surface immunoglobulin of human thymus cells and its *in vitro* biosynthesis. *Proc. Natl. Acad. Sci. (U.S.A.) 70:* 3716-20.

Moroz, C. and Lahat, N. 1974. *In vitro* biosynthesis and molecular arrangement of surface immunoglobulin of mouse thymus cells. In *The Immune System: Genes, Receptors, Signals,* ed. E. E. Sercarz, A. R. Williamson and C. F. Fox, pp. 233-46. New York: Academic Press.

Müller-Eberhard, H. J. 1968. Chemistry and reaction mechanisms of complement. *Adv. Immunol. 8:* 1-80.

Murakawa, S. 1968. Studies on transplantation immunity in the Japanese newt, *Cynops pyrrhogaster. Sabco J. 4:* 17-32.

Naff, G. B., Pensky, J., and Lepow, I. H. 1964. The macromolecular nature of the first component of human complement. *J. Exp. Med. 119:* 593-613.

Nakamuro, K., Tanigaki, N., and Pressman, D. 1973. Multiple common properties of human $\beta_2$ microglobulin and the common portion fragments derived from H2-A antigens. *Proc. Natl. Acad. Sci. (U.S.A.) 70:* 2863-65.

Nash, D. R. and Mach, J. P. 1971. Immunoglobulin classes in aquatic mammals. Characterization by serologic cross-reactivity, molecular size and binding of human free secretory component. *J. Immunol. 107:* 1424-30.

Nash, D. R., Vaerman, J. P., Bazin, H., and Heremans, J. F. 1969. Identification of IgA in rat serum and secretions. *J. Immunol. 103:* 145-48.

Natvig, J. B. and Kunkel, H. G. 1973. Human immunoglobulins: classes, subclasses, genetic variants, and idiotypes. *Adv. Immunol. 16:* 1-59.

Natvig, J. B., Kunkel, H. G., and Litwin, S. D. 1967. Genetic markers of the heavy chain subgroups of human $\gamma$G globulin. *Cold Spring Harbor Symp. Quant. Biol. 32:* 173-80.

Nieuwkoop, P. D. and Faber, J. 1967. *Normal Table of Xenopus laevis*, second edition. Amsterdam: North Holland Publishing Company.

Nisonoff, A., Reichlin, M., and Margoliash, E. 1970. Immunological activity of cytochrome c. II. Localization of a major antigenic determinant of human cytochrome c. *J. Biol. Chem. 245:* 940-46.

Noguchi, H. 1903. A study of immunization—haemolysins, agglutinins, precipitins and coagulins in cold-blooded animals. *Centralbl. f. Bakt. Abt. Orig. 33:* 353-62.

Nolan, C. and Margoliash, E. 1968. Comparative aspects of primary structures of proteins. *Annu. Rev. Biochem. 37:* 727-90.

Nossal, G. J. V. 1967. Mechanisms of antibody production. *Annu. Rev. Med. 18:* 81-96.

Nossal, G. J. V. 1974. Principles of immunological tolerance and immunocyte receptor blockade. *Adv. Cancer Res., 20:* 93-130.

Nossal, G. J. V. and Ada, G. L. 1971. *Antigens, Lymphoid Cells and the Immune Response.* N.Y.: Academic Press.

Nossal, G. J. V., Ada, G. L., and Austin, C. M. 1965. Antigens in immunity. IX. The antigen content of single antibody-forming cells. *J. Exp. Med. 121:* 945-54.

Nossal, G. J. V., Austin, C. M. and Ada, G. L. 1965. Antigens in immunity. VII. Analysis of immunological memory. *Immunology 9:* 333-48.

Nossal, G. J. V., Warner, N. L., and Lewis, H. 1971. Incidence of cells simultaneously secreting IgM and IgG antibody to sheep erythrocytes. *Cell. Immunol. 2:* 41-53.

Nossal, G. J. V., Szenberg, A., Ada, G. L., and Austin, C. M. 1964. Single cell studies on 19S antibody production. *J. Exp. Med. 119:* 485-502.

Nossal, G. J. V., Warner, N. L., Lewis, H., and Sprent, J. 1972. Quantitative features of a sandwich radioimmunolabeling technique for lymphocyte surface receptors. *J. Exp. Med. 135:* 405-28.

Nuttal, G. 1888. Experimente über die bacterienfeindlichen Einflüsse des thierischen Körpers. *Z. Hyg. 4:* 353-94.

O'Daly, J. A. and Cebra, J. J. 1971. Chemical and physicochemical studies of the component polypeptide chains of rabbit secretory immunoglobulin A. *Biochemistry 10:* 3843-50.

Ogilvie, B. M. and Jones, V. E. 1969. Protective immunity in helminth diseases. *Proc. R. Soc. Med. 62:* 298-301.

Oka, H. 1970. Colony specificity in compound ascidians. In *Profiles of Japanese Science and Scientists*, ed. H. Yukawa, pp. 196-205. Tokyo: Kodanska.

Oka, H. and Watanabe, H. 1960. Problems of colony specificity in compound ascidians. *Bulletin of the Marine Biological Station of Asamushi 10:* 153-59.

Olins, D. E. and Edelman, G. M. 1964. Reconstitution of 7S molecules from L and H polypeptide chains of antibodies and γ-globulins. *J. Exp. Med. 119:* 789-815.

Onoue, K., Grossberg, A. L., Yagi, Y., and Pressman, D. 1968. Immunoglobulin M antibodies with 10 combining sites. *Science 162:* 574-76.

Orlans, E. 1967. Fowl antibody VIII. A comparison of natural, primary and secondary antibodies to erythrocytes in hen sera; their transmission to yolk and chick. *Immunology 12:* 27-37.

Overton, J. 1966. The fine structure of blood cells in the ascidian, *Perophora viridis. J. Morphol. 119:* 305-26.

Owen, R. D. 1945. Immunogenetic consequences of vascular anastomoses between bovine twins. *Science 102:* 400-401.

Padlan, E. A., Segal, D. M., Cohen, G. H., and Davies, D. R. 1974. The three dimensional structure of the antigen binding site of the McPc 603 protein. In *The Immune System: Genes, Receptors, Signals*, ed. E. E. Sercarz, A. R. Williamson, and C. F. Fox, pp. 7-17. New York: Academic Press.

Papermaster, B. W., Condie, R. M., Finstad, J., and Good, R. A. 1964.

Evolution of the immune response. I. The phylogenetic development of adaptive immunologic responsiveness in vertebrates. *J. Exp. Med. 119:* 105-30.

Parish, C. R. and Marchalonis, J. J. 1970. A simple and rapid acrylamide gel method for estimating the molecular weights of proteins and protein subunits. *Anal. Biochem. 34:* 436-50.

Parkhouse, R. M. E., Askonas, B. A., and Dourmashkin, R. R. 1970. Electron microscopic studies of mouse immunoglobulin M; structure and reconstitution following reduction. *Immunology 18:* 575-84.

Patten, W. 1912. *The evolution of the vertebrates and their kin.* Philadelphia: Blakiston.

Pauling, L. 1940. A theory of the structure and process of formation of antibodies. *J. Am. Chem. Soc. 62:* 2643-57.

Pellegrino, M. A., Ferrone, S., Mittal, K. K., Götze, D., Terasaki, P. I., and Reisfeld, R. A. 1974. Cross-reactivity between human and murine lymphocyte antigens: III. Reactivity of H-2 Allo and zenoantisera with human lymphoid cells. *Immunogenetics 1:* 158-73.

Perlmann, P. and Holm, G. 1969. Cytotoxic effects of lymphoid cells *in vitro. Adv. Immunol. 11:* 117-93.

Pernis, B., Forni, L., and Amante, L. 1971. Immunoglobulins as cell receptors. *Ann. N.Y. Acad. Sci. 190:* 429-31.

Peterson, P. A., Cunningham, B. A., Berggard, I. and Edelman, G. M. 1972. $\beta_2$-microglobulin—a free immunoglobulin domain. *Proc. Natl. Acad. Sci. (U.S.A.) 69:* 1697-1701.

Phillips, J. H. 1966. Immunological processes and recognition of foreignness in the invertebrates. In *Phylogeny of Immunity,* ed. R. T. Smith, P. A. Miescher, and R. A. Good, pp. 133-39. Gainesville: University of Florida Press.

Pickering, R. J., Wolfson, M. R., Good, R. A., and Gewurz, H. 1969. Passive hemolysis by serum and cobra venom factor: a new mechanism inducing membrane damage by complement. *Proc. Natl. Acad. Sci. (U.S.A.) 62:* 521-27.

Pillemer, L., Schoenberg, H., Blum, L., and Wurz, L. 1955. Properdin system and immunity. II. Interaction of the properdin system with polysaccharides. *Science 122:* 545-49.

Pollara, B., Finstad, J. and Good, R. A. 1966. The phylogenetic development of immunoglobulins. In *Phylogeny of Immunity*, ed. R. T. Smith, P. A. Miescher and R. A. Good, pp. 88-97. Gainesville: University of Florida Press.

Pollara, B., Suran, A., Finstad, J., and Good, R. A. 1968. N-terminal amino acid sequences of immunoglobulin chains in *Polyodon spathula. Proc. Natl. Acad. Sci. (U.S.A.) 59:* 1307-12.

Pollara, B., Cain, W. A., Finstad, J., and Good, R. A. 1969. The amphibian as a key step in the evolution of lymphoid tissue and diverse immunoglobulin classes. In *Biology of Amphibian Tumors,* ed. M. Mizell, pp. 177-83. New York: Springer-Verlag.

Porter, R. R. 1959. The hydrolysis of rabbit γ-globulin and antibodies with crystalline papain. *Biochem. J. 73:* 119-26.

Prager, E. M. and Wilson, A. C. 1971. The dependence of immunological cross-reactivity upon sequence resemblance among lysozymes. I. Microcomplement fixation studies. *J. Biol. Chem. 246:* 5978-89.

Premkumar, M., Shoyab, M., and Williamson, A. R. 1974. Germ line basis for antibody diversity: Immunoglobulin $V_H$ and $C_H$ gene frequences measured by hybridization. *Proc. Natl. Acad. Sci. (U.S.A.) 71:* 99-103.

Press, E. M., Fleet, G. W. J., and Fisher, C. E. 1971. Affinity labeling of rabbit antibodies with ε-4-azido-2-nitrophenyl lysine. In *Progress in Immunology,* ed. B. Amos, pp. 233-41. New York: Academic Press.

Putnam, F. W. 1957. Aberrations of protein metabolism in multiple myeloma. Interrelationships of abnormal serum globulins and Bence-Jones proteins. *Physiol. Rev. 37:* 512-38.

Putnam, F. W., Shimizu, A., Paul, C., and Shinoda, T. 1972. Variation and homology in immunoglobulin heavy chains. *Fed. Proc. 31:* 193-205.

Putnam, F. W., Florent, G., Paul, C., Shinoda, T., and Shimizu, A. 1973. Complete amino acid sequence of the mu heavy chain of a human IgM immunoglobulin. *Science. 182:* 287-91.

Rabellino, E., Colon, S., Grey, H. M., and Unanue, E. R. 1971. Immunoglobulins on the surface of lymphocytes. I. Distribution and quantitation. *J. Exp. Med. 133:* 156-67.

Raff, M. C. 1971. Surface antigenic markers for distinguishing T and B lymphocytes in mice. *Transplant. Rev. 6:* 52-80.

Rapaport, F. T. 1972. The biological significance of cross reactions between histocompatibility antigens and antigens of bacterial and/or heterologous mammalian origin. In *Histocompatibility Antigens: Markers of Biological Individuality,* ed. B. D. Kahan and R. A. Reisfeld, pp. 181-208. New York: Academic Press.

Ray, A. and Cebra, J. J. 1972. Localization of affinity-labeled residues in the primary structure of anti-dinitrophenyl antibody raised in strain 13 guinea pigs. *Biochemistry 11:* 3647-57.

Reinisch, C. L. and Bang, F. B. 1971. Cell recognition: reactions of the sea star (*Asterias vulgaris*) to the injection of amebocytes of sea urchin (*Arbacia punctulata*). *Cell. Immunol. 2:* 496-503.

Reisfeld, R. A. and Kahan, B. D. 1971. Extraction and purification of soluble histocompatibility antigens. *Transplant. Rev. 6:* 81-112.

Ridgway, G. J., Hodgins, H. O., and Klontz, G. W. 1966. The immune response in teleosts. In *Phylogeny of Immunity*, ed. R. T. Smith, P. A. Miescher, and R. A. Good, pp. 199-207. Gainesville: University of Florida Press.

Rieber, E. P. and Riethmuller, G. 1974a. Surface immunoglobulin on thymus cells. I. Increased immunogenicity of heterologous anti-Ig bound to thymus cells. *Z. Immunitaetsforsch. 147:* 262-75.

Riethmuller, G., Rieber, E. P., and Seeger, I. 1971. Suppression of graft-versus-host reaction by univalent antiglobulin antibody. *Nature 230:* 248.

Robert, L., Fayolle, J., Derouette S., and Zabriskie, J. 1972. Homology of amino acid composition of structural glycoproteins, transplantation antigens, cell wall glycoproteins, and streptococcus A cell membrane. *Transplant. Proc. 4:* 415-18.

Roelants, G. E., Forni, L. and Pernis, B. 1973. Blocking and redistribution ("capping") of antigen receptors on T and B lymphocytes by anti-immunoglobulin antibody. *J. Exp. Med. 137:* 1060-77.

Roelants, G. E., Ryden, A., Hagg, L. B., and Loor, F. 1974. Active synthesis of immunoglobulin receptors for antigen by T lymphocytes. *Nature 247:* 106-09.

Roitt, I. M., Greaves, M. F., Torrigiani, G., Brostoff, J., and Playfair, J. H. L. 1969. The cellular basis of immunological responses. *Lancet 2:* 367-71.

Rollinghoff, M., Wagner, H., Cone, R. E., and Marchalonis, J. J. 1973. Syngeneic tumor immunity *in vitro:* Release of antigen-specific immunoglobulin from cytotoxic effector cells. *Nature [New Biol.] 243:* 21-23.

Rolley, R. T. and Marchalonis, J. J. 1972. Release and assay of antigen-binding immunoglobulin from the surface of lymphocytes of unsensitized mice. *Transplantation 14:* 734-41.

Romano, E. L., Geczy, C. L. and Steiner, L. A. 1973. Reaction of frog antiserum with guinea pig complement. *Immunochemistry 10:* 655-57.

Romer, A. S. 1962. *The Vertebrate Body*. Philadelphia: W. B. Saunders.

Romer, A. S. 1966. *Vertebrate Paleontology*. Chicago: University of Chicago Press.

Romer, A. S. 1967. Major steps in vertebrate evolution. *Science 158:* 1629-37.

Rose, M. E. and Orlans, E. 1962. Fowl antibody: III. Its haemolytic activity with complements of various species and some properties of fowl complement. *Immunology 5:* 633-41.

Rosenquist, G. L. and Hoffman, R. Z. 1972. The production of anti-DNP antibody in the bullfrog, *Rana catesbeiana. J. Immunol. 108:* 1499-1505.

Ross, G. D. and Jensen, J. A. 1973a. The first component (C in) of the complement system of the nurse shark (*Ginglymostoma cirratum*). I. Hemolytic characteristics of partially purified C in. *J. Immunol. 110:* 175-82.

Ross, G. D. and Jensen, J. A. 1973b. The first component (C in) of the complement system of the nurse shark (*Ginglymostoma cirratum*). II. Purification of the first component by ultracentrifugation and studies of its physicochemical properties. *J. Immunol. 110:* 911-18.

Roth, S., McGuire, E. J., and Roseman, S. 1971. Evidence for cell surface glycosyltransferases. Their potential role in cellular recognition. *J. Cell Biol. 51:* 536-47.

Roubal, W. T., Etlinger, H. M., and Hodgins, H. O. 1974. Spin-label studies of a hapten combining site of rainbow trout antibody. *J. Immunol. 113:* 309-15.

Rouse, B. T. and Warner, N. L. 1972. Suppression of graft-versus-host reactions in chickens by pretreatment of leucocytes with anti-light chain sera. *Cell. Immunol. 3:* 470-77.

Rowlands, D. T. Jr. 1969. General mechanisms and principles of immunity in cold blooded vertebrates. In *Immunity to Parasitic Animals*, Vol. I. G. J. Jackson, R. Herman, and I. Singer, pp. 231-48. New York: Appleton-Century-Crofts.

Rowlands, D. T. Jr. and Dudley, M. A. 1968. The isolation of immunoglobulins of the adult opossum (*Didelphys virginiana*). *J. Immunol. 100:* 736-46.

Rowlands, D. T. Jr., Blakeslee, D., and Lin, H. H. 1972. The early immune response and immunoglobulins of opossum embryos. *J. Immunol. 108:* 941-46.

Rowlands, D. T. Jr., La Via, M. F., and Block, M. H. 1964. The blood-forming tissues and blood of the newborn opossum (*Didelphys virginiana*) II. Ontogenesis of antibody formation to flagella of *Salmonella typhimurium. J. Immunol. 93:* 157-64.

Ruben, L. N. 1970. Immunological maturation and lymphoreticular cancer transformation in larval *Xenopus laevis*, the South African clawed toad. *Dev. Biol. 22:* 43-58.

Ruben, L. N. 1975. Ontogeny, phylogeny and cellular cooperation. *Amer. Zool. 15:* 93-106.

Ruben, L. N., van der Hoven, A., and Dutton, R. W. 1973. Cellular cooperation in hapten-carrier response in the newt, *Triturus viridescens. Cell. Immunol. 6:* 300-14.

Rubin, B. and Wigzell, H. 1973. Hapten-reactive helper lymphocytes. *Nature 242:* 467-69.

Rutishauser, U. and Edelman, G. M. 1972. Binding of thymus- and bone marrow-derived lymphoid cells to antigen-derivatized fibers. *Proc. Natl. Acad. Sci. (U.S.A.) 69:* 3774-78.

Sailendri, K. Ph. D. Thesis, Maduri University, Madurai, India, 1973. Cited by Cohen, N. Phylogenetic emergence of lymphoid tissues and cells. In *The Lymphocyte: Structure and Function*, ed. J. J. Marchalonis, New York: Marcel Dekker, Inc., forthcoming.

Salt, G. 1963. The defence reactions of insects to metazoan parasites. *Parasitology 53:* 527-642.

Saluk, P. H., Drauss, J., and Clem, L. W. 1970. The presence of two antigenically distinct light chains ($\kappa$ and $\lambda$) in alligator immunoglobulins. *Proc. Soc. Exp. Biol. Med. 133:* 365-69.

Sanders, B. G., Travis, J. C., and Wiley, K. L. 1973. Chicken low molecular weight immunoglobulin heavy chains: a comparison with the gamma chain of man. *Comp. Biochem. Physiol. (B) 45:* 189-95.

Sarma, V. R., Silverton, E. W., Davies, D. R., and Terry, W. D. 1971. The three-dimensional structure at 6Å resolution of a human $\gamma$G1 immunoglobulin molecule. *J. Biol. Chem. 246:* 3753-59.

Sarvas, H. and Mäkelä, O. 1970. Haptenated bacteriophage in the assay of antibody quantity and affinity: maturation of an immune response. *Immunochemistry 7:* 933-43.

Scatchard, G. 1949. The attractions of proteins for small molecules and ions. *Ann. N.Y. Acad. Sci. 51:* 660-72.

Scharff, M. D. 1967. The assembly of gamma globulin in relation to its synthesis and secretion. In *Nobel Symposium 3, Gammaglobulins: Structure and Control of Biosynthesis*, ed. J. Killander, pp. 385-98. Stockholm: Almgrist and Witsell.

Schierman, L. W. and Nordskog, A. W. 1963. Influence of the B blood group histocompatibility locus in chickens on a graft-versus-host reaction. *Nature 197:* 511-12.

Schmid, K., Kaufmann, H., Isemura, S., Bauer, F., Emura, J., Motoyama, T., Ishigeiro, M., and Nanno, S. 1973. Structure of $\alpha$-acid glycoprotein. The complete amino acid sequence, multiple amino acid substitutions, and homology with the immunoglobulins. *Biochemistry 12:* 2711-24.

Scott, M. T. 1971. Recognition of foreignness in invertebrates. Transplantation studies using the American cockroach (*Periplaneta americana*). *Transplantation. 11:* 78-86.

Scott, M. T. 1972. Partial characterization of the hemagglutinating activity on hemolymph of the American cockroach (*Periplaneta americana*). *J. Invertebr. Pathol. 19:* 66-71.

Seaman, G. R. and Robert, N. L. 1968. Immunological response of male cockroaches to injection of *Tetrahymena pyriformis. Science 161:* 1359-61.

Segall, M., Schendel, D. J., and Zur, K. S. 1973. MLC workshop summary report. *Tissue Antigens 3:* 353-57.

Segrest, J. P., Jackson, R. L., Andrews, E. P., and Marchesi, V. T. 1971. Human erythrocyte membrane glycoprotein: a reevaluation of the molecular weight as determined by SDS polyacrylamide gel electrophoresis. *Biochem. Biophys. Res. Commun. 44:* 390-94.

Sell, S. and Sheppard, H. W. Jr. 1973. Rabbit blood lymphocytes may be T cells with surface immunoglobulins. *Science 182:* 586-87.

Shelton, E. and Smith, M. 1970. The ultrastructure of carp (*Cyprinus carpio*) immunoglobulin: a tetrameric macroglobulin. *J. Mol. Biol. 54:* 615-17.

Shortman, K. 1972. Physical procedures for the separation of animal cells. *Annu. Rev. Biophys. Bioengin. 1:* 93-130.

Shreffler, C. D., David, C. S., Passmore, H. C., and Klein, J. 1971. Genetic organization and evolution of the mouse H-2 region: a duplication model. *Transplant. Proc. 3:* 176-79.

Shumway, W. 1940. Stages in the normal development of *Rana pipiens.* I. External Form. *Anat. Rec. 78:* 139-47.

Shuster, J., Warner, N. L., and Fudenberg, H. H. 1969. Cross-reactivity of primate immunoglobulins. *Ann. N.Y. Acad. Sci. 162:* 195-201.

Sidky, Y. A. and Auerbach, R. 1968. Tissue culture analysis of immunological capacity of snapping turtles. *J. Exp. Zool. 167:* 187-96.

Sigel, M. M. and Clem, L. W. 1965. Antibody response of fish to viral antigens. *Ann. N.Y. Acad. Sci. 126:* 662-77.

Sigel, M. M. and Clem, L. W. 1966. Immunologic anamnesis in elasmobranchs. In *Phylogeny of Immunity,* ed. R. T. Smith, P. A. Miescher, and R. A. Good, pp. 190-97. Gainesville: University of Florida Press.

Sigel, M. M., Ortiz-Muniz, G., Lee, J. C., and Lopez, D. M. 1972. Immunobiological reactivities at the cellular level in the nurse shark. In *L'étude Phylogenique et Ontogenique de la Réponse Immunitaire et son Apport a la Théorie Immunologique,* ed. P. Liacopoulos and J. Panijel, pp. 113-19. Boulogne: Inserm.

Simonsen, M. 1957. The impact on the developing embryo and newborn animal of adult homologous cells. *Acta Pathol. Microbiol. Scand. 40:* 480-500.

Simpson, G. G. 1953. *The Major Features of Evolution.* New York: Columbia University Press.

Singer, S. J. and Doolittle, R. F. 1966. Antibody activity sites and immunoglobulin molecules. *Science 153:* 13-25.

Sirotinin, N. N. 1960. A comparative physiological study of the mechanism of antibody formation. In *Mechanism of antibody formation*, ed. M. Holub and L. Jaroskova, pp. 113-17. Prague: Academia.

Sledge, C., Clem, L. W. and Hood, L. 1974. Antibody structure: amino terminal sequences of nurse shark light and heavy chains. *J. Immunol. 112:* 941-48.

Smith, A. M. and Potter, M. 1969. Teleost anti-dinitrophenyl bovine serum albumin (DNP-BSA) precipitation antibody having cross reactivity with purine and pyrimidine substituted BSA and deoxyribonucleic acid (DNA). *Fed. Proc. 28:* 819.

Smith, G. P. 1973. *The Variation and Adaptive Expression of Antibodies.* Cambridge: Harvard University Press.

Smith, R. T. and Bridges, R. A. 1958. Immunological unresponsiveness in rabbits produced by neonatal injection of defined antigens. *J. Exp. Med. 108:* 227-50.

Smith, W. I., Ladoulis, C. T., Misra, D. N., Gill, T. J. III, and Bazin, H. 1975. Lymphocyte plasma membranes. III. Composition of lymphocyte plasma membranes from normal and immunized rats. *Biochem, Biophys. Acta 382:* 506-25.

Smithies, O. 1967. The genetic basis of antibody variability. *Cold Spring Harbor Symp. Quant. Biol. 32:* 161-66.

Smithies, O. 1970. Pathways through networks of branched DNA. *Science 172:* 574-77.

Smithies, O. and Poulik, M. D. 1972. Dog homologue of human $\beta_2$ microglobulin. *Proc. Natl. Acad. Sci. (U.S.A.) 69:* 2914-17.

Smithies, O., Gibson, D. M., Fanning, E. M., Percy, M. E., Parr, D. M. and Connell, G. E. 1970. Deletions in immunoglobulin polypeptide chains as evidence for breakage and repair in DNA. *Science 172:* 574-77.

Snell, G. D., Cherry, M., and Demant, P. 1971. Evidence that H-2 private specificities can be arranged in two mutually exclusive systems possibly homologous with two subsystems of HL-A. *Transplant. Proc. 3:* 183-86.

Sprent, J. F. A., 1959. Parasitism, immunity, and evolution. In *The Evolution of Living Organisms*, ed. G. W. Leeper, pp. 149-165. Melbourne: Melbourne University Press.

Stanton, T. C., Sledge, C., Capra, J. D., Woods, R., Clem, W., and Hood, L., 1974. A sequence restriction in the variable region of immunoglobulin light chains from sharks, birds, and mammals. *J. Immunol. 112:* 633-40.

Stavnezer, J. and Huang, R. C. C. 1971. Synthesis of a mouse immunoglobulin light chain in a rabbit reticulocyte cell-free system. *Nature [New Biol.] 230:* 172-76.

Stephens, J. M. 1959. Immune responses of some insects to some bacterial antigens. *Can. J. Microbiol. 5:* 203-28.

Stephens, J. M. and Marshall, J. H. 1962. Some properties of an immune factor isolated from the blood of actively immunized wax moth larvae. *Can. J. Microbiol. 8:* 719-25.

Steward, M. W., Todd, C. W., Kindt, T. J., and David, G. S. 1969. Low molecular weight mercaptoethanol-sensitive antibody in rabbits. *Immunochemistry. 6:* 649-58.

Stobo, J. D. and Tomasi, T. B., Jr. 1967. A low molecular weight immunoglobulin antigenically related to 19S IgM. *J. Clin. Invest. 46:* 1329-37.

Suran, A. A., Tarail, M. H., and Papermaster, B. W. 1967. Immunoglobulins of the leopard shark. I. Isolation and characterization of 17S and 7S immunoglobulins with precipitating activity. *J. Immunol. 99:* 679-86.

Svasti, J. and Milstein, C. 1972. The complete amino acid sequence of a mouse κ light chain. *Biochem. J. 128:* 427-44.

Szenberg, A. and Warner, N. L. 1962. Dissociation of immunological responsiveness in fowls with a hormonally arrested development of lymphoid tissue. *Nature 194:* 146-47.

Szenberg, A., Cone, R. E., and Marchalonis, J. J. 1974. Isolation of surface immunoglobulins from lymphocytes of chicken thymus and bursa. *Nature 250:* 118-20.

Szenberg, A., Lind. P., and Clarke, K. 1965. IgG and IgM antibodies in fowl serum. *Aust. J. Exp. Biol. Med. Sci. 43:* 451-54.

Tabel, H. and Ingram, D. G. 1972. Immunoglobulins of mink. Evidence for five immunoglobulin classes of 7S type. *Immunology 22:* 933-42.

Tanaka, K. 1973. Allogeneic inhibition in a compound ascidian, *Botryllus primigenus* Oka. II. Cellular and humoral responses in "nonfusion" reaction. *Cell. Immunol. 7:* 427-43.

Taniguchi, M. and Tada, T. 1974. Regulation of homocytotropic antibody formation in the rat. X. IgT-like molecule for the induction of homocytotropic antibody response. *J. Immunol. 113:* 1757-69.

Taylor, A. C. and Kollros, J. J. 1946. Stages in the normal development of *Rana pipiens* larvae. *Anat. Rec. 94:* 7-23.

Theis, G. A. and Thorbecke, G. J. 1973. Suppression of delayed hypersensitivity reactions in bursectomized chickens by passively administered antiimmunoglobulin antisera. *J. Immunol. 110:* 91-97.

Thoenes, G. H. and Hildemann, W. H. 1970. Immunological responses of pacific hagfish. II. Serum antibody production to soluble antigen.

In *Developmental Aspects of Antibody Formation and Structure*, pp. 711-22. Prague: Czech. Acad. Sci.

Thomas, L. 1959. Discussion. In *Cellular and Humoral Aspects of the Hypersensitive State*, ed. H. S. Lawrence, pp. 529-32. New York: Hoeber-Harper.

Thomas, W. R., Turner, K. J., Eadie, M. E., and Yadav, M. 1972. The immune response of the quokka (*Setonix brachyurus*). The production of a low molecular weight antibody. *Immunology 22:* 401-16.

Thompson, R. A. and Rowe, D. S. 1968. Reactive hemolysis—a distinctive form of red cell lysis. *Immunology, 14:* 745-62.

Thorbecke, G. J., Warner, N. L., Hochwald, G. M., and Ohanian, S. H. 1968. Immune globulin production by the bursa of Fabricius in young chickens. *Immunology 15:* 123-34.

Thorsby, E. 1971. A tentative new model for the organization of the mouse H-2 histocompatibility system: Two segregant series of antigens. *Eur. J. Immunol. 1:* 57-59.

Tillet, W. S. and Francis, T. Jr. 1930. Serological reactions in pneumonia with a nonprotein somatic fraction of pneumococcus. *J. Exp. Med. 52:* 561-71.

Tiselius, A. 1937. Electrophoresis of serum globulin. I. *Biochem. J. 31:* 313-17.

Tonegawa, S., Steinberg, C., Dube, S., and Bernardini, A. 1974. Evidence for somatic generation of antibody diversity. *Proc. Natl. Acad. Sci. (U.S.A.) 71:* 4027-31.

Tournefier, A. 1972. Les reactions immunitaires chez les amphibians urodeles. III. Rôle du thymus dans l'immunité de transplantation. Capacité d'immunisation aux antigènes particuliares chez le urodele et le triton alpestre adultes. In *L'étude Phylogenique et Ontogenique de la Réponse Immunitaire et son Apport a la Théorie Immunologique*, ed. P. Liacopoulos et J. Panijel, pp. 105-12. Boulogne: Inserm.

Trnka, Z. and Franek, F. 1960. Studies on the formation and characteristics of antibodies in frogs. *Folia Microbiol. (Praha) 5:* 374-80.

Trump, G. N. 1970. Goldfish immunoglobulins and antibodies to bovine serum albumin. *J. Immunol. 104:* 1267-75.

Trump, G. N. and Hildemann, W. H. 1970. Antibody responses of goldfish to bovine serum albumin. Primary and secondary responses. *Immunology 19:* 621-27.

Turner, R. J. and Manning, M. J. 1974. Thymic dependence of amphibian antibody responses. *Eur. J. Immunol. 4:* 343-46.

Turpen, J. B., Volpe, E. P., and Cohen, N. 1973. Ontogeny and peripheralization of thymic lymphocytes. *Science 182:* 931-33.

Tyler, A. and Scheer, B. T. 1945. Natural heteroagglutinins in the serum of the spiny lobster, *Panulirus interruptus.* II. Chemical and antigenic relation to blood proteins. *Biol. Bull. 89:* 193-200.

Uhr, J. W. 1964. The heterogeneity of the immune response. *Science, 145:* 457-64.

Uhr, J. W. and Moller, G. 1968. Regulatory effect of antibody on the immune response. *Adv. Immunol. 8:* 81-127.

Uhr, J. W., Finkelstein, M. S., and Franklin, E. C. 1962. Antibody response to bacteriophage $\phi$X174 in nonmammalian vertebrates. *Proc. Soc. Exp. Biol. (Med.) 111:* 13-15.

Underdown, B. J., Simms, E. S., and Eisen, H. N. 1971. Subunit structure and number of combining sites of the immunoglobulin A myeloma protein produced by mouse plasmacytoma MOPC-315. *Biochemistry 10:* 4359-68.

Vaerman, J. P. and Heremans, J. F. 1968. The immunoglobulins of the dog. I. Identification of canine immunoglobulins homologous to human IgA and IgM. *Immunochemistry 5:* 425-32.

Vaerman, J. P. and Heremans, J. F. 1971. IgA and other immunoglobulins from the European hedgehog. *J. Immunol. 107:* 201-11.

Vaerman, J. P., Heremans, J. F., and van Kerckhoven, G. 1969. Identification of IgA in several mammalian species. *J. Immunol. 103:* 1421-23.

Vaughn, P. P. 1962. The paleozoic microsaurs as close relatives of reptiles, again. *Amer. Midl. Nat. 67:* 79-84.

Vitetta, E. S. and Uhr, J. W. 1973. Synthesis, transport, dynamics and fate of cell surface Ig and alloantigens in murine lymphocytes. *Transplant. Rev. 14:* 50-75.

Vitetta, E. S., Bianco, C., Nessenweig, V. and Uhr, J. W. 1971. Cell surface immunoglobulin. IV Distribution among thymocytes, bone marrow cells, and their derived populations. *J. Exp. Med., 136:* 81-93.

Voisin, G. A. 1971. Immunological facilitation, a broadening of the concept of the enhancement phenomenon. *Prog. Allergy 15:* 328-485.

Volpe, E. P. 1971. Immunological tolerance in amphibians. *Am. Zool. 11:* 207-18.

Volpe,, E. P. and Earley, E. M. 1970. Somatic cell mating and segregation in chimeric frogs. *Science 168:* 850-52.

Volpe, E. P. and McKinnell, R. G. 1966. Successful tissue transplantation in frogs produced by nuclear transfer. *J. Hered. 57:* 167-74.

Waldenström, J. 1952. Abnormal proteins in myeloma. *Adv. Intern. Med. 5:* 398-440.

Walker, A. D. 1972 New light on the origin of birds and crocodiles. *Nature 237:* 257-63.

Wang, A. C., Pink, J. R. L., Fudenberg, H. H., and Ohms, J. 1970. A variable region subclass of heavy chains common to immunoglobulins G, A, and M and characterized by an unblocked amino-terminal residue. *Proc. Natl. Acad. Sci. (U.S.A.) 66:* 657-63.

Warner, N. L. 1964. The immunological competence of thymic cell suspensions. *Aust. J. Exp. Biol. Med. Sci. 42:* 401-16.

Warner, N. L. 1972. Differentiation of immunocytes and the evolution of immunological potential. In *Immunogenicity,* ed. F. Borek, pp. 467-537. Amsterdam: North-Holland.

Warner, N. L. 1974. Membrane immunoglobulins and antigen receptor on B and T lymphocytes. *Adv. Immunol. 19:* 67-216.

Warner, N. L. and Marchalonis, J. J. 1972. Structural differences in mouse IgA myeloma proteins of different allotypes. *J. Immunol. 109:* 657-61.

Warner, N. L., Byrt, P., and Ada, G. L. 1970. Blocking of the lymphocyte antigen receptor site with anti-immunoglobulin sera *in vitro. Nature 226:* 942-43.

Warner, N. L., Szenberg, A., and Burnet, F. M. 1962. The immunological role of different lymphoid organs in the chickens. I. Dissociation of immunological responsiveness. *Aust. J. Exp. Biol. Med. Sci. 40:* 373-87.

Warner, N. L., Uhr, J. W., Thorbecke, G. J., and Ovary, Z. 1969. Immunoglobulins, antibodies, and the bursa of Fabricius: Induction of agammaglobulinemia and the loss of all antibody-forming capacity by hormonal bursectomy. *J. Immunol. 103:* 1317-30.

Wasserman, R. L., Kehoe, J. M., and Capra, J. D. 1974. The $V_H III$ subgroup of immunoglobulin heavy chains: phylogenetically associated residues in several avian species. *J. Immunol. 13:* 954-57.

Watts, R. L. and Watts, D. C. 1968. Gene duplication and the evolution of enzymes. *Nature 217:* 1125-30.

Webster, R. O. and Pollara, B. 1969. Isolation and partical characterization of transferrin in the sea lamprey, *Petromyzon marinus. Comp. Biochem. Physiol. 39:* 509-27.

Weigle, W. O. 1973. Immunological unresponsiveness. *Adv. Immunol. 16:* 61-122.

Weinbaum, F. I., Gilmour, D. G., and Thorbercke, G. J. 1973. Immunocompetent cells of the chicken. III. Cooperation of carrier-sensitized T cells from agammaglobulinemic donors with hapten-immune B cells. *J. Immunol. 110:* 1434-36.

Wetherall, J. D. 1969. *Immunoglobulins of the lizard* Tiliqua rugosa. Doctoral Dissertation, the University of Adelaide, Adelaide.

Widal, F. and Sicard, E. 1897. Influence de l'organisme sur les propriétés acquises par les humeurs du fait de l'infection. (L'agglutination chez quelques animaux à sang-froid). *C. R. Soc. Biol. (Paris) 4:* 1047-50.

Wilde, C. E. III, and Koshland, M. E. 1972. The role of J chain in immunoglobulin assembly. *Fed. Proc. 31:* 755 Abs.

Williamson, A. R. 1971. Antibody isoelectric spectra. Analysis of the heterogeneity of antibody molecules in serum by isoelectric focusing in gel and specific detection with hapten. *Eur. J. Immunol. 1:* 390-94.

Witschi, E. 1957. *Development of Vertebrates.* Philadelphia: W. B. Saunders Co.

Wolfe, H. R., Amin, A., Mueller, A. P., and Aronsen, F. R. 1960. The secondary response of chickens given a primary inoculation of bovine serum albumin at different ages. *Int. Arch. Allergy Appl. Immunol. 17:* 106-15.

World Health Organization 1964. Nomenclature of human immunoglobulin. *Bull. W. H. O. 30:* 447-50.

World Health Organization 1969a. Cell-mediated immune responses. *W. H. O. Tech. Rep. Ser.* No. 423.

World Health Organization 1969b. An extension of the nomenclature for immunoglobulins. *Bull. W. H. O. 41:* 975-78.

Wu, T. T. and Kabat, E. A. 1970. An analysis of the sequences of the variable regions of Bence-Jones proteins and myeloma light chains and their implications for antibody complementarity. *J. Exp. Med. 132:* 211-50.

Yadav, M., Stanley, N. F., and Waring, H. 1972a. The thymus glands of a marsupial, *Setonix brachyurus* (Quokka), and their role in immune responses. Structure and growth of the thymus glands. *Aust. J. Exp. Biol. Med. Sci. 50:* 347-56.

Yadav, M., Stanley, N. F., and Waring, H. 1972b. The thymus glands of a marsupial, *Setonix brachyurus* (Quokka), and their role in immune responses. Effect of thymectomy on somatic growth and blood leucocytes. *Aust. J. Exp. Biol. Med. Sci. 50:* 357-64.

Yocum, D., Cuchens, M., and Clem, L. W. 1975. The hapten-carrier effect in teleost fish. *J. Immunol. 114:* 925-27.

Young, J. Z. 1962. *The Life of Vertebrates.* Oxford: Clarendon Press.

Yphantis, D. A. 1964. Equilibrium ultracentrifugation of dilute solutions. *Biochemistry 3:* 297-317.

Zabriskie, J. B. 1967. Mimetic relationships between group A. streptococci and mammalian tissues. *Adv. Immunol. 7:* 147-88.

Zimmerman, B., Shalatin, N., and Grey, H. M. 1971. Structural studies on the duck 5.7S and 7.8S immunoglobulins. *Biochemistry. 10:* 482-88.

Zuckerkandl, E. 1965. The evolution of hemoglobin. *Sci. Am. 212* (5): 110-18.

Zuckerkandl, E. and Pauling, L. 1965. Evolutionary divergence and convergence in proteins. In *Evolving Genes and Proteins*, ed. V. Bryson and H. J. Vogel, pp. 97-166. New York: Academic Press.

# Index